Urban Design

Christa Reicher

Urban Design

Springer

Christa Reicher
Chair of Urban Planning and Design
Rhenish-Westphalian Technical University
Aachen, Nordrhein-Westfalen, Germany

ISBN 978-3-658-34369-9 ISBN 978-3-658-34370-5 (eBook)
https://doi.org/10.1007/978-3-658-34370-5

This book is a translation of the original German edition „Städtebauliches Entwerfen“ by Reicher, Christa, published by Springer Fachmedien Wiesbaden GmbH in 2017. The translation was done with the help of artificial intelligence (machine translation by the service DeepL.com). A subsequent human revision was done primarily in terms of content, so that the book will read stylistically differently from a conventional translation. Springer Nature works continuously to further the development of tools for the production of books and on the related technologies to support the authors.

This Springer imprint is published by the registered company Springer Fachmedien Wiesbaden GmbH part of Springer Nature.
The registered company address is: Abraham-Lincoln-Str. 46, 65189 Wiesbaden, Germany

Preface

Today, urban design is a core competence of every planner and architect training. This has been repeatedly confirmed in teaching and planning practice. In planning, monitoring and implementation of urban development projects, I have experienced how important the connection between an individual house and its context as well as the level of the neighbourhood and the city is. Public space is an important link in this process—as the backbone and business card of a city. Built and open space must be examined and planned with equal intensity, because a quality neighbourhood can only be created through a balanced dialogue. I have compiled my experiences from planning practice and as a university lecturer at various architecture and planning faculties over the last twenty years in this handbook. The fifth edition is supplemented by two current thematic focuses: informal urban development and regional urban planning. This thematic and spatial orientation of urban design has gained importance in recent years and is therefore given appropriate consideration in the updated edition.

Urban design, urban planning and urban development are complex processes involving not only architects, urban planners and engineers, but also politicians, institutional and private investors, creative people, sociologists, climate researchers, lighting designers, event managers and, more than ever before, the urban community itself. Urban design requires an examination of the whole complex city and must define qualities both at its edges and in its centre. In doing so, it is necessary to listen to the history of the city, understand its origins, connect to it or—where the substance no longer corresponds to the rules of the present—develop a new layer. The modern city today is one of these layers; our actions today will also be one of these layers tomorrow! But urban design is always a process of discovery and research that makes the historical foundations of a place comprehensible and advances history(ies).

Architects and planners who will design cities in the future and continue to build existing cities must learn to deal with this new complexity. They must be empowered to define the framework for possible developments with the necessary technical know-how and their own position. The handbook aims to meet this demand by providing assistance in reading about cities and a tool for designing urban structures. The layer method presented here is the result of many design processes and is intended to do justice to the complexity of the city.

In the past few years, I have worked with various colleagues on urban development design projects at the interface of architecture and urban planning. Prof. Anne Klasen-Habeney and Prof. Klaus Köpke, with whom I have often worked together in the past on courses, excursions and projects, contributed to the conception of the book. Prof. Ulrike Beuter for the open space, Päivi Kataikko-Grigoleit for the social and educational infrastructure and Prof. Dr. Angela Million for trade and supply have contributed as authors to the "urban components" of the city. They are all experts in the individual subject areas.

The chapter "Urban design" and the layer method presented here was developed in discussion and cooperation with my colleagues Fabio Bayro Kaiser, Peter Empting, Päivi Kataikko-Grigoleit, Dr. Ilka Mecklenbrauck, Dr. Jan Polivka, Anne Söfker-Rieniets and Yasemin Utku. Prof. Dr. Frank Roost has contributed as author to the chapter on image and branding strategies, Stephan Willinger has contributed as author to the chapter on informal urban development and Dr. Ilka Mecklenbrauck has contributed as author to the chapter on activation of the inner city.

I would like to thank Dirk Haas and Klaus Köpke for their editorial support. The illustrations were created by Amir-Reza Rahnama, unless they were taken from the respective draft authors of the projects. Amir-Reza Rahnama and Marcus Lumma are responsible for the layout. Fabio Bayro Kaiser and Mladen Milinovic assisted with the English autotranslation.

So this book was only possible with the participation of many colleagues. For this I would like to thank everyone very much!

Aachen, Germany Christa Reicher

Contents

Introduction 1

1 Tasks and Challenges for Architects, Urban and Landscape Planners

Today, in the first "urban century" of history, cities and urban design are facing a multitude of new challenges. This does not mean that the guidelines and principles of urban design will now be completely different from those of the nineteenth or twentieth century. However, the dynamics and complexity of urban development, as seen, for example, in the spatial and temporal coexistence of growth, shrinkage and stagnation, requires an even more processual understanding of urban planning than was necessary in the phases of the city in the industrial age. Cities, urban districts and urban regions do not develop in only one direction; contemporary urban design concepts must therefore take more than just one possible reality into account, which must be designed and shaped.

On the one hand, such a processual understanding of urban design requires a convincing overall concept that determines the necessary constants in the form of leitmotifs, goals and urban design rules. On the other hand, it must also take into account changeability and variability, grant the groups of actors involved in urban design leeway and demand an active discourse with urban society. Such robust yet flexible cities and urban spaces are created on the basis of integrated thinking and action. Therefore, urban design—more than ever before—encompasses a multitude of economic, cultural, aesthetic, social, technical and last but not least political aspects, the complexity of which has grown with the city, which has also become increasingly complex.

> Urban design is not a linear, finite trace from program to implementation, but a basically endless process of developing, conceiving, rejecting and replacing spatial proposals, a permanent exploration of the situation (Erik Pasveer)

Urban design is therefore an extremely complex, creative process that can become a challenging experiment with structures, forms, landscape, topography and much more, especially for becoming architects, urban and landscape planners. The design

C. Reicher, *Urban Design*, https://doi.org/10.1007/978-3-658-34370-5_1

of a garden, for example, is no less important than the design of a building, the form no less than the programme, open space deserves the same attention as the built city, and the usability of a space is just as important as its aesthetic quality.

Fig. 1.1 Tower of Babel, Pieter Breughel, 1563

2 Urban Design of Product and Process

The process of design is no longer limited to the development of a spatial product, a physical space ("hardware"), but increasingly also includes the development of spatial processes and the associated strategies ("software"): these can be methods of communication and participation, marketing and branding strategies or even novel planning instruments that enable and implement high-quality urban development. Following this understanding of the task, this guide to urban design differentiates between product and process.

Depending on the specific tasks, there are numerous interfaces between product and process. In this book on urban design, both dimensions are dealt with in detail, even if the design of physical spaces is the main focus. Ultimately, urban design is a spatial choreography of structures and fillings, which also includes the uncertain and surprising—qualities that are inherent in both the city as a topos and the design process.

Urban design as a **product**:

A. Spatial structure (of the physical space)
   - Building
   - Free space

B. Space filling
   - Use
   - Programme

Urban design as a **process**:

C. Spatial organisation
   - Planning procedures, methodology
   - Formal and informal rules
D. Space activation
   - Branding and Imagination
   - Event, staging
   - Participation

> Every form is the frozen snapshot of a process. So the work is a stop of becoming and not a frozen goal. (El Lissitzky)

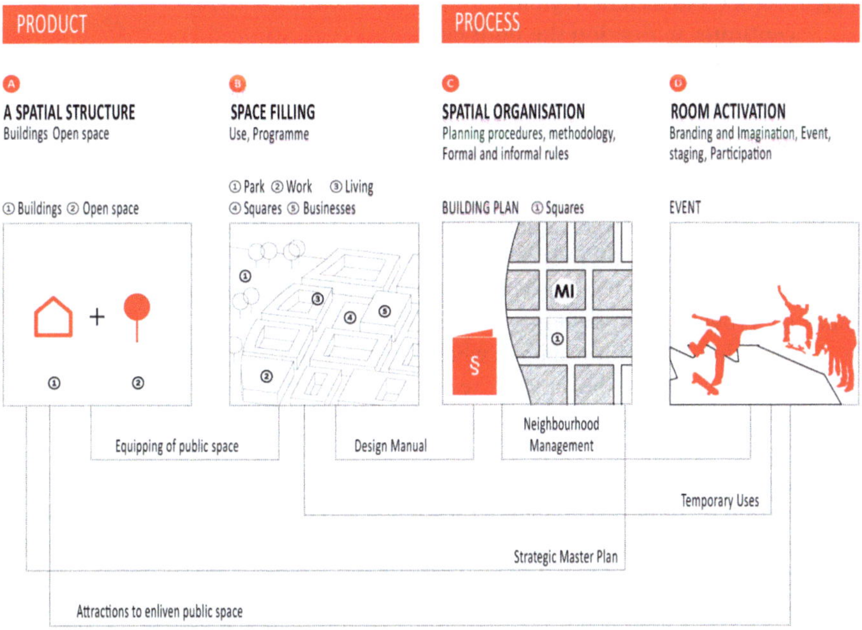

Fig. 1.2 Urban design of product and process

3 Definition and Understanding: What Is Urban Design?

Urban design deals with the spatial organisation and design of the environment, in both urban and rural contexts. Even though the discipline of urban design only emerged in the second half of the nineteenth century, urban design has existed since planning processes for the construction of cities have existed. The origins of the discipline therefore go back to the history of the city, especially since the construction of settlements or cities is understood from the very beginning as a field of action with cultural, social, economic, technical and design aspects: After all, cities would not exist if they had been planned and built permanently against the needs of their inhabitants, against basic economic and ecological principles, and without appropriate infrastructures.

Over the centuries, urban design has had to develop solutions for very different requirements, because the specific challenges and the associated urban design tasks have changed in and with the respective eras. Today's cities, urban districts and urban regions with their complex and sometimes contradictory development trends, in turn, present urban design with numerous new challenges and require a correspondingly multifaceted understanding of its tasks.

. . .is urban and regional planning in the third dimension.

While urban and spatial planning is often limited to two-dimensional representations of the city and its functional areas, urban design adds the third dimension of the city and the design of space. In this sense, urban design is a shaping urban planning, up to the architectural planning of space—from the city neighbourhood to the city region.

. . .continues to develop the built city.

In contrast to phases of rapid urban growth, which were characteristic of the industrialisation of the nineteenth and twentieth centuries, urban design in Europe is currently increasingly concerned with the development conditions and possibilities of the existing city, i.e. the already built city. However, even here the first impression of an already built city is deceptive, since the rate of structural change—depending on the location—is between 2 and 5% per year, so that statistically speaking, cities will be completely rebuilt over the course of 2–3 generations. Urban redevelopment and urban renewal are therefore constant tasks in urban design, regardless of whether cities are growing or shrinking.

In shrinking cities, for example, which are characterized by abandonment of use and empty buildings, the largely new task is to design decreasing density or emptiness or to develop temporary solutions for the preservation of valuable building fabric in the cities.

Cities are reading books of history. Each stage of the history of urban development has left its mark, which must be respected and developed further. The specific "fingerprint" determines the identity of a city or neighbourhood.

. . .designs new neighbourhoods and quarters.

The growth spurts of past eras have slowed down. Nevertheless, there is enormous development pressure, especially in the growing cities, which is being answered either by building new districts and neighbourhoods on vacant land or by planning urban expansion areas. Such urban expansions are, however, generally on a smaller scale than the "new cities" of the 1950s and 1960s.

New urban neighbourhoods can often react more easily than existing parts of the city to new demands for living and working, to changing lifestyles and new technical or design possibilities of urban design. They are therefore also a specific feature of our current epoch and its social and aesthetic practice.

Urban design describes a vision of future living conditions (Peter Zlonicky)

…**is an agreement between public and private interests.**

The construction of cities has always been an achievement of public and private actors, albeit with different and changing weight.

Urban design cannot be considered and carried out exclusively as a private or public task. With the diminishing financial resources of cities and municipalities, private actors have greater scope for action. The many public-private partnership (PPP) projects, but also the numerous civil society initiatives at local, regional and international level are evidence of the changing balance of power and new alliances in urban development and urban design.

…**directs spatial development.**

Urban design has the task of steering spatial and especially structural development. This includes the planning of the various land uses, the determination of the boundaries and heights of the buildings as well as the planning for the development of future areas of use and their building structures. The statements of urban design can be understood as a hypothesis for the future development of physical space. This hypothesis can be verified and at the same time changed. Urban design thus shows a future reality without specifying it in detail.

> By urban design we understand the control of spatial, especially structural development in the municipal area. The field of activity ranges from the long-term spatial disposition of land use and infrastructure investments to the design of the three-dimensional framework for the structural design. Urban design finds its legal expression in the urban land-use plans with their two main types of plan, the land use plan and the development plan. (Gerd Albers)

…**provides the conditions for a space to function.**

Urban design provides a structure that sets the course and at the same time sets a framework for the functional character of space. The compatibility of uses, especially with regard to mixed use, is essentially determined by how the corresponding structures are designed in terms of scale, technical infrastructure, zoning, etc. The liveliness of a place is significantly influenced by the interaction of the individual uses. This interplay in turn requires rules and specifications to enable use and appropriation on the one hand and to minimize or avoid conflicts on the other.

> Urban design as a shaping task comprises the spatial arrangement and the structural formulation of urban living areas. Order results from the perceptible relationships between the nature of the site (surroundings, ground relief, vegetation, climate, weather) and the objects artificially built into it by man, such as houses, paths, fences or monuments, which as products of human art and technology stand for what is conceivable and feasible here. (Jürgen Hotzan)

…**determines the space between the building and its context.**

Urban design formulates the physical framework for the allocation and form of buildings, facilities and public, semi-public and private open spaces. The focus is on the formulation and implementation of spatial contexts and meanings. In this respect, urban designs fulfil an intermediary role between the individual object and its context: establishing such interrelationships between buildings and urban space, between neighbourhood and city as a whole, between city park and cultural landscape is a core concern of urban design.

. . .provides the foundation for aesthetic qualification and value adding.

More than ever, the high-quality design of urban spaces has become an important factor in urban development. Urban design shapes "addresses" and thus creates the conditions for the settlement of sophisticated businesses with interesting jobs, as well as for high-quality residential neighbourhoods that promise long-term stability of value and well-being.

. . .is the prerequisite and result of attractive urban planning culture.

The future of a city will depend to a large extent on whether it succeeds in creating its own, in the best case unmistakable urban profile. This does not so much mean the individual, possibly iconographic building, but rather aims at the urban context: the city square, the park, the neighbourhood, the campus or the city centre. The way buildings are joined together and urban spaces are designed determines the face of a city. In the competition for inhabitants, purchasing power and attractiveness, such architectural qualities of cities are becoming increasingly important. However, qualified debates on urban design and building culture are no less important, as they form the social basis for the constant development of the spatial and architectural identity of a city or city district.

. . .is the ordering of social relationships.

The built structures represent the framework on which people's communication and actions are based. They can be designed to promote joint activities and dialogues, but they can also stand in the way of them. Thus, through urban design, social relations are ordered and the opportunities for joint trade are increased.

Urban design is the forward planning of the habitat of local communities. . . (Gerd Albers)

Fig. 1.3 The City as a Container of Life, woodcut by Frans Masereel

To Understand the Basic Terms 2

1 Sustainability

The term sustainability is understood as a balanced relationship between three basic factors: ecology, economy, and social responsibility.

Ecological action pursues the goal of preserving natural capital through the economical use of natural resources (such as water, air, and soil). The economic factor, the economy, is relevant for the material satisfaction of basic human needs as well as for the creation and assurance of quality of life. The societal and social dimension of sustainability demands that all people have equal access to resources and optimal conditions for the development of their lives. In essence, this means a right to a dignified life for all, but also social justice between the present and future generations. Sustainable action should, if possible, strike a balance between these three dimensions. If one of these dimensions is neglected, a positive and sustainable development is jeopardised. However, this does not mean that all dimensions must be treated equally in every case.

The term "sustainable development" has a history that goes back to the sixteenth century. The principle of sustainability was used in forestry at the time and referred to the form of management in which, in the long term, no more wood is felled than what grows back.

Four centuries later, in 1987, the concept of sustainable development was presented by the World Commission on Environment and Development in the so-called "Brundtland Report". This report called for the objectives of economic and social development to be examined and defined in terms of their permanence. According to this, development is sustainable if it "meets the needs of the present without compromising the ability of future generations to meet their own needs and to choose their own lifestyle".

The principle of sustainable development later formed the basis for the UN Conference on "Environment and Development" in Rio in 1992, which gave a decisive impetus for action in the countries involved. Under the title "Agenda 21", that is, sustainable action for the twenty-first century, concrete concepts and programmes have

C. Reicher, *Urban Design*, https://doi.org/10.1007/978-3-658-34370-5_2

since then been developed from the national to the local level. The "Agenda 21" identifies various procedures and technologies that must be developed to ensure that human needs can be met through the prudent management of natural resources. The first objective is to improve and transform the decision-making process in order to achieve sustainable results.

Relevance of Sustainable Urban Development

With increasing urbanisation, suburbanisation, and the spatial expansion of cities into urban regions, which are usually associated with high environmental pollution, the importance of sustainable urban development has also increased. Cities are the focal points of social and economic development. They offer a wide range of opportunities for social interaction and communication. A city's ecological, economic, and social performance also determines whether and how people feel "at home" there. This becomes relevant in times of declining population in Europe and competing cities. Urban development influences the quality of life of people. In terms of sustainability, urban development can constructively influence the negotiation process between economic stakeholders, investors, and different social interest groups on the one hand and environmental concerns on the other.

Apart from economic aspects, central components of sustainable urban development and thus also of sustainable urban design are, for example:

- the consideration of environmental protection (nature conservation, landscape management, etc.),
- the promotion of environmentally sound energy supply and transport systems,
- the use of efficient and environmentally friendly local transport systems,
- the promotion of environmentally friendly construction using local building materials and locally available natural resources,
- institutionalising the participation of local communities (dialogue with citizens, participation models, etc.).

The consideration of these concerns in an appropriate balance is the prerequisite for a convincing urban development concept.

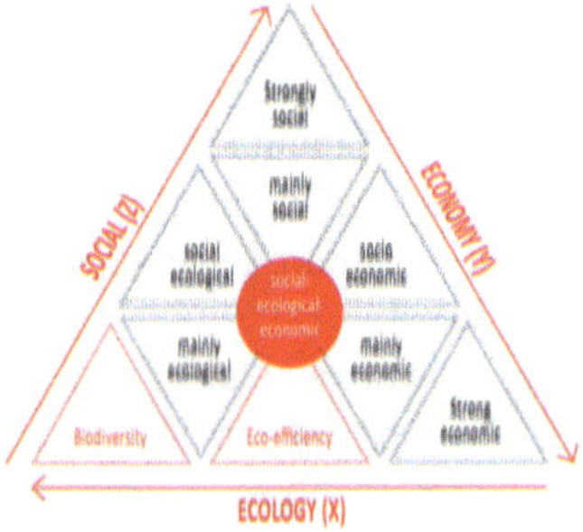

Fig. 2.1 The three-pillar model of sustainability

The integrating sustainability triangle allows for continuous merging of the three dimensions of ecology, economy, and social affairs.

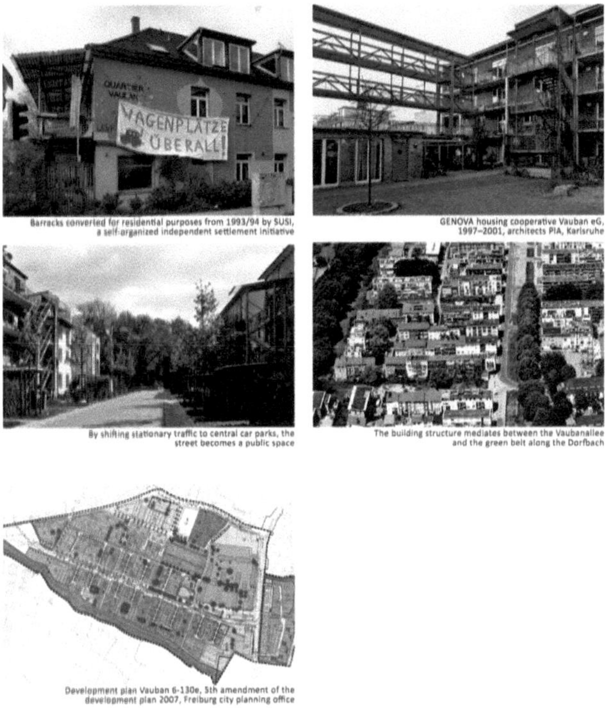

Fig. 2.2 Vauban district, Freiburg

With the withdrawal of the French troops in 1992, an area of approximately 38 hectares in the south of the city of Freiburg became free. In 1993, the municipal council decided to apply the planning law instrument of urban development measures according to the building code. The district Vauban with about 2000 residential units was created. All buildings were constructed at least in accordance with the Freiburg low-energy standard. Today about 5000 people live in the Vauban district.

Ecological Quality Preservation of the old stock of trees and the biotope in the Dorfbach, greening of facades and roofs, reuse of existing buildings, energy optimisation, low-energy house standard for all new buildings, low-energy, passive house or plus-energy house standard, partially thermal solar systems, local heating supply, CHP with wood chips, solar settlement on the Schlierberg, rainwater seepage on site, combined vacuum drainage, greening of roofs, low-waste construction initiative.

Economic Quality
Low construction costs, diversified offer: social housing, ownership, hire purchase. Mixed use: commercial, mixed, and residential construction areas, different housing typologies, good technical and social infrastructure.

Social Quality
Primary schools and kindergartens, initiative for shops in the neighbourhood, neighbourhood social work, neighbourhood centre in the former officers' casino, delivery service for food, square for farmers' market, square at the neighbourhood entrance, traffic-free residential streets, mix of social groups.

2 Urbanity

Urbanity comes from the Latin word urbs (city), which originally referred to Rome as the centre of the world. "Urban" therefore means "belonging to the city" (Durth and Sigel 2009, p. 773). The term urbanity refers to a way of life that can arise in a large and cosmopolitan city. Urbanity has become a key concept in the discussion about the future of the city. Whereas urbanity was once understood more as a social way of life and concept, today's understanding is often reduced to the urban structure and primarily refers to the dense city, which is equated with the city of the nineteenth century.

In the late 1960s, architects and urban planners hoped to create urbanity through density. This proved to be an illusion. The concept of urbanity has been misunderstood. The urban development model of the 1970s, "urbanity through density," reduced the understanding of urbanity to an enlargement of the building scale, introduced vertical urban planning without taking into account the aspect of lived urbanity.

In the built city, the indications of urbanity—enclosed squares and streets, markets, and festivals—often intertwine. The misunderstanding that urbanity can only be created by a dense building structure and manageable spaces is nevertheless widespread. The distinction made by Thomas Sieverts between built and lived urbanity is a helpful differentiation for understanding and designing urban spaces.

The concept of the urban city cannot only be defined by the compact building structure and the numerical population density, but also includes cultural aspects:

- Mixed uses and functions
- Social integration capacity
- Possibilities of appropriation of space
- Well-proportioned public spaces
- Urban architecture
- Typological traditions

Can Urbanity Be Planned?
The understanding of and discussion about urbanity, which is an important basis for urban design, can be traced back to various criteria:

Building Density

Building density is one of the basic characteristics of the city and is, therefore, an essential condition for urbanity. Building density is a quantitatively measurable criterion. It is defined as the extent to which a property is used for construction or as a measure of the volume of buildings on a built-up area. It implies the ratio of developed to undeveloped area. In urban land use planning, the building density is determined by the coverage, the floor area ratio (FAR), or the volume area ratio. The coverage indicates the proportion of a property covered by buildings. The floor area ratio (FAR) refers to the ratio of the floor area (corresponding to the external dimensions of the building) to the area of the plot. The volume area ratio introduces a third dimension and relates the building mass to the land area. The listed parameters do not affect urbanity in the same way. The coverage indicates the measure of urban density more strongly than the FAR. It is an indicator of land use which, in conjunction with development and open spaces, has a decisive influence on the character of an urban appearance. Building density is only one of the prerequisites of urbanity. Density alone, however, is—as has been stated—not sufficient to create urbanity.

Social Density

In addition to building density, there is also social density, which refers to the number of people and their activities in a neighbourhood.

As a rule, the population density of a city or district is also associated with the building density. However, the two need not be identical. In Wilhelminian style neighbourhoods, the density of buildings has remained the same, while the population density has decreased due to the increasing demand for living space and the progressing suburbanisation. Likewise, the ratio of population and available space can vary greatly: In Manhattan there are about 25,000 inhabitants per square kilometre, in New York City about 9400, in Hamburg 2150, in Kassel 1800, in Weimar 1150.

Besides population density, there are other density types such as employment density and interaction density. However, these two density types are only of limited significance for the urbanity of urban neighbourhoods, as their values can also be high in suburban locations. Similarly, a high density of buildings does not necessarily imply a high density of interaction. Social density, especially in conjunction with population density, is an important criterion for urbanity.

Mixed Use

The heterogeneity and thus the liveliness of urban structures is decisively influenced by the structure and composition of its uses. The mixture of compatible uses decisively determines the urban character of a neighbourhood. The term mixed use refers to the mixture of the functions of living, working, and recreation, which used to be a natural part of urban life, but has been increasingly lost in the course of the separation of functions—accompanied by industrialisation and, in a distinctly pragmatic form, by modernity.

The different uses can be interwoven in different ways—in a single building as well as within a single area. The functions can be separated vertically and horizontally. The following categories are relevant for urban design:

- the small-scale mix of uses at the neighbourhood level (horizontal mix)
- the mixture of uses within the building structure (vertical mixture).

Both categories can be meaningfully linked with each other. The advantages of mixed use at the neighbourhood level—apart from strengthening an urban character—are manifold:

- more compact urban structures
- accesebility and shorter traffic routes
- more efficient use of space
- economic use of the infrastructure
- greater possibilities for conversion and adaptation

These advantages are now increasingly being recognised by investors and developers, resulting in a paradigm shift in urban development, in addition to an increased mix of uses at neighbourhood level.

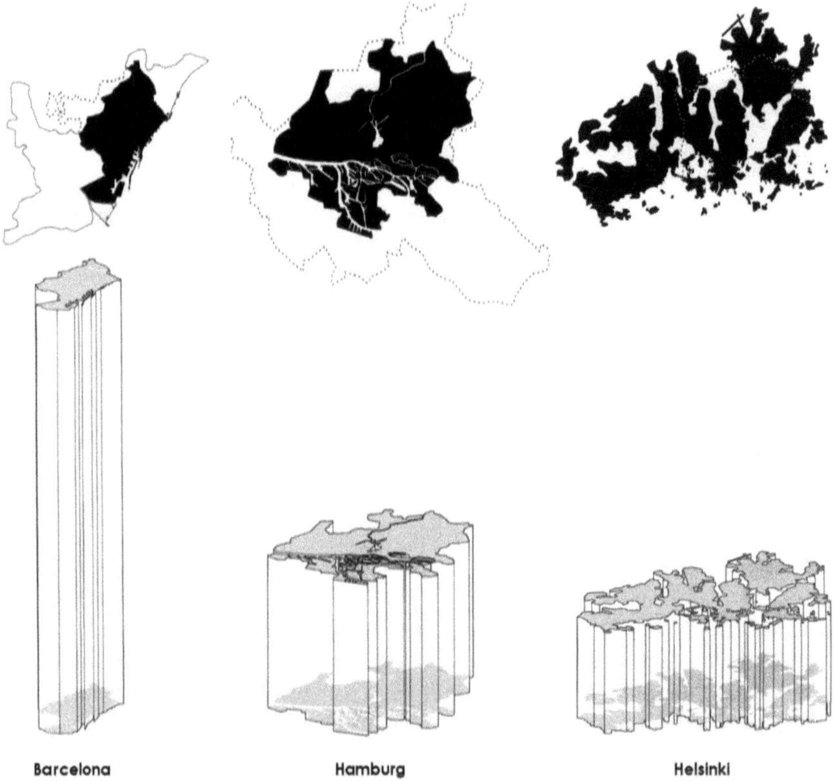

Fig. 2.3 Density: Barcelona, Hamburg, Helsinki, El Khafif. Three-dimensional density profiles illustrate the differences in population and employment density in European metropolitan areas

Urban Form

A central aspect of urbanity is the urban design of buildings and public spaces. This aspect, which is subjectively perceived in very different ways, plays a decisive role in determining whether a development or urban space has a more metropolitan, small-town, or village-like appearance. Thus the architecture of the buildings, both their architectural style and their degree of representation, strongly determines the urban effect of the built structure. The character of the public space is essentially determined by the scale, the built design elements, and also the greenery.

The role that the urban form plays in the urbanity of a neighbourhood can be illustrated using the example of the "townhouses". In terms of density, these townhouses are comparable to the classic terraced houses. Nevertheless, the two building typologies differ greatly in their spatial effect, which is not least due to the design. In addition to a frequently raised ground floor, the façade of a townhouse has a much more sophisticated architectural design. Regular "house personalities" line up next to each other and, in the interplay of unity and height development, shape the desired urbanity with this design requirement.

Taking into account a more complex understanding of urbanity, urban spaces can also be planned to a certain extent. Physical density, for which there are clear parameters, can be combined with a density of events and experiences. A prerequisite for planning is that the architectural imprint takes into account the human scale of experience and creates a framework for diverse uses and possibilities of appropriation.

Fig. 2.4 Townhouses Berlin

3 Identity

Since the 1980s, in addition to the concept of image, another concept has entered the discussions about urban design: urban identity. One of the fundamental works in this field was published by George Herbert Mead in 1973 under the title "Spirit, Identity and Society." According to Mead's theory, identity is not an innate quality of man. It is acquired by an individual by reflecting his or her process of experience and activity within a society. Identity is formed through communication with society.

The city itself cannot have an identity in the original context of use of the word. However, the city can very well have an identity-forming effect. What is meant is the uniqueness of a city. Cities can be distinguished by numerous characteristics such as their history, culture, political leadership, and urban appearance. The greatest importance is attached to features that are considered to be particularly characteristic and typical, such as the sequence of places, materials, etc. This makes it possible to differentiate a city from other cities. The identity frees them from their anonymity. In other words: the more concise the city identity is, the more clearly it can be distinguished from its competitors and made recognisable again.

> Identity means conformity with all characteristics or emotional attachment to persons, concrete or ideal objects (Klaus Humpert)

Bearer of Identity
Identity is expressed both in the built signs, in striking landscape elements, and in the urban context, on the level of the neighbourhood.

Buildings and ensembles that have the character of landmarks of a city can become identity bearers for a city and its surroundings just as much as a specific assignment of buildings to each other. In the same way, a landscape can be so striking and unique that specific features such as topography, water, or forest create unmistakable situations with a high recognition effect.

The neighbourhood can also assume a function as a carrier of identity. As a spatial unit with special reference points such as church, care centre, etc., it becomes a frame of reference for the residents at the neighbourhood level.

An object in the sense of both a striking building and a landscape element can take on such a role, but of course only if it is deliberately placed in its environment and given a high-quality appearance. The same applies to a larger unit in an urban planning context.

Source of Identity
Particularly in the fragmented urban landscape, reference is repeatedly made to the importance of the individual, special building as a source of identity. Today, identity is a key concept in the constructive handling of our urban condition. Everything that constitutes the identity of a city, an urban space, including its symbolic value, deserves first priority. We need elements that make the city unmistakable. The typical characteristics of a city are the most valuable things that cities can and should "grow"

with. The identity-forming element must compete against the mass product, stand out from the masses without claiming that the connection to the environment is unimportant.

The meaning of identity must not stop at this spatial dimension, even if social relations, which play an important role in the formation of identity, are not dealt with more strongly here. Identity has something to do with the possibility of appropriation of spaces by humans, the building of trust through the recognisability of familiar images.

Even Kevin Lynch came to the conclusion that only a memorable image of the city, that is, an image that is clearly remembered by the inhabitants, can produce an identity that in turn implies identification of the inhabitants (see also, Lynch 1965, p. 18). The image of the city contains three components that are essential for the memorability of the image: identity, structure, and meaning. All three elements are connected with each other. Identity in this context means a clear recognition of the place. The place is detached from the context by the aspect of individuality and is thus identified. Furthermore, the viewer must be able to establish a spatial and/or structural relationship to the place, which in turn is linked to a sense of meaning. The existing structure, i.e. the physical space, determines the identity of the place through its recognizability, meaning, and sense.

Identity and Identification

In his studies, Kevin Lynch describes the meaning of identity with the inherent identification of the inhabitant and establishes the connection between identity, structure, and meaning as factors for a possible image production.

The appearance of the city is, therefore, not only dependent on the design but also on whether the population identifies with the city. The identification with a place depends on its usability and the possibilities of appropriation. The connotation of a place, which Lynch addresses, can result from experiences, which are in turn influenced by the structural, can be caused by the physical space.

> A usable image first requires the identification of an object, which makes it possible to distinguish it from other objects and to recognize it as a separate "being". We call this "identity"—not in the sense of agreement with anything else, but in the sense of "individuality". Secondly, the image must contain a spatial or structural relationship of the object to the observer and to other objects. And finally, the object must have some meaning to the observer—either practical or emotional. (Kevin Lynch)

Lynch integrates the users as active components of the built urban space. They are actors on the public stage and thus part of the action, part of the spatial structure, and observers of the spectacle. Through the processes of identity formation, the action in the urban space thus presents itself as a multiple-relevant synthetic achievement.

All in all, an influence on identity seems to be possible only to a limited extent because—in addition to the living conditions of the inhabitants—it is also based on many changeable influences such as the geographical location, the climate, and the natural environment. Identity is rather an impression of the city built up over a longer period of time, which is the basis for the inhabitants' identification with their city. In contrast, the image is more dynamic, can be influenced and can sometimes be short-lived.

Fig. 2.5 The Dortmunder U—Center for art and creativity

4 Aesthetics

Aesthetics (from Greek aisthesis: perception) means, from its origin, the doctrine of perceptible beauty as well as the harmony in art and nature.

The concept of aesthetics, which today is widely discussed in architecture and urban planning, was already used in antiquity by the philosopher Aristotle. For the visual arts and architecture, since the Renaissance, there has been an attempt, still recognized by art history today, to derive the sense of beauty anthropomorphically from the proportions of the human body. The "golden section" was used as basis for design. The ideal proportion is that a person perceives the division of an area or distance in a ratio of about 3:5 as particularly harmonious. From Leonardo da Vinci and Palladio to Schinkel and Camillo Sitte, architects and urban planners used this design principle. Le Corbusier also used this principle in the development of the MODULOR, in order to give architecture a mathematical order based on the human measure. In this way, the beauty, the aesthetics of proportions made manageable, could be objectified in a certain way.

In the eighteenth century, the philosopher Alexander Gottlieb Baumgarten developed his own theory of aesthetics. He defined aesthetics as the theory of "sensual cognition" and understood it to mean above all sensual perception and the comprehension of impressions.

The philosopher and writer Umberto Eco sees a close interweaving of the beautiful and the good:

> 'Beautiful' is—besides 'graceful', 'pretty' or even 'sublime', 'wonderful', 'splendid' and similar words—an adjective we often use to describe something we like. It seems, then, that what is beautiful is identical with what is good, and indeed there has been a close connection between the beautiful and the good in different periods of history. However, if we judge by our everyday experience, we tend to call good not only what we like, but also what we would like to have. (Umberto Eco)

In this sense Eco considers the similarity between the beautiful and the good from the aspect of the desire for what one likes.

In architecture and urban planning, the discussion about beauty and aesthetics is often dismissed with reference to questions of taste, although this aspect has a major influence on the atmosphere of urban space, its quality, and acceptance.

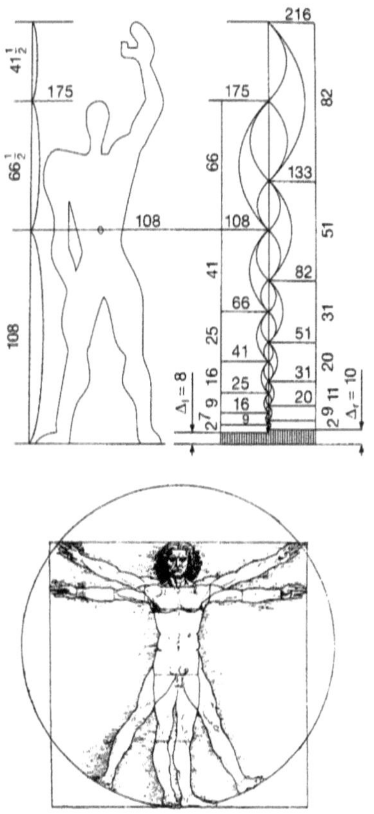

Fig. 2.6 Human dimensions. Modulor, Le Corbusier (above), Leonardo DaVinci

5 Building Culture

The history and culture of a society are always reflected in its cities, its buildings, and its public spaces. At the same time, every new building is an expression of social esteem and wishes for the future. Buildings and spaces build bridges between generations and, over long periods of time, still reveal the wishes, efforts, responsibilities, and also the failures of past realities (Durth and Sigel 2009, p. 9). The way we use these bridges—our buildings and urban spaces—and how we deal with them expresses our own self-image. This is called building culture.

Since 2001, the concept of building culture has again been increasingly addressed by politicians and experts in the field of planning and building. The debate about the necessity, the characteristics and the benefits of building culture is not fundamentally new. In his work "Baukultur," Durth presents various developments in building culture in the twentieth century, shows how the term was used, strained, misunderstood, and strategically related, and how the built environment often presented the concept of building culture as a distorted image.

Today, the subject of building culture is increasingly being discussed as a factor that determines architectural quality in addition to external and internal design and the inclusion of the spatial context: flexible usage options, economic viability, and the consideration of ecological concerns. In addition, contemporary debates are attempting to involve the general public to a greater extent in the discussions, which up to now have mostly been conducted among experts, in order to achieve a social understanding of the qualities of the built environment. The aim of these efforts is to increase the appreciation of living spaces as a whole and to develop and raise awareness of the built environment. The shape of the built environment makes a major contribution to how the inhabitants of a city or region identify with it, how intensively they deal with their city and its transformation and how responsibly they deal with the spaces they use (see also Burgdorff 2006).

A high level of building culture is the result of dynamic and constantly evolving quality criteria. However, progress in building culture is not to be understood as a goal-oriented development process toward the best solution, but rather as a successful search for the appropriate solution in each case, which must be determined by the changing tasks and social needs (see also, Durth 2009, p. 10). Designing and creating spaces is a creative process in which a conscious decision is made as to which form or appearance a building, an ensemble or even an urban space. Everything that surrounds us has been thought and designed by someone. In the built environment, design is the key to many decision-making processes. At the strategic level, it generates visions for new spaces and structures. On the operational level, it defines how our environment functions, looks and feels.

PROCESS	Concept, Planning, Realization, Operation, receive, redesign, demolition, Waste disposal/recycling	PRODUCT	built, infrastructure, landscape/open space, salt, functions
INDICATORS PROCESS		INDICATORS PRODUCT	
	professional execution		measured introduction into the context
	appropriate participation		respectful treatment of the history of the place
	high transparency		aesthetic salt
	interdisciplinary cooperation		high authenticity and identification potential
	creative process design		flexible usage possibilities
	Quality demanding procedures		nutritious innovation
	economic viability		economic viability
	deadline and cost-safe procedure		good marketability
			resource efficiency, ecological issues

Fig. 2.7 Fields of observation and indicators of building culture

In the concrete handling of building culture, the consideration fields "process" and "product" offer possibilities of for understanding. The entire life cycle (planning, realization, preservation, demolition, disposal) and all relevant scale levels (object, environment, region) are addressed.

> Building culture is product and process: buildings, city, built landscape are just as much the subject of the concept of building culture as their production and the way we deal with them. (Klaus Selle)

This high value of the process concept leads Durth to the thesis that building culture is invisible in essential parts, but that its quality is measured by the visible result—the product. The process concept refers to the entire spectrum of planning, realization, and redesign. Next to it is the concrete object, that is, the product that is created by the process. The product is also based on a comprehensive consideration of the built environment, infrastructure, and landscape.

Questions of taste and fashion are often at the centre of discussions about building culture. Tastes vary and fashions change frequently. Something new and unfamiliar is usually accepted as beautiful as time passes and familiarity grows. However, the foundations for quality design are clearly defined and outlast fashions and personal tastes. Quality design in the sense of building culture is based on three basic principles:

- Robustness + durability;
- Usefulness + efficiency and,
- Beauty + the ability to please people.

If these basic principles are adhered to, good conditions for an increase in urban building culture are given. The focus should be on an urban design that places great value on suitability for everyday use in a comprehensive sense, on sustainability and durability, on their usability and social acceptance and adaptability, on economic efficiency and, last but not least, on the quality of design appropriate to the specific location.

6 The Interaction of the Basic Concepts

How do these terms—sustainability, urbanity, identity, aesthetics, and building culture—interact in the sense of a design claim in urban planning?

The terms presented play a central role in urban design. They can—each considered individually—be an objective in the development of urban planning concepts. However, it is much more important that they produce quality in their interaction.

Thus, sustainable aesthetics means that—free from architectural fashions—a high quality of design is produced over a long period of time and with a minimized use of resources. Likewise, the identity-creating building culture, which is created through communication with the actors and participants, is a central building block of the city.

It may not always be about the spectacular answer, but sometimes simply about the modest, appropriate answer. To find or make the usual and ordinary interesting is just as much a task:

> For this is the amazing thing, that despite all the ugly buildings, despite the noise, despite all that can be blamed on it, to whom whoever wants to see it, the great city is a miracle of beauty and poetry, a fairy tale, more colourful, more varied, more varied than any that any poet has ever told... (AugustEndell)

Literature

1. Sustainability

Agenda 21, Deutsche Fassung 1997: Bericht der Weltkommission für Umwelt und Entwicklung 42/87, 1987 (Originaltext in Englisch).

2. Urbanity

Albers, Gerd; Hoerkens, Beate (Hrsg.) 1991: Urbanität in Deutschland. Stuttgart, Berlin, Köln: Kohlhammer.
Frech, Siegfried (Hrsg.) 2010: Urbanität neu planen. Stadtplanung, Stadtumbau, Stadtentwicklung. Schwalbach/TS: Wochenschau Verlag.
Lampugnani, M. Vittorio (Hrsg.) 2007: Städtische Dichte. Zürich: Verlag Neue Zürcher Zeitung.

3. Identity

Lynch, Kevin 1965: Das Bild der Stadt. Berlin: Ullstein.

4. Aesthetics

Eco, Umberto (Hrsg.) 2010: Die Geschichte der Hässlichkeit. München: dtv.
Henckmann, Wolfhart; Lotter, Konrad (Hrsg.) 2004: Lexikon der Ästhetik. München.

5. Building Culture

Burgdorff, Frauke 2006: Gestalt geben. In: 5 Jahre Landesinitiative StadtBauKultur NRW. Gelsenkirchen, Düsseldorf.
Durth, Werner; Sigel, Paul 2009: Baukultur. Spiegel gesellschaftlichen Wandels. Berlin: jovis.

A Journey: From History to Theory

3

1 The History of Urban Development

Urban design requires knowledge of the historical processes of urban development: cities and urban layouts are reading books of history and contain important empirical knowledge about the characteristics and laws of the city. This knowledge is indispensable because many current urban design tasks can only be adequately dealt with in knowledge of the history of a city and the history of ideas of urban design.

The First Forms of Urban Organization

For about 5000 years, people have lived in urban forms of organization. In the course of history, they have created diverse living spaces for themselves, which are an expression of their respective social, economic and cultural conditions. The city, as a comparatively young spatial entity, has differentiated itself in very different ways during these past 5000 years: Not only have different (developmental) stages emerged, but also very different cultures of cities.

The basic **triggering factors** for the development of spatial and building structures can basically be divided into two categories: natural and social conditions.

1. **Natural conditions**:
   - Climate conditions
   - Topography
   - Availability of resources (e.g. building materials)
2. **Social conditions**:
   - Social forms of organization
   - Forms of governance
   - Protection and safety
   - Claims for use

C. Reicher, *Urban Design*, https://doi.org/10.1007/978-3-658-34370-5_3

While the climate (rain, heat...) and the available resources (stone, clay...) can lead to a strong differentiation of building typologies, social conditions tend to find expression in special urban spatial concepts in connection with public buildings or representative buildings. In addition, the different demands on the use of space result from economic and technological developments, which in the past have primarily led to a differentiation of zones of use within a city. For example, the spatial-functional separation of residential and workplaces has enabled the development of a wide variety of building typologies (residential, commercial, utility and transport buildings) in the course of settlement history.

In the settlements of primitive peoples, it is mainly the natural conditions that can be read, but also the need for a protected interior area. The huts of many indigenous peoples are uniform spaces in which domestic weaving takes place. There is hardly any zoning of different areas of activity here.

The village complex of the Yanomami Indians separates itself from the jungle, which is perceived as hostile, with closed back walls. It is arranged in a circle around a village square. This settlement is set up for a limited time, for the duration of the settlement of its inhabitants.

In the African settlement in Cameroon, which is also grouped around a circular centre, both functional specialization and social differentiation are visible.

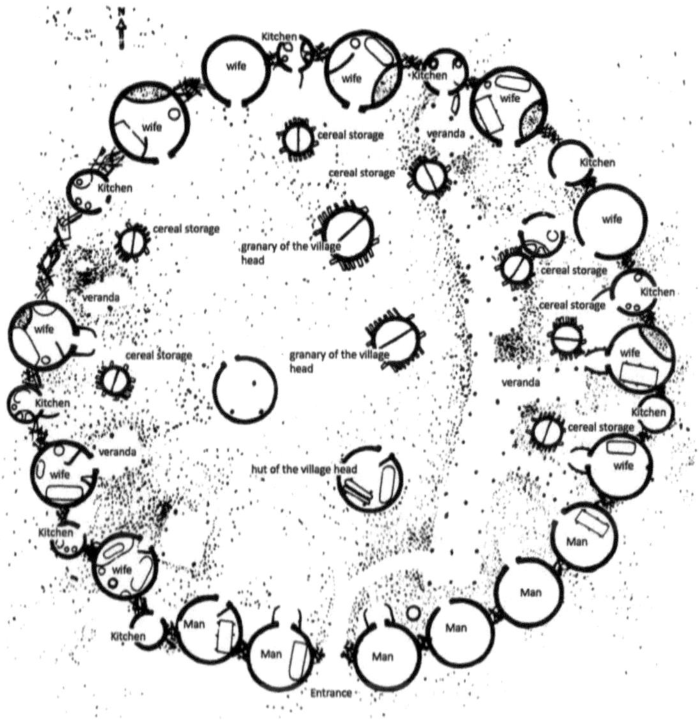

Fig. 3.1 Settlement in present-day Cameroon, Africa

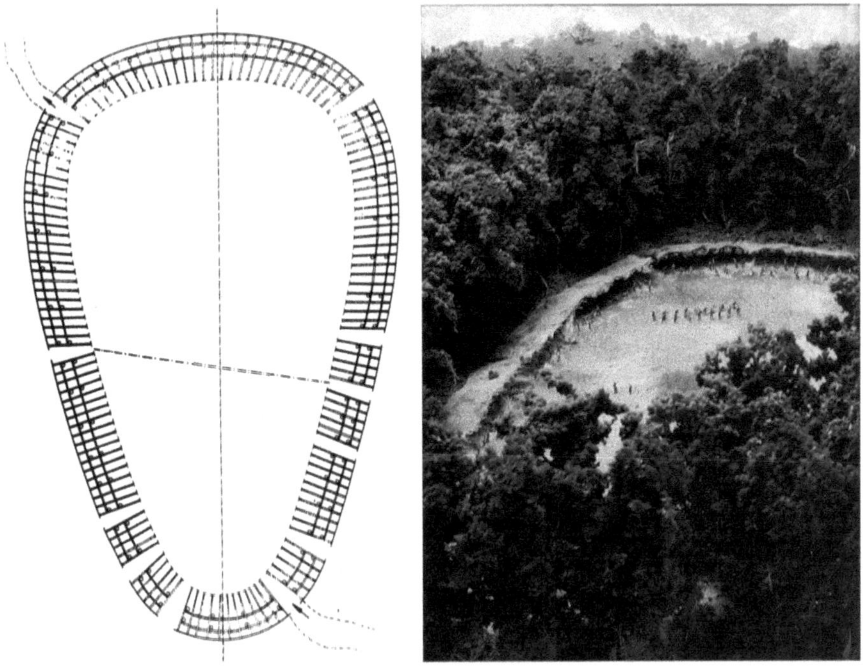

Fig. 3.2 Weather roof settlement of the Yanomami Indians, Brazil

1.1 The Origin of the City

The city of Jericho on the west bank of the Jordan River is considered one of the oldest continuously inhabited cities in the world. It was founded as a settlement around 8000 BC and most likely originated as a trading centre and not as an agricultural town. The Turkish settlement of Catal Huyuk is one of the first known cities besides Jericho. Here, also at the beginning of the eighth millennium BC, a new form of organization of social coexistence was formed. In the following millennia, the focus of the settlements in the Near East shifted to the southern river valleys between the Euphrates and Tigris. Here, since the fourth millennium BC, steadily growing urban centres have developed, with cities such as Ur, Erbil or Babylon.

The ground plan of the excavated city of Ur already shows features that justify the term "city" spatially and functionally:

Temonos = political power
Port and canal = trade and traffic
House and palace = social structure
City wall = protection and safety

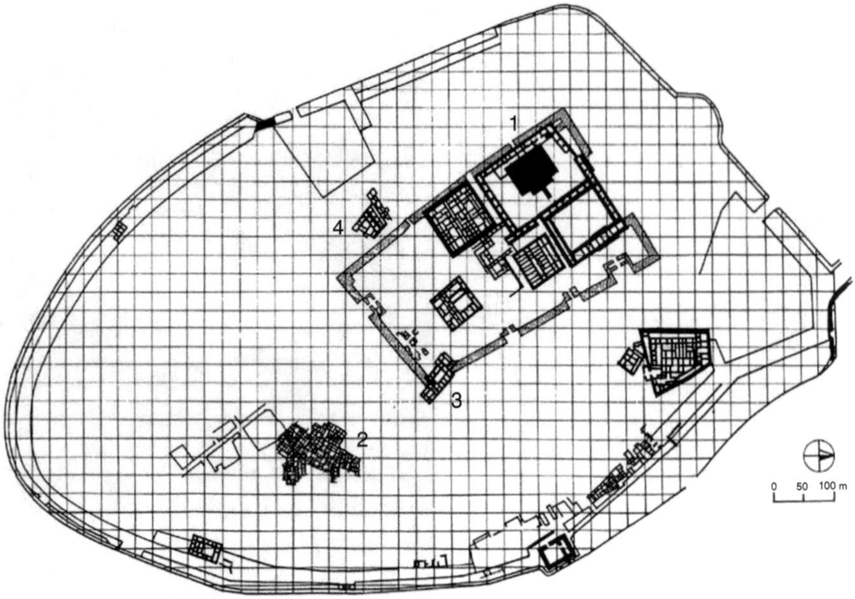

Fig. 3.3 Location of buildings and districts in the city of Ur, Iraq

Fig. 3.4 Aerial photograph of the city of Erbil, Iraq

Fig. 3.5 Catal Huyuk excavation area within the 17 m high and 13 ha large earth mound of the former city (above), excavated buildings (below)

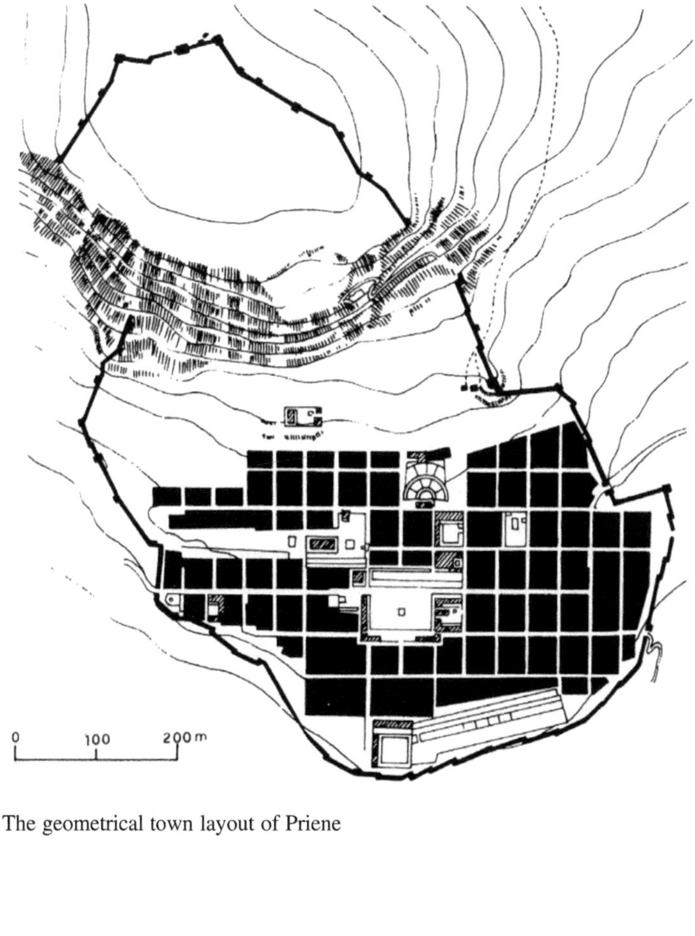

Fig. 3.6 The geometrical town layout of Priene

Fig. 3.7 Warendorf

The free city in Greece, on the other hand, represents a special feature because it already represents a novel character of social coexistence. From its basic structure, the city forms a spatial unit that is clearly demarcated from the outside. The actual city area is divided into three areas: the private area of the residential buildings, the sacred area with the temples and the public area for meetings and events. Hippodamus of Miletus is considered the inventor of the "equal division of the city", in which "one part was intended for culture, the second for public purposes and the third for private property". The plans for the expansion of Piraeus, Miletus and Olynth originate from him. These cities were planned according to a strictly geometrical principle, into which the individual houses had to fit.

The geometrical layout of the city was a means of enabling efficient use of the area, expanding the built-up areas step by step by adding blocks of houses, and thus systematically controlling urban development.

Even in these first cities, it is clear that the principle of spatial, economic and social division of labour was of central importance for the development of the city. It formed the foundation of urban culture and made the concentration of the population in a confined space possible and necessary.

City Foundations in Germany

Around 100 AD, social groups living in present-day Germany were mainly divided into tribes with a culture that was predominantly oriented towards hunting, agriculture and military technology. In these natural or agrarian cultural landscapes, the Saxon settlement of Warendorf in Westphalia was an "island of civilization". The settlement existed until 800 AD and was then abandoned. Surprising was the multitude of different house layouts. Rural settlements like these formed the islands of civilization, which either already existed since the Germanic and Celtic early times or were created in the course of clearing forest areas.

Roman city foundations in present-day Germany at the turn of the century, which had no urban culture, were therefore technical and civilising innovations. Urban lifestyles were therefore initially hardly able to assert themselves. It was only with the construction of armies from which cities gradually developed, the construction of residential and religious buildings made of stone and the application of new technologies, for example in road and bridge construction, that the social and political forms of organization in present-day Germany gradually changed.

The most striking urban structural feature of Roman town foundations is their strict geometric grid, which determined the architecturally closed and extensive city. However, the decline of the Roman Empire did not stop at its city foundations: the city of Trier, one of the four capitals of the Roman Empire, fell into disrepair after its heyday between 250 and 400 AD due to the destruction that began in the fifth century. In the course of the following centuries, an irregular town layout, typical of medieval town foundations, developed on the formerly rectangular street grid.

1.2 The Medieval City

The foundations of many cities were laid in the Middle Ages; the German urban system in particular can be traced back to the numerous medieval city foundations. Important features of medieval towns are the clear separation between town and country (inside and outside), the functional division of the urban area (merchants' neighbourhood, craftsmen's neighbourhood etc.) and a frequently irregular urban structure compared to the grid-shaped Roman town foundations. Seen from the outside, the cities with their closed ramparts and walls appeared as compact, almost object-like structures in the middle of an otherwise open and extensive rural environment. At first glance, the vertical cityscape with its large number of church and defence towers seemed even more impressive than the close-knit inner city of small squares and narrow streets.

Most medieval town layouts are based on irregular development and building patterns and therefore appear unplanned; nevertheless, they did not emerge by chance, but follow local requirements and special features. However, there were also "planned cities" with regular urban ground plans, mostly in the case of completely new towns or large-scale urban expansions.

Despite this, until the end of the Middle Ages, "city" was, from a design point of view, primarily an arrangement of individual buildings. In the process, different spatial forms of organization of the individual houses appeared: The house presents itself as an individual element and approaches the street with its facade, its "face".

> In the Middle Ages, a decisive and formative basic element of all conceptions of the city was the house, namely the individual house, the 'house personality,' which stood out from its neighbours, differed and yet was arranged in the row by form and detail and material, the rowed individual house, but not the row house. (Wolfgang Rauda)

The example of the Place des Vosges in Paris (started in 1605) shows how a square front is formed by individual buildings. The individual building can still be read, it has its own hipped roof, and a wider arcade support separates the houses from each other. Unmistakable is the creative idea that the individual house becomes a square wall. In the centre of the square stands the equestrian monument of the king; the facades of the buildings form the framework for its spatial staging.

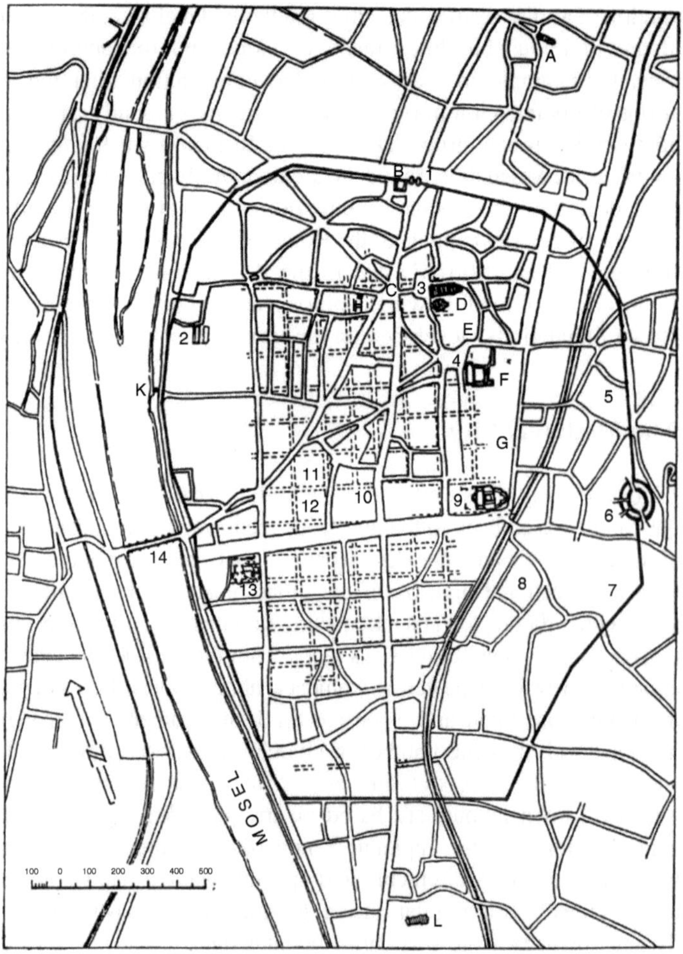

Fig. 3.8 Trier, plan of the regional capital of the late Roman Empire

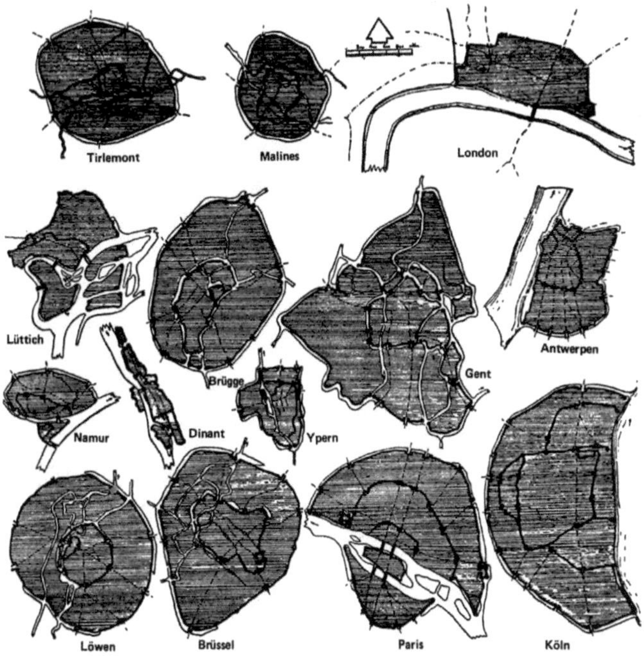

Fig. 3.9 Plans of northern European cities with the erected wall rings

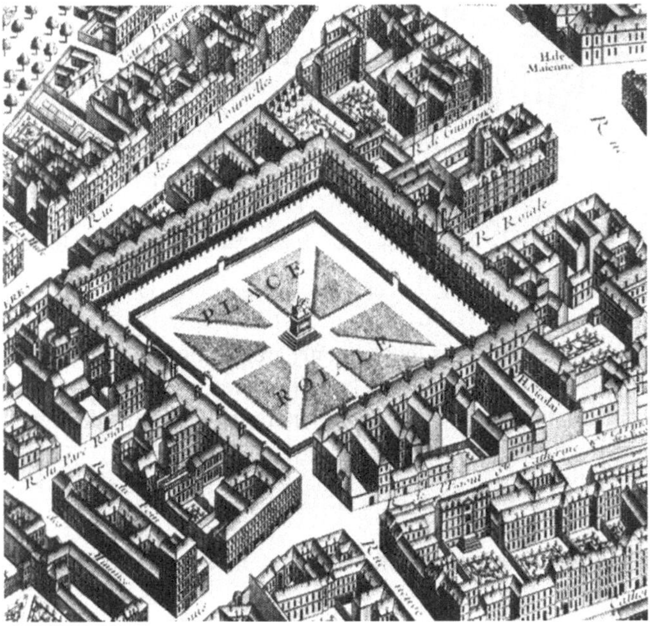

Fig. 3.10 Place de Vosges, Paris, 1650

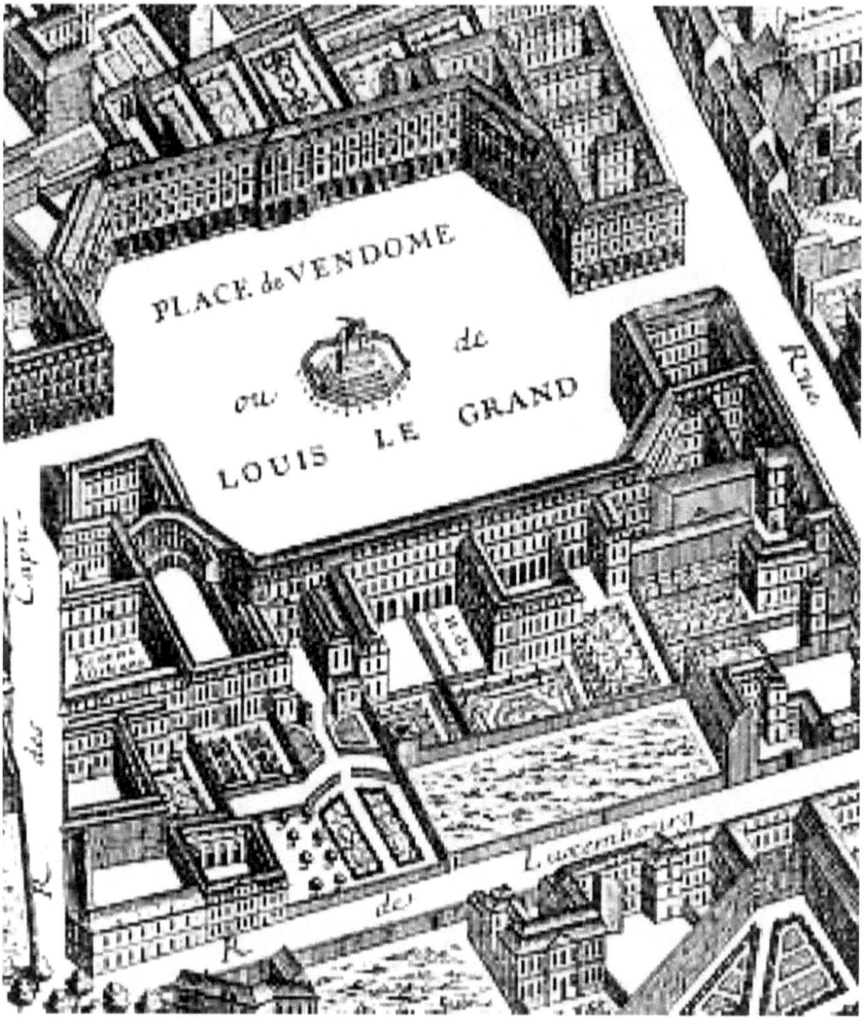

Fig. 3.11 Place de Vendome, Paris, part of a plan by Turgot from 1731

The urban design concept is thus not only created by the repetition or addition of volumes, but also by the position the building takes in relation to the public space. This insight is extremely important for urban design tasks that deal with integration into existing structures.

1.3 The City in Absolutism

In the early fourteenth century, the art historical epoch of the Renaissance spread throughout Europe, starting in northern Italy. The concept of the Renaissance (rebirth) ties in with the rational teachings and values of antiquity and places human dignity at the centre. With regard to urban design and architecture, the Renaissance concept is made visible by the rediscovery of the teachings of Vitruvius (88–26 BC) and the "Ten Books on Architecture" ("Zehn Bücher über die Baukunst") by Leon Battista Alberti, who coined the term "city as a work of art". In this epoch, which is characterized above all by the artistic idealization of the city, new urban structures, but above all urban utopias, have emerged.

The ideal city for Albrecht Dürer's "City of the King" follows the idea of a square fortress city with a castle in the middle, surrounded by residential buildings and a surrounding moat. Regularity, geometry and formal principles of order are developed as a deliberate break with the prevailing traditions. Ideal cities are characterized by the schematic regularity of the urban layout; the defensive aspect determines the outer form.

The Renaissance did not produce a city type of its own, but the additive principle of medieval cities is replaced by a designed greater order in which each building takes its place as part of a final whole.

In the Baroque period, these design principles expanded to a new cultural achievement: the spatial perspective of the Baroque. It first appeared as stage design; some of the new urban planners were also stage designers (e.g. Bernini). The new city, according to Lewis Mumford, was the "attempt of a stage design" for the staging of power:

> In the new city, the building is a frame for the avenue, but the avenue is first and foremost a parade ground—a place where spectators gather on the sidewalks or at the windows to watch the parade. . . .to watch the army's triumphal marches, to be duly awed and intimidated. (Lewis Mumford).

Since the idea of the baroque city comes close to a three-dimensional large-scale sculpture, the creative scope for the individual buildings to take on a life of their own is very limited. Symmetry is one of the most important design rules in the Baroque. This also applies to rooms and elements shaped by the landscape: they are subject to the same strict ideas of order and design repertoire. Versailles and Karlsruhe are particularly striking examples.

While in Versailles, the open spaces were designed according to Baroque principles, in Karlsruhe, the radial aisles leading to the castle follow a different but similarly rigid geometry.

1.4 The Gründerzeit City

The next major surge of change for European cities began with industrialization and the urban expansions of the Wilhelminian period, particularly in the period from the middle of the nineteenth century. The enormous growth of the economy and population led to extreme structural densification in all places. Cities expanded mainly through multi-storey building blocks on a grid-like ground plan, making maximum use of the land. Especially in overcrowded, poorly lit and ventilated tenement blocks in big cities, this spatial densification is accompanied by miserable living and housing conditions. The planning philosophy of those years is characterized by an engineering-like, rational-technical understanding of urban design: The "Systematic Urban Expansion" in Barcelona or the "Inner Urban Expansion" in Vienna are such rational, large-scale, almost anonymous settlement patterns that hardly show any specific relationship between context and object or urban space and singular building.

Fig. 3.12 Versailles, view around 1668 by P. Patell

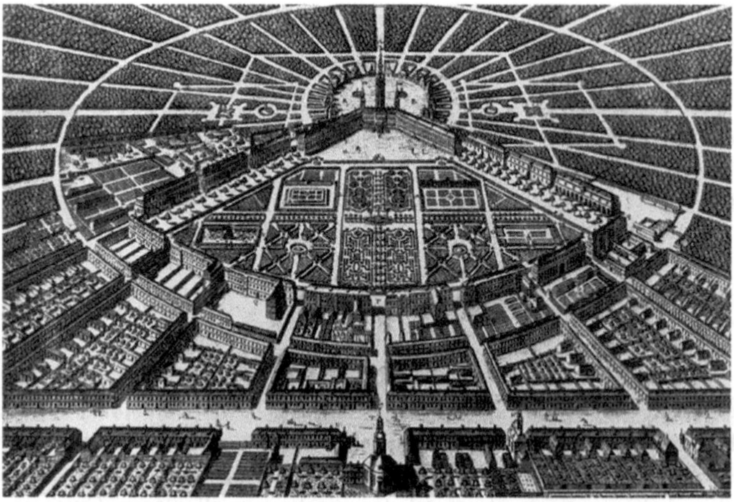

Fig. 3.13 Karlsruhe, city complex with castle, engraving by Christian Thran, 1739

At the end of the nineteenth century, Camillo Sitte took up this theme of the interplay between context and object in his book "Urban design according to its artistic principles" ("Der Städtebau nach seinen künstlerischen Grundsätzen"; 1889). He refers to the Middle Ages and the Renaissance as "... there was still a lively practical use of city squares for public life and, in this context, there was also a correspondence between these and the adjacent public buildings...". Just as Camillo Sitte analyses the layout of squares in terms of their spatial impact and draws conclusions for urban design, he makes recommendations for the position of buildings. Camillo Sitte emphasizes public space as an important connecting component, which decisively shapes the relationship between context and object.

> In modern urban construction, the relationship between built-up and empty floor space is just about reversed. In the past, the empty space (streets and squares) was a closed whole of calculated form; today, the building plots are distributed as regularly closed figures, what remains in between is street or square. (Camillo Sitte).

1.5 Modern Age Urban Design

With the beginning of the twentieth century, urban design debate has focused on improving housing conditions and reforming urban life. Classic Modernism and its idea of functional urban design was—just like Ebenezer Howard's garden city—an immediate counter-design to the dense metropolis of the fading nineteenth and beginning twentieth centuries. Leading architects and urban planners saw a completely "new city" as the key to the realization of a modern, contemporary way of life. These visions of modernity were discussed in 1933 in the International Congress of Modern Architecture (Congrès Internatiaux d'Architecture Moderne, CIAM) and summarized by Le Corbusier in a paper known as the "Charter of Athens" ("Charta von Athen"). Among other things, this paper propagated the separation of the urban functions of living, working, leisure and transport, and formulated the idea of a "structured and loosened-up city"—in other words, those principles that had a decisive influence on urban design and development in Germany during the reconstruction phase after the Second World War.

Conceptually, the city model of modernity is tantamount to a reversal of the previously prevailing ideas of the city. It manifests itself in the image of a dispersed collection of autonomous and solitary buildings in a uniform, open spatial continuum. Public space primarily serves the increased mobility requirements within and between the various functional areas of the city; traditional elements of urban space such as streets are also deconstructed along functional requirements or mutate into "parkways" following the ideal of open space.

The claim and reality of the city of modernity are often appreciated, but at least as often vehemently criticized:

> ... in its present form, which has not disappeared, the city of modern architecture, as an accumulation of strikingly incompatible objects, has become as problematic as the traditional city it sought to replace. (Colin Rowe)

1.6 Reconstruction and New Models

After the Second World War and its far-reaching destruction of many cities, the necessary reconstruction is both an occasion and an opportunity for the development and realization of new urban design models. The post-war period can be divided into two important phases:

1. **Reconstruction**: according to old and new concept (ca. 1945–1955)
2. **Implementation of new models**: loosened-up city models (ca. 1955–1970)

However, the reconstruction of cities is determined by a number of factors, which often leave little room for a completely different urban layout:

- Existing roads, with their underground technical infrastructure, represent the most important resources for reconstruction, as in many cases they have been preserved despite the above-ground war damage.
- Existing land tenure conditions decisively determine the possibilities for the reconstruction of buildings and urban districts.

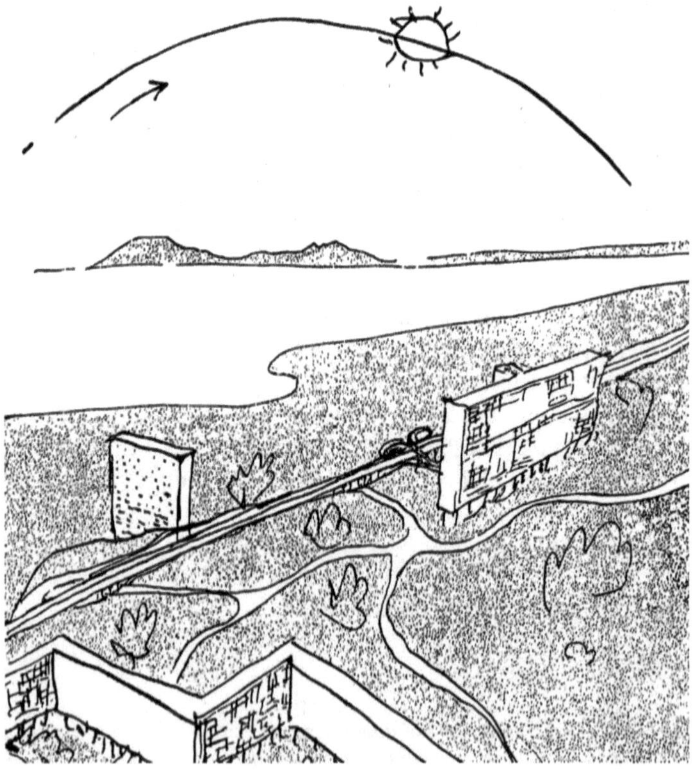

Fig. 3.14 Sketch for residential units, Le Corbusier, 1942

- The undestroyed building stock is an important determinant in all reconstruction plans. This applies in particular to buildings that are classified as cultural heritage worthy of preservation.

The possibilities to create completely new urban structures with the reconstruction are therefore not available everywhere. Characteristic is rather the ambivalence between the historical town layout and new urban architecture, when many traces of the old town are preserved, but the town has to meet new ideas and new demands at the same time.

The "structured and loosened-up city" formulated by Göderitz, Rainer and Hoffmann in 1957 is the most important urban design model in this phase. It paints the picture of a city divided into individual settlement areas, loosened up by green spaces and closely linked to the surrounding recreational landscape. The high-density city centre is abandoned in favour of several urban units; the model goes hand in hand with the spatial distribution of the various functions, as provided for in the Athens Charter.

Other models of that era, such as the "organic urban architecture" or the "car-friendly city" (Hans Bernhard Reichow), try to formulate a new model of city and traffic ("man and car") in view of growing mobility and associated effects of car traffic.

As a counter-design to the structured and loosened-up city with its flowing spaces in so-called cityscapes, which are criticized as less urban, the urban design model of "urbanity through density"—also known as "vertical urban design"—became more and more established in the 1960s. High density is seen as a necessary precondition for urbanity. The corresponding planning often leads to new satellite towns or large housing estates, which, however, already a generation later gained a reputation of being urban design mistakes: Above all, the promise of urbanity has rarely been fulfilled.

With the "European Year of Monument Conservation" in 1975, the attitude towards existing building stock changes because in urban development and architectural planning, much more attention is paid to the historical context of planning and building projects. Extensive demolitions in the context of urban redevelopment projects, as they took place in the 1960s, are now hardly conceivable. Decidedly urbanistic thinking returned to urban design practice in the 1980s at the latest—supported by the programme of the International Building Exhibition in Berlin (1984–1987). The inner city plan in Berlin, which was launched at the end of the 1990s, is the best-known example in this country of an attempt to orientate the further construction of the city on the historical dimensions of urban spaces and on "urban components"—instead of individual buildings. At the same time, at the end of the 1990s, the USA urban design reform movement of the New Urbanism also made its way into Europe. New Urbanism criticizes an urban design that is solely committed to functional and economic interests and that has led to "urban sprawl" in the USA. This reform movement contrasts the sprawling urban landscape with an understanding of urbanity that is based on the model of the small or medium-sized city with well-proportioned urban spaces and functional neighbourhoods. The projects realized in Europe with reference to the reform movement, however, often bear witness not only to their socio-spatial exclusivity, but also to nostalgic, scene-like design repertoires that trace the cityscape of small historic towns.

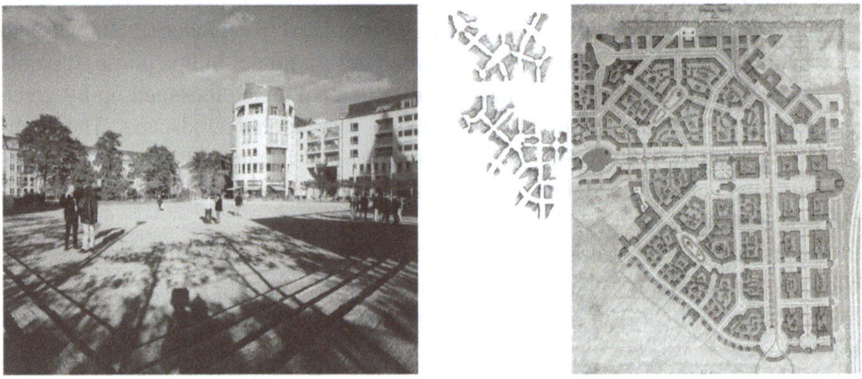

Fig. 3.15 Kirchsteigfeld, Potsdam, Krier Kohl Architects

Fig. 3.16 Inner city plan of Berlin

2 Urban Design Theory

Just as knowledge of the history of the city's origins is an important basis for urban design, so is knowledge of urban design theories: Theories and scientific or artistic positions associated with them expand the necessary thinking space for a profound urban design practice. The theoretical approaches selected here deal with the built and spatial organization as well as the structure of the city. They analyse the problems and tasks of the respective epoch and use them as a starting point for urban design concepts for the future, with which—in critical retrospect—new problems and tasks often went hand in hand.

2.1 Ildefonso Cerdá

Ildefonso Cerdá (1815–1876) is considered the old master of urban design theory in the industrial age. Based on his practical experience with urban design projects, he arrived at a theory of urban design and even at the conviction that urban design is a science. "And I therefore see the need to explore, justify and consolidate the foundations and principles upon which this science must be built" (Cerdá 1867, Volume II, p. 1).

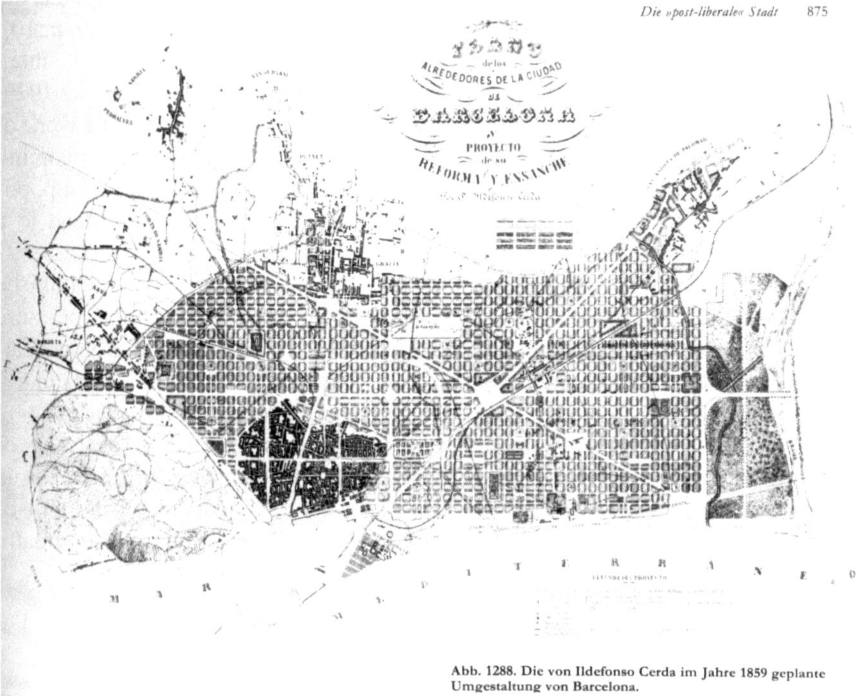

Fig. 3.17 Barcelona urban expansion, Ildefonso Cerdá, 1859

Cerdá became internationally famous mainly because of its 1859 urban expansion plan for Barcelona, which is structured by succinct diagonal streets.

In Cerdá's planning and theory, basic needs of people (sunlight, ventilation, transport, etc.) are placed at the centre of the considerations. In addition to this social orientation of urban design, which he regarded as a "great social necessity", he was one of the first theorists to deal with the networking of urban structures. He saw establishing relationships and connections between different parts of the "action space city", i.e. between the city centre, the suburbs, the periphery and the region, as an important functional and spatial challenge. His proposals for streets that were optimized for pedestrians and light railways that were unknown at the time reflect this then innovative approach to the networking of urban spaces. Cerdá's concern was to develop a "perfect" urban design, whose strict theoretical principles, however, must be verified in practice.

2.2 Le Corbusier

Le Corbusier (1887–1965) is considered the founder of the functional city. In his "Principles of Urbanism" and in the Charter of Athens, which he was instrumental in shaping, the criticism of the traditional city and its inability to adapt to technical achievements and future requirements is strongly expressed. His plea for "perfect order" is underpinned by his criticism of existing structures: "The curved road is the way of the donkey, the straight road is the way of man" (Le Corbusier 1929, p. 10). He bases his desire for order on human needs: "The more perfect order is, the more comfortable he (man) feels. . . . The human work is called order" (Le Corbusier 1929, p. 20). "At the highest levels of creation, we strive for the purest order: the work of art" (Le Corbusier 1929, p. 21). Le Corbusier drew the foundations for his planning and his theory of urban development from statistical forecasts, especially on population development and the increase in private transport, which would lead to a "congestion" of the inner cities.

For him, modern urban planning was based on the idea that living should be separated from all other areas of life and that the city as a whole should be more strongly divided into its four areas—living, leisure, work and transport. The radicalism with which Le Corbusier rethought and planned the city was particularly remarkable. His "Plan Voisin" (1925) was based on the idea of making the existing city disappear completely to create the new city centre for Paris, consisting of skyscrapers, on the new "Tabula rasa". Le Corbusier was an advocate of the utopia of perfect planning. For him, urban design was an experiment with the aim of remedying the existing shortcomings of the city, but at the price of the equally existing urban spatial and atmospheric qualities.

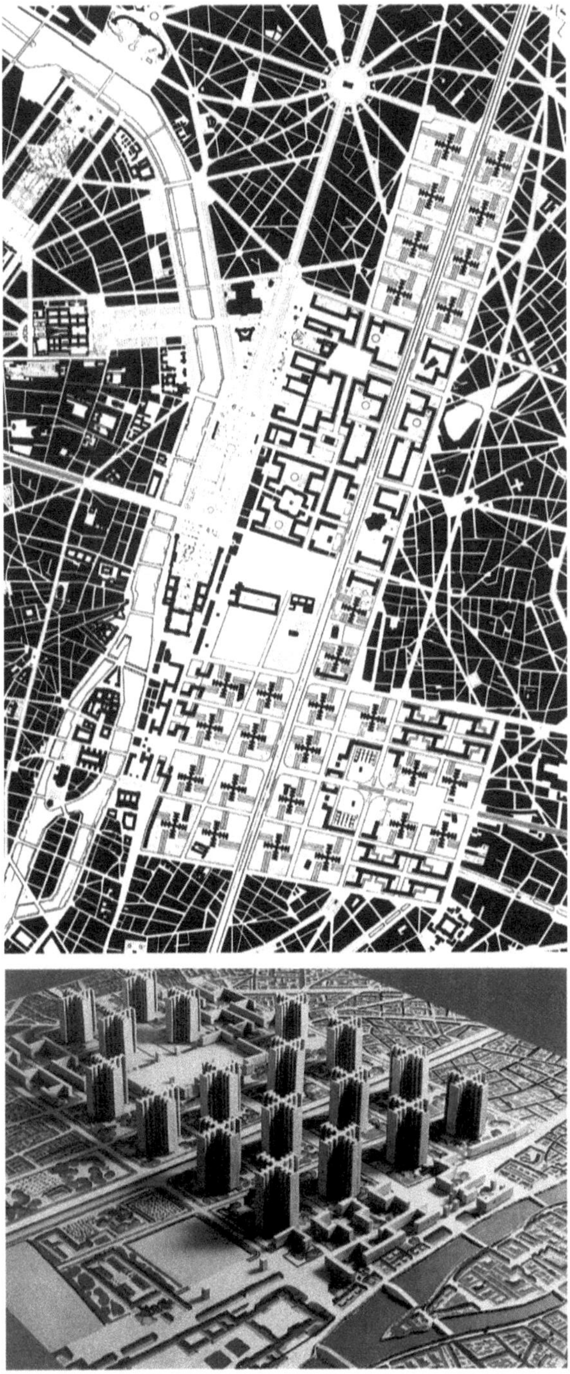

Fig. 3.18 Plan Voisin for Paris, Le Corbusier, 1925

2.3 Kevin Lynch

In 1960, Kevin Lynch (1918–1984) published “The Image of the City”, a standard work for urban design, which deals primarily with the meaning of physical space in terms of its visual recognizability. With the subsequent “Theory of Good City Form” (1981), he places these findings from his urban research and observation in a new context. His concept of space is mainly based on anthropological research results and urban design features.

In Lynch’s view, understanding the city is related to assessing how it should be: “. . .it is impossible to explain how a city should be without understanding how it is” (Lynch 1981, p. 39). Elsewhere he emphasizes that a good theory must use concepts and methods in such a way that they correspond to their object, i.e. the form of settlement.

With regard to the spatial form of the city, Kevin Lynch distinguishes five dimensions of action in his urban design theory. The best possible form should essentially:

– secure life,
– make the city experienceable,
– have a framework plan for action and behaviour,
– provide general accessibility and
– regulate the disposal of the land.

In his statements, Lynch refers to the physical form of the city, but without neglecting that the quality of an urban space or settlement depends on much more than its form.

2.4 Aldo Rossi

Aldo Rossi (1931–1997), in his book “L’Architeturra della Città” (1973), “The Architecture of the City” (1975), introduced a “sketch of a fundamental theory of the urban” into the discussion on urbanism. In doing so, he refers to two central systems of the city, and thus also of urban design: “One regards the city as the architectural product of functions, the other sees in it a spatial structure” (Rossi 1975, p. 15). While the functionalist system of the city starts out from an analysis of social, political and economic facts and looks at the city from the perspective of these disciplines, the other system tends to take its geographical and architectural character as its starting point.

In his preoccupation with the city, Aldo Rossi focuses on the figure, on what can be experienced concretely, which in turn leaves specific impressions on people. “Individual and community meet and permeate each other in the city” (Rossi 1975, p. 13). His theory of the urban, with its analytical method and the systematic attempt to differentiate the city into the public and private spheres, or even into primary elements and residential buildings, contributes to a view of urban design that is interdisciplinary even in its basic approach.

2.5 Christopher Alexander

Christopher Alexander (*1936) developed his Pattern Language in 1977, a language of spatial patterns in which he describes the city, the buildings and the construction. He was looking for a language of archetypal forms and relationships that correspond to the timeless architectures and relationships of traditional communities. The found derivations of the total of 253 patterns are based on observations of proven building structures and architectures, such as old English universities or Italian villages, which exemplarily satisfy the spatial requirements and elementary human needs.

Fig. 3.19 Pattern 60 “Accessible green” (“Erreichbare Grünfläche”), Christopher Alexander

Christopher Alexander has tried to transfer the regularity of a language to spatial organization and design. The individual building blocks can be combined "like words in a sentence . . ." and lead to differentiated statements. So, all patterns are based on the principle of relationships.

On the one hand, the patterns should serve to define and make architectural and spatial elements comprehensible, and on the other hand, they should be an aid for planning, including the design of urban development concepts. By means of arrangement of physical parts and laws, a design problem, be it of a technical, aesthetic or social nature, can be solved or a corresponding design goal can be achieved.

Each pattern first describes the empirical background of a problem that occurs repeatedly in our environment and then explains the empirical background the core of the solution of this problem, and in such a way that this solution can be applied in many ways. Finally, the pattern is related to all the smaller patterns of language, with the help of which it can be supplemented or even improved.

This structure has two main intentions. The first is to show each pattern in conjunction with other patterns, so that the collection of all 253 patterns can be understood as a whole, as a language in which infinite combinations are created. Second, the problem and the solution of a pattern should be presented in such a way that one can judge and modify it oneself without losing the central idea.

It is significant that Christopher Alexander makes no distinction between the structural-spatial and social organization. This attempt at systematization addresses both the demands of use and the existing context. Ultimately, Christopher Alexander has tried to identify simple generic forms and relationships that are usually found in most communities, have grown out of a social structure and are easy for most people to understand. Each pattern can be regarded as a scientific hypothesis. In this sense, each pattern represents the best current assumption about which arrangement of the physical environment will work best in solving the problem at hand.

2.6 Gerd Albers

Gerd Albers (*1919) gained his urban design theory from his many years of work on the history of ideas of urban design (1975) and his findings from practical experience. In his view, practice looks for theories that accompany action or facilitate decisions, and at the same time, he expresses his reservations, observation from practice, about a higher-level theory that guides it.

Gerd Albers does not make a clear distinction between the disciplines of urban design and urban planning; he uses both terms synonymously. He concedes two paths to urban planning and urban design, which are also relevant in the theoretical debate: the theoretically founded concrete objective for a planning area and the perfection of the planning process. He has repeatedly stressed the importance of continuity, caution and perseverance in the design of cities and that these in turn depend on the interaction between urban design and the use of space.

In his theory of spatial planning, Albers emphasizes three fields of action:

- A typological order of planning problems or of developments in society and economy that lead to changes in space,
- Composition rules for spatial elements that exist on different spatial scale levels up to the design level,
- Strategies for translating spatial concepts into reality.

For the success of spatial planning—and thus also of urban design—he considers research that spans different branches of science and an interdisciplinary approach to planning problems to be indispensable, almost essential.

2.7 Thomas Sieverts

Thomas Sieverts (*1934), with his book "Zwischenstadt. Between place and world, space and time, city and country" ("Zwischenstadt. Zwischen Ort und Welt, Raum und Zeit, Stadt und Land"), has made an apt analysis of the existing urban planning conditions by speaking of the "urbanization of the city" and the "urbanization of landscape". The Zwischenstadt is not to be understood—as is often interpreted—as an urban planning model, but rather as the variety of settlement and development forms visible in our environment, whose edges have become fluid to the landscape.

In Sieverts' view, the "Zwischenstadt" requires new instruments to read the existing structures and to intervene in their design. "The essential contribution to the 'readability' of an urban region must be made through the process of small creative 'interventions' and the allocation of meanings because readability first of all means—as we have tried to work out—conscious perception, memorization, memory" (Sieverts 1997).

According to Sieverts, this process of intervention can happen in different ways: Previously imperceptible elements can be made visible and accessible and thus experienced or become memorable through a "cultural charge". Elements that initially appear unattractive are aesthetically charged. Likewise, elements with negative emotional connotations can be reinterpreted by connecting them to positive events and, in combination with other elements, become chains of sight and experience.

With Zwischenstadt, Thomas Sieverts has opened up a new perspective on urban and landscape structures and thus made an important contribution to the legibility of an urban region. The potential for such readability is different in each Zwischenstadt and must be tracked down separately. To this end, it provides numerous approaches for planning action, but above all an informative theoretical framework.

Literature

1. History of Urban Development

Albers, Gerd 1975: Entwicklungslinien im Städtebau. Ideen, Thesen, Aussagen (1875–1945). Düsseldorf: Bertelsmann.
Benevolo, Leonardo 1993: Die Geschichte der Stadt. Frankfurt, New York: campus.
Klotz, Heinrich 1991: Von der Urhütte zum Wolkenkratzer, München: Prestel.
Mumford, Lewis 1987: Die Stadt. Geschichte und Ausblick. Band 1 + 2. dtv.

2. Theory of Urban Development

Cerdá, Ildefonso 1867: Teoria general de la urbanizacion, Bd. I–II. Madrid.
Frick, Dieter 2008: Theorie des Städtebaus, Tübingen, Berlin: Wasmuth.
Le Corbusier 1929: Städtebau. Berlin, Leipzig: Deutsche Verlagsanstalt.
Lynch, Kevin 1981: A Theory of Good City Form. Cambridge: MIT Press.
Rossi, Aldo 1975: Die Architektur der Stadt – Skizze zu einer grundlegenden Theorie des Urbanen. Düsseldorf.
Sieverts, Thomas (1997): Zwischenstadt. Zwischen Ort und Welt, Raum und Zeit, Stadt und Land. Braunschweig: Vieweg.

The "Grammar" of the City 4

1 Understanding the Place and the Code: Transformation of the Museum Quarter Vienna

A place can only be understood if not only the present conditions are analyzed but also the historical changes are traced. Using the example of the Museum Quarter in Vienna, the transformation process of a place up to the present situation can be presented systematically. The mutual interplay of planning and realization, of buildings and open space, of built structures and uses, of architecture and appropriation of space has changed the place over centuries and ultimately shaped its "code."

C. Reicher, *Urban Design*, https://doi.org/10.1007/978-3-658-34370-5_4

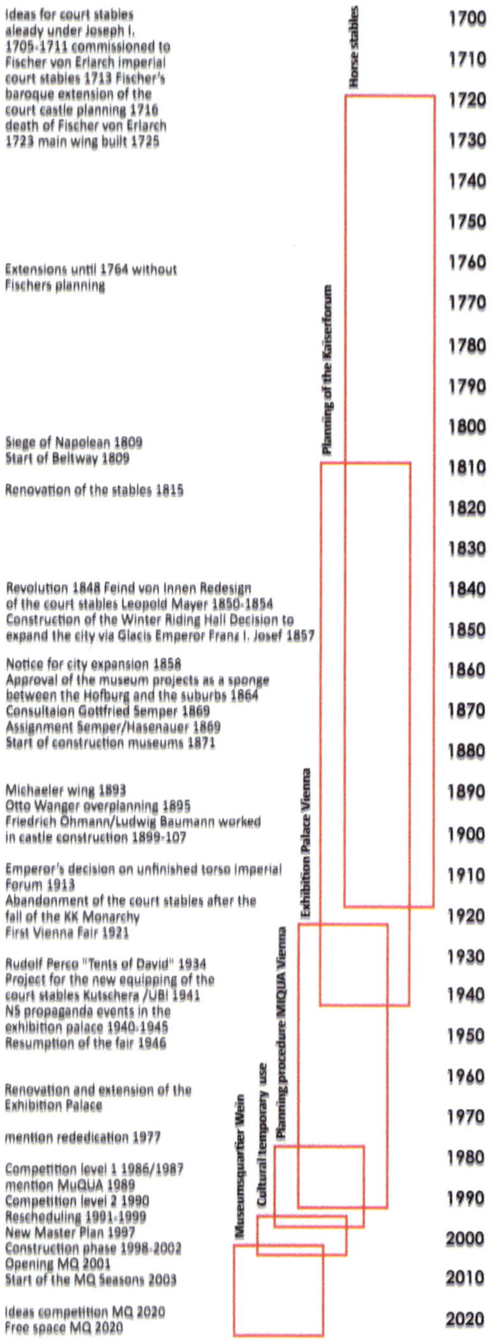

Fig. 4.1 Timeline about the transformation of the location. Phases of transformation—from the imperial court stables to the Museum Quarter Vienna (right)

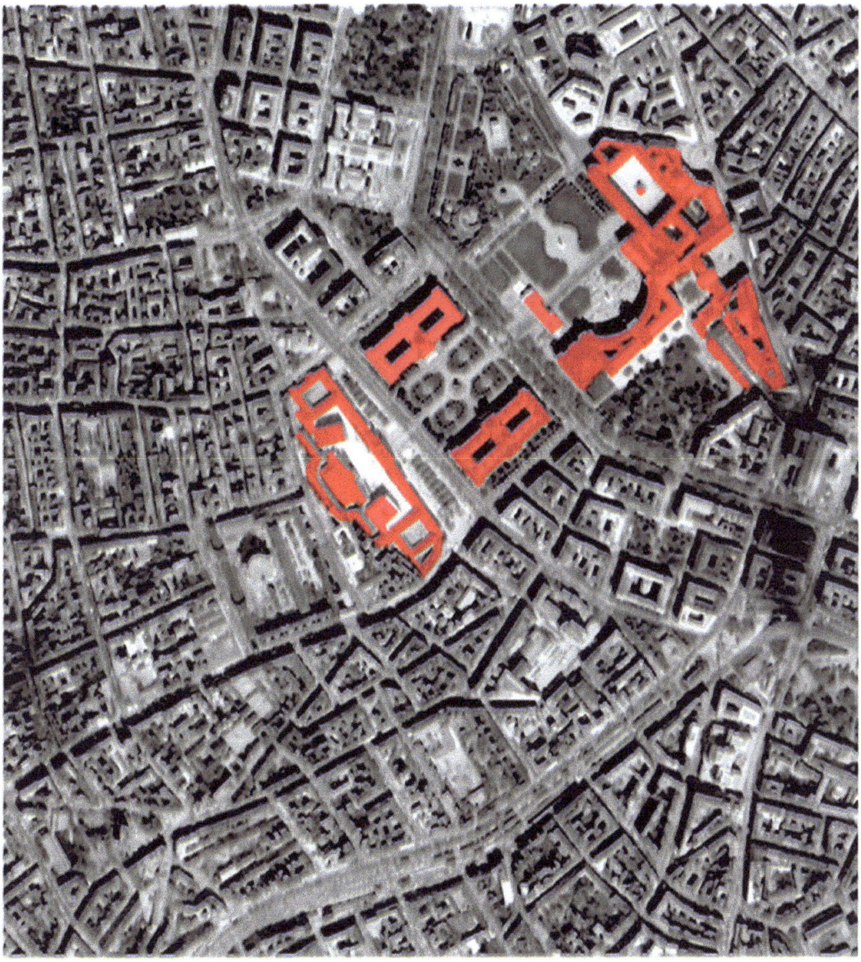

Fig. 4.2 Aerial photograph of the Museum Quartier Vienna. Imperial Hofburg, Semper Museums, and Museum Quarter (below)

The First Transformation: Imperial Court Stables

Today, the Museum Quarter is defined in its external appearance by the 350 m long façade of the baroque old building designed by architect Fischer von Erlach. In 1718, plans were drawn up for the construction of the imperial court stables for approximately 600 horses on the outskirts of the Viennese suburb of Mariahilf opposite the Hofburg on the site of a former poultry farm. The new construction of the Court Stables represented the first extension of the inner residential city that extended beyond the belt of fortifications. This form of baroque urban expansion, which was a matter of course in other European cities, was considered a novelty for Vienna. In the redesign of the Hofburg, carried out between 1719 and 1723 according to the plans of Fischer von Erlach, one thing stands out before one: The main façade—despite planning freedom—is not positioned parallel to the Leopold Wing, which at that time was the only new building that formed a distinctive part of the residence. The orientation of the façade reacts to the medieval core of the Hofburg. This spacious overall concept planned by Fischer von Erlach, which can still be felt today, was never completed. In the rear area, outbuildings were erected until 1764—without any recognizable principle of order—to cover the space required. The semicircular end had got stuck in its beginnings.

From an urban design point of view, the imperial court stables represent a remarkable fragment. Fischer's plans fell into oblivion at the end of the eighteenth century. They became relevant again in the middle of the nineteenth century, when the development of the Ring Road was pushed forward.

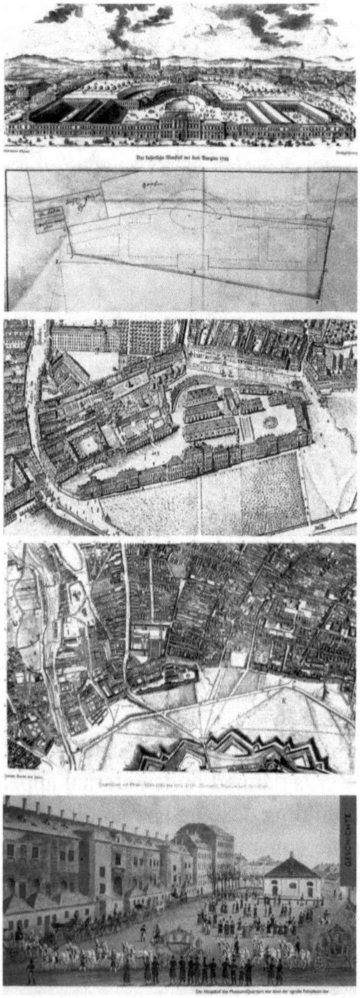

Fig. 4.3 The first transformation: Imperial court stables. Ideal plan "Design of a Historical Architect" Fischer von Erlach, 1721. Site plan in the area of the military "demolition," 1718. Court stables 1770 by Joseph Daniel von Huber. Orientation of the court stables toward the medieval core of the Hofburg, 1769. Main courtyard as driving area, 1834 Balthasar Wiegand

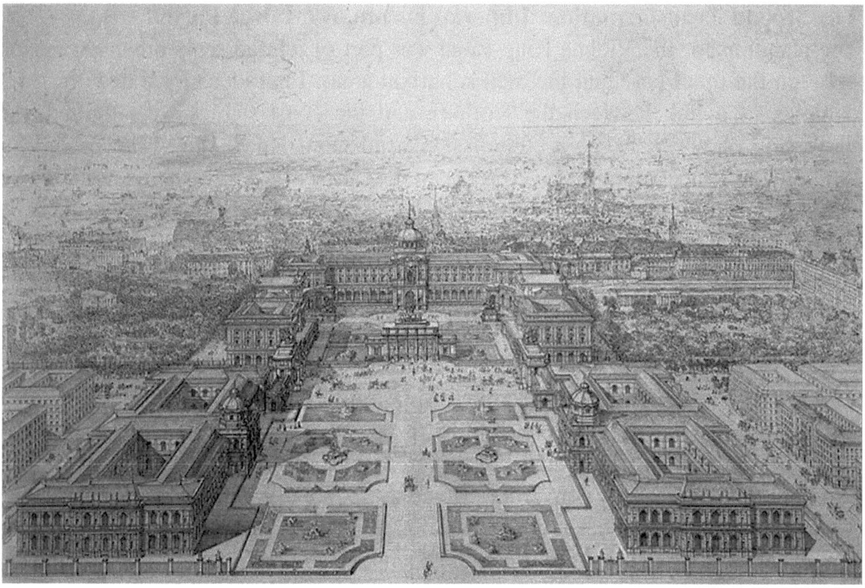

Fig. 4.4 The second transformation: Imperial forum as a large-scale form. Gottfried Semper and Karls Hassenauer, first draft for the Vienna Imperial Forum 1869

Fig. 4.5 The third transformation: Vienna Trade Fair Palace. Exhibition hall G built after the Second World War. Reconstruction of the Trade Fair Palace 1961. Fair poster Vienna

The Second Transformation: Imperial Forum as a Large Form

The planning for the Vienna Ring Road was part of a large-scale urban expansion between the inner core and the then suburban areas. First ideas for a new museum complex as a link between the Hofburg and the court stables were drafted, the discussion of which led to a commission from Gottfried Semper. The design for the imperial shape submitted in 1869 served as the basis for the realization, which was subsequently accompanied by various master builders with different ideas on the integration of the Court Stables in their faulty alignment and lack of representation. Shortly before the First World War the work was stopped for economic and design reasons. As a result, the second wing of the castle remained unfinished, leaving the Imperial Forum as the torso of a major urban development project.

The Third Transformation: Vienna Trade Fair Palace

The first Vienna Fair, which opened in 1921, was spread over several locations in the city. The Hofburg and the grounds of the old court stables functioned as one of these locations and were thus converted into inner city exhibition areas, the so-called Vienna Trade Fair Palace. Between 1940 and 1945 the grounds were used for events organized by the Nazi regime. When the Vienna Trade Fair resumed its activities in 1946, necessary conversions were carried out and new halls were added. With the withdrawal of trade fair activities and the new master plan for Vienna Trade Fair, which concentrated all activities on the Prater grounds, discussions about the conversion of the grounds began in the late 1970s. When the holdings of the contemporary Ludwig collection were shown at the House of Artists in Vienna in spring 1977, the desire for permanent access to the works of art arose. This was the beginning of a discussion lasting almost 20 years on the concept for the conversion of the Trade Fair Palace, which also addressed the unfinished plans for the Imperial Forum.

The Fourth Transformation: Museum Quarter

In 1993, the decision was made to build a cultural forum on the grounds of the Trade Fair Palace in the future. This was to concentrate the various art and cultural institutions that previously had no fixed location in Vienna. The spatial programme designed for the competition procedure included non-museum uses as an important part of the new concept, in order to achieve a mixture of uses at the location. The architects Ortner and Ortner won the competition with a project that included the Kunsthalle Vienna, the Museum of Modern Art, and a multifunctional event hall. As a signal to the outside world, the architects proposed two towers: The Art Library as a "reading tower" in the main courtyard and an office tower. The project was not without controversy. A citizens' initiative acted against any form of contemporary architecture in the centre of Vienna and demanded instead a horse museum in the old court stables. The critical discussion about the concept of the reading tower finally resulted in a new dimensioning of the building masses, which were reduced by half and reduced in height. The concept realized on this basis enters into a symbiosis with the existing building stock. The exterior appearance is dominated by the Baroque old building; it is only in the interior that a new urban space unfolds, which is also characterized by the new building.

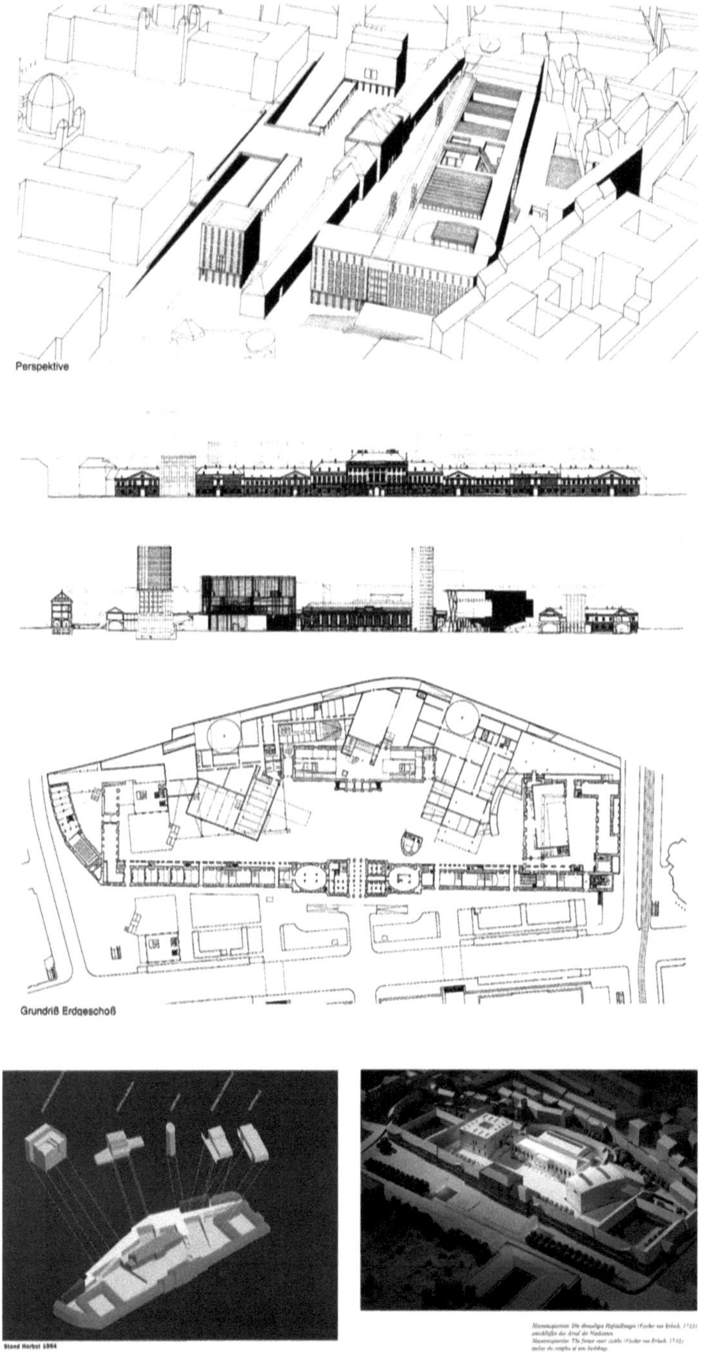

Fig. 4.6 The fourth transformation: Museum Quarter. Competition MUQUA level 1, Ortner & Ortner. Competition MUQUA level 2, sections, floor plan, Ortner & Ortner. Implants 1994, Ortner & Ortner. Plan status 1995 ARGE Ortner & Ortner/Wehdorn

The Fifth Transformation: Intermediate Cultural Uses

The old exhibition halls were used for temporary cultural purposes during the long planning process. From 1992 to 2000, numerous installations, cultural displays, and performances were staged. The basis for this was an "MQ Settlement Concept," which envisaged a colourful mix of uses and temporary actions. Thus, the Museum Quarter was opened to the public in an "Open House" clearing out campaign. In this way, the users and the future public transformed the location and parts of the existing buildings, thus creating an emotional bond with the location. The temporary cultural use not only opened up the area to visitors and cultural actors, but also had a decisive influence on the present and future character of the area through various forms of programmatic and architectural intermediate states.

Fig. 4.7 The fifth transformation: Intermediate cultural uses

The Museum Quarter is an example of transformation of a place that has changed over the centuries due to external conditions, changing demands for use and a struggle to find the right urban design and architectural approach. In its current state it represents a "code" which on the one hand reveals references to the most diverse historical phases, but on the other hand also makes clear its different dimensions: the symbiosis of built structure and open space, the interaction of the ground plan and the design of public space, the architectural form and scale, the forms of appropriation and programming of space.

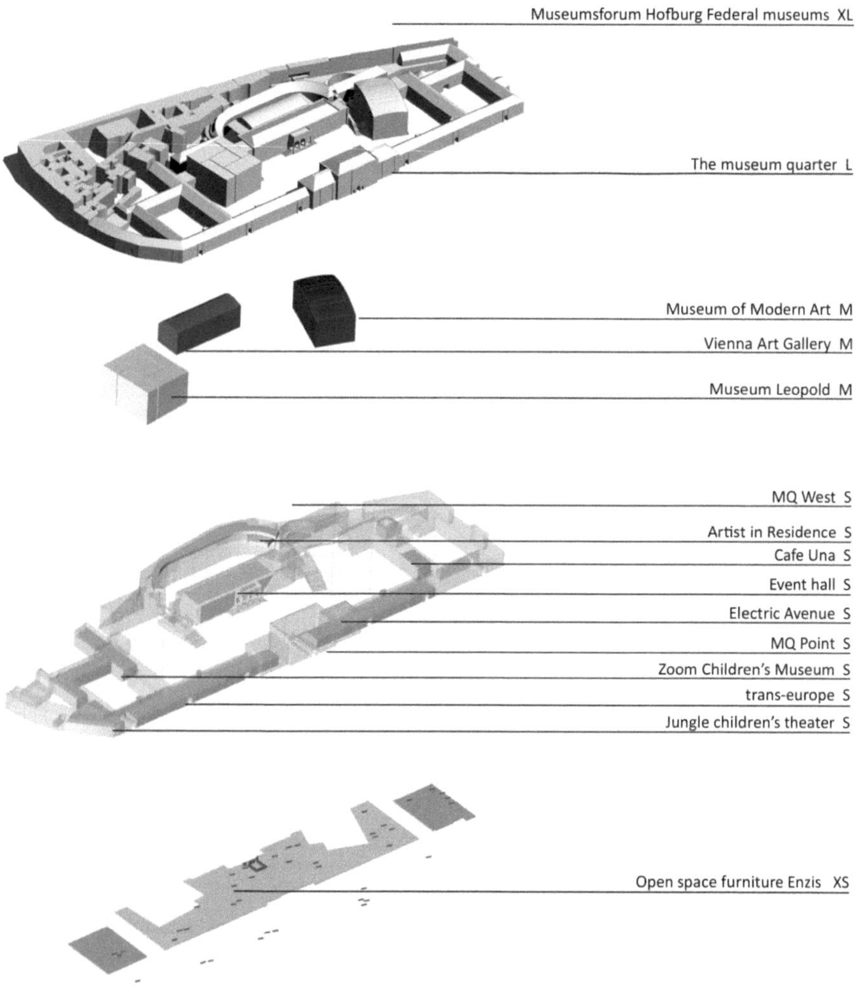

Fig. 4.8 Scale composition Museum Quarter Vienna

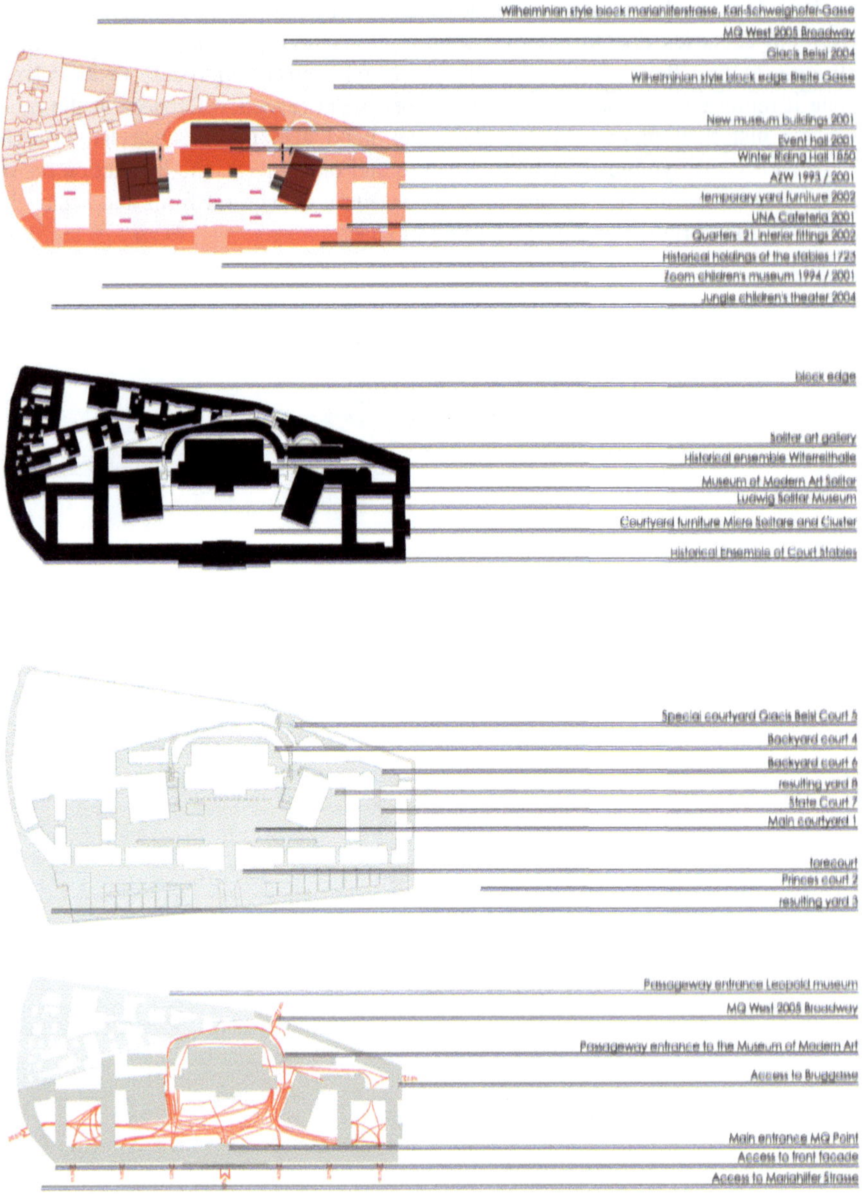

Fig. 4.9 Urban typology: Clusters and neighbourhoods, open spaces, and networking

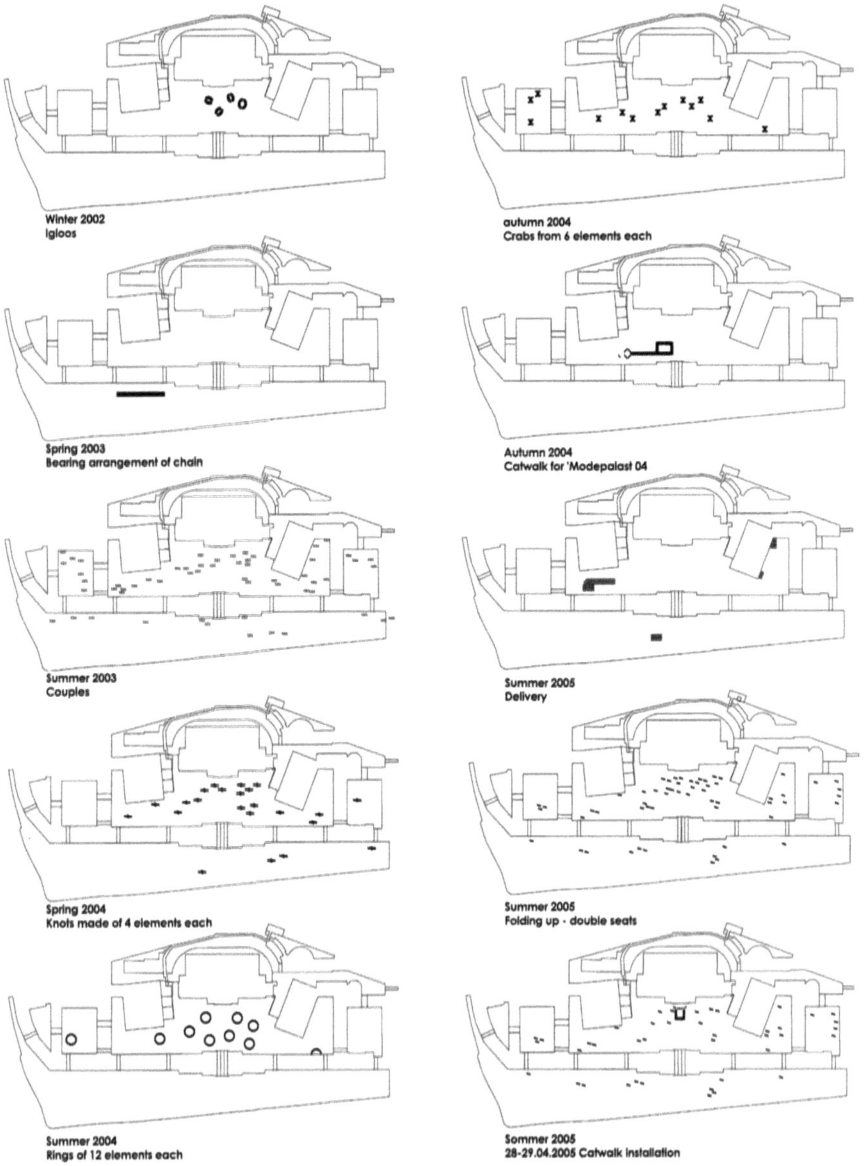

Fig. 4.10 Inventories of Court furniture 2002–2006

2 Typologies and Layers of the City

2.1 Structural Levels and Structural Models

Structural Levels

When looking at the city and its surroundings, different levels can be identified, each of which builds on the other and between which interrelationships exist:

- the city, embedded in the region,
- the district as part of the city,
- the neighbourhood or quarter as a unit within the district,
- the plot of land and the building as a basic building block.

The boundaries between these defined units are not only determined by structures or spatial breaks, but also by functional laws.

Structural Models

The structure of a city is characterized by its elements and their arrangement in relation to each other. Since in cities of similar size the same elements—albeit in different quantities and sizes—usually occur, their difference is mainly determined by the arrangement of the elements in relation to each other. Essential here are those elements that have a lasting effect on the functional process. These include:

- The transport networks
- Utilities (commercial, educational, social, cultural institutions)
- Larger outdoor areas
- Commercial and industrial sites.

To be able to depict such a complex structure as a city, schematizations and simplifications are necessary. For this purpose, the so-called structural models are used, which represent the essential connections in an abstract way.

The structures of cities can be traced back to three basic types, which correspond to the settlement elements "point, line, area":

- the concentric urban structure
- the linear urban structure
- the area of the city structure

However, these basic types are rarely found in their pure form, but mostly in mixed forms. Cities, for example, correspond to the concentric structural model in terms of their basic type, but within this structure they have, for example, band-like elements. In a similar way, the principles of arrangement can also be considered at regional and district level.

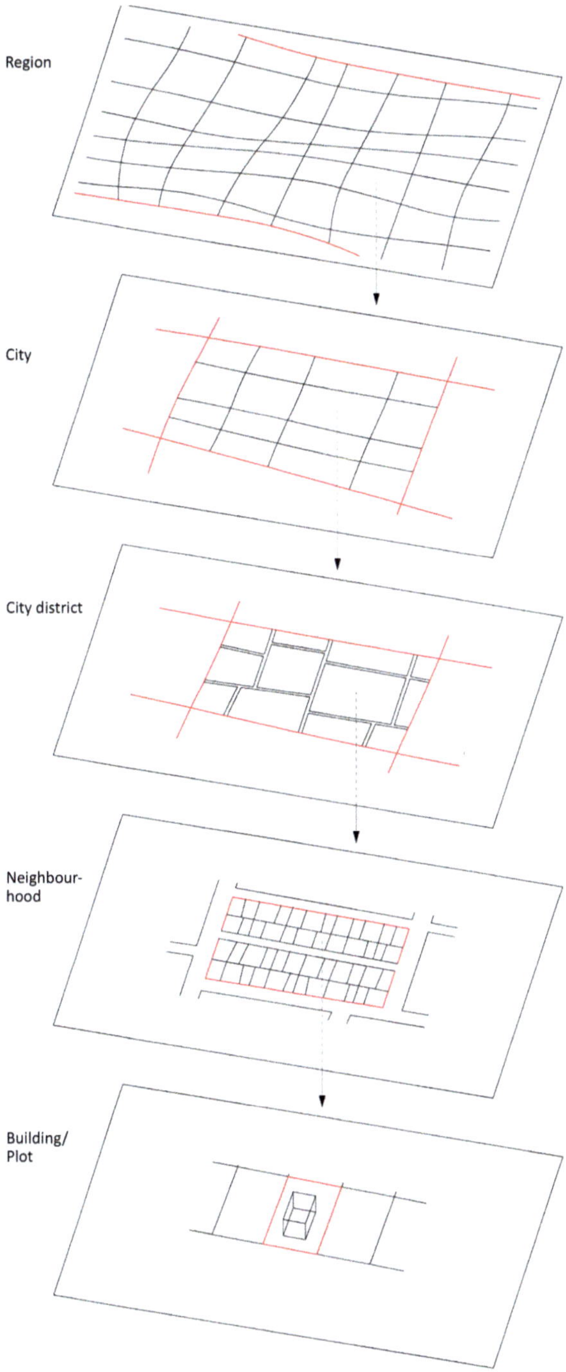

Fig. 4.11 Structural levels

The Concentric Structural Model

Concentric urban structures are characterised by the following features:

- a pronounced centre with a clear structural focus
- transport network oriented radially toward the centre
- the density of meaning and functions decreases toward the edges

This type of structure is common, as most of the larger cities today have developed in more or less large rings starting from the historical core. This is due to the aim of having the shortest possible route to the centre on the one hand and the economic management of the infrastructure (roads, water supply, canal, etc.) on the other. Typical problems are the traffic congestions in the centre, a hardly possible expansion of the centre in line with the rest of the city's growth and the frequent undersupply of recreational and open spaces close to the centre.

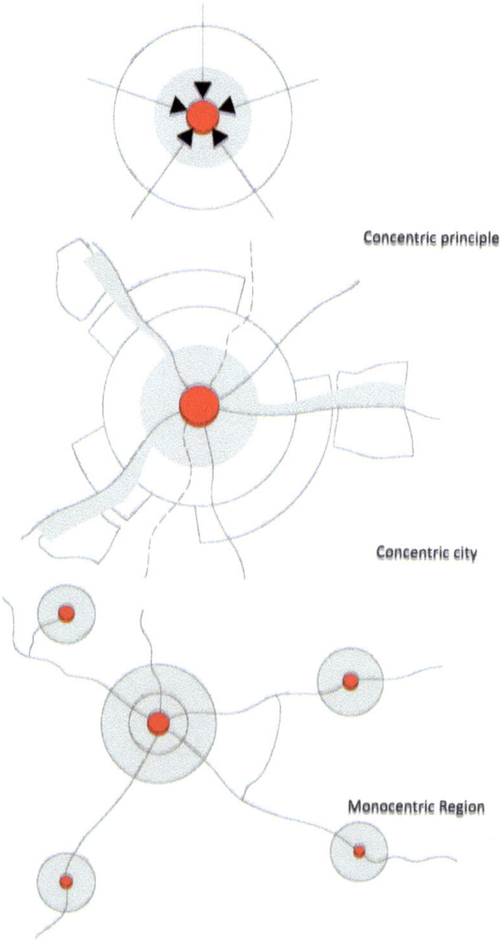

Fig. 4.12 Concentric structural model

Fig. 4.13 Medieval town complex. Urban layout of Arles within the walls of the ancient amphitheatre

The Linear Structural Model

Linear urban structures have the following characteristics:

- central supply structure arranged like a band, without a clear centre
- linear transport infrastructure in the form of a transport corridor
- linearly distributed density of meaning and function

The backbone of a band city is usually an efficient transport corridor (main road, railway line, river, canal, etc.) from which the individual settlement or neighbourhood units are accessed. The central supply and service facilities are usually located directly in or along the corridor. Behind them are the residential areas.

The band- or line-shaped arrangement of appropriately dimensioned urban units offer some advantages in the allocation of residential, working, supply, and recreation areas along with efficient infrastructure, but on the other hand only allows a relatively small width of the settlement zone, so that they are usually supplemented by the star- or comb-shaped bands.

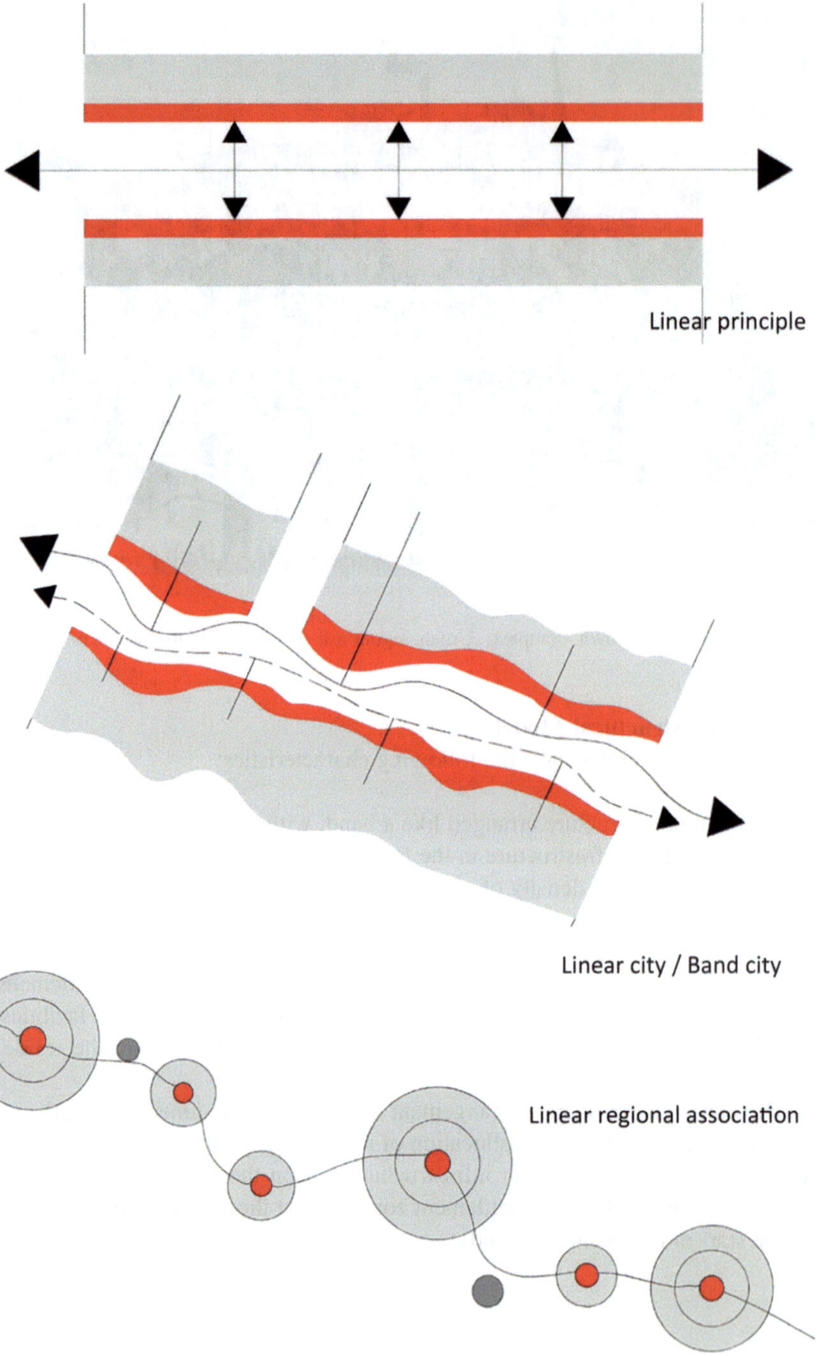

Fig. 4.14 Linear structural model

Fig. 4.15 Band city, Arturo Soria y Mata, 1894. Completed section of the band city around Madrid

The Extensive Structural Model

Spatial urban structures have the following characteristics:

- no pronounced centre
- extensive transport network
- homogeneous distribution of the functions over the entire surface

An even distribution of urban functions over the entire urban area is usually only achieved approximately, as the various functions have specific requirements for their locations. The model is, therefore, to be understood in such a way that the individual functional units are not concentrated but rather arranged as decentral as possible, combined with a largely nonhierarchical, grid-like traffic network. This offers advantages for the formation of viable sub-centres, which in turn serve as the basis for network-like models or models of decentralized concentration.

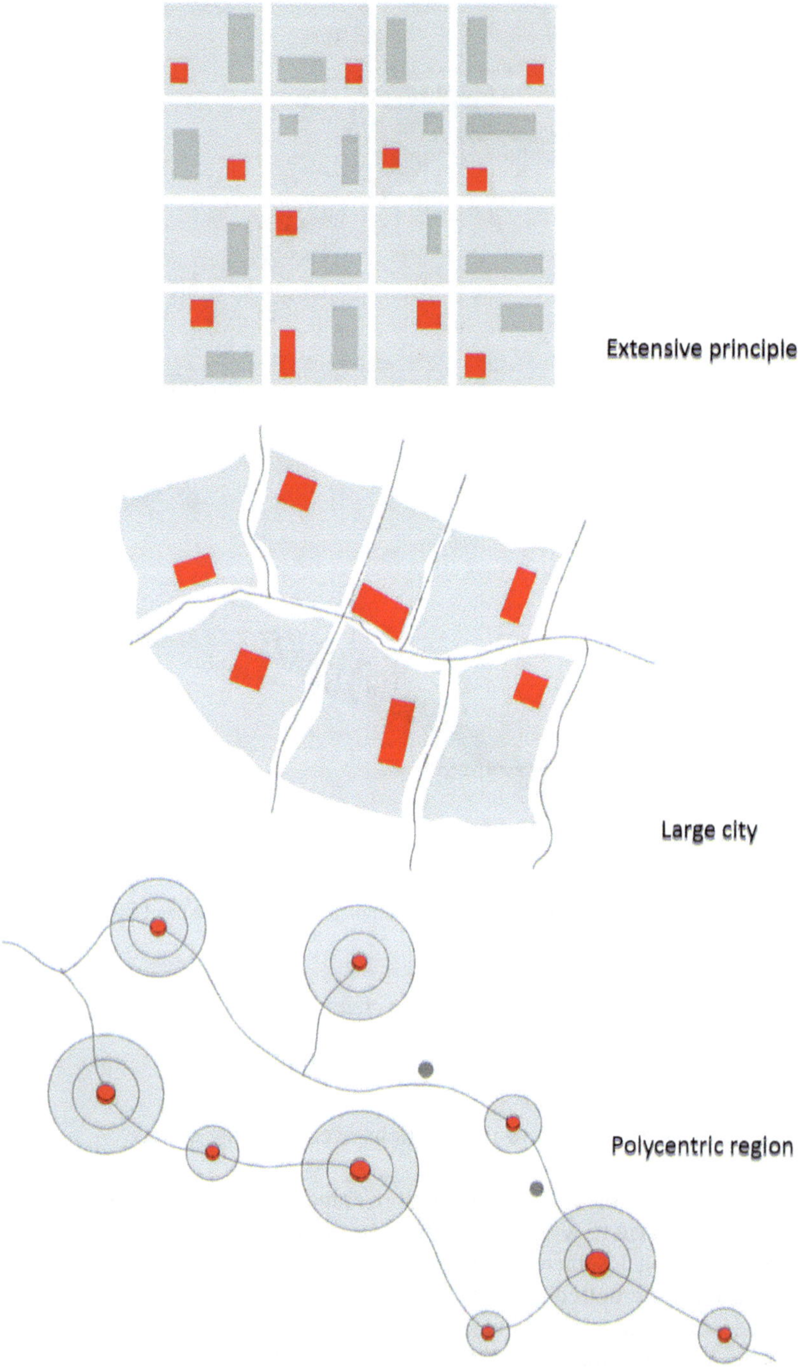

Fig. 4.16 Extensive structural model

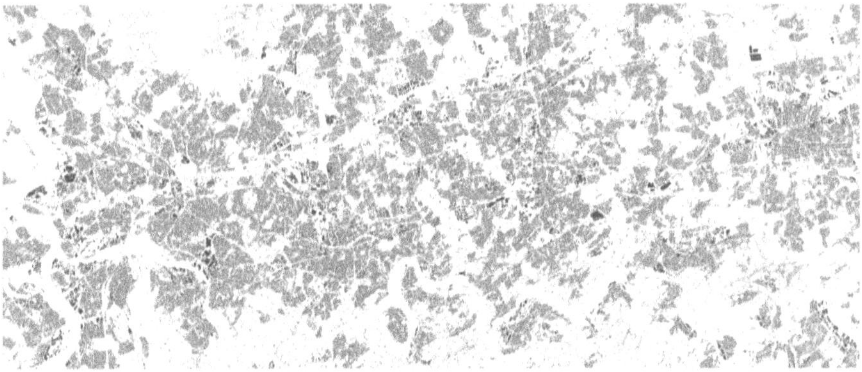

Fig. 4.17 The Building Structure of the Ruhr Area, 2010

2.2 Urban Structure

The physical city structure is the decisive organizational pattern of coexistence of the inhabitants in one place, in one city. The physical structure of the city is not to be understood as a rigid system but as a dynamic one. The visible structure of the city, its buildings, and the spaces they shape are subject to permanent change.

The structure of a city, the so-called urban structure, creates a framework that influences many areas of life of its inhabitants in the long term. It reacts to changing framework conditions, such as social values, laws, political, or economic decisions. The reactions to change are complex and, measured against human activity, usually proceed rather slowly. The structure of a city is thus both the basis and the result of human action.

Hardly any of the other factors are as strongly marked by historical references as the development of our cities: Financial crises and building booms, the effects of war that destroyed grown urban structures and created space for new developments, parties with their ideas on transport policy and urban planning—all these have left their mark and will continue to influence life in and the character of cities in the future.

The city structure determines the distribution of central functions: residential and commercial areas, shopping miles, cultural sites, etc. The transport system ensures that these points are more or less easily accessible: Streets with cars and taxis, footpaths and cycle paths, trams, bus lines, railway lines, airports. Depending on the size, location, and historical development of the city, the individual modes of transport can be of very different importance.

Changes in the urban structure are reflected in the ecological, social, and economic field. The speed at which such changes take place is reflected in the quality of the urban structure. Radical, rapid changes can lead to ruptures and distortions that destroy established structures and regularities.

Definition of Terms

Urban structure is the arrangement and interaction of the individual parts of a city to form an overall structure, including the principles, ideas, and laws underlying this arrangement. A spatial image of the urban structure is the ground plan of the city. In the ground plan of the city many individual laws can be read. The ground plan can also be described as the "genetic code" of the city, which shows specific patterns of an arrangement of the city's building blocks (building and open space structures), which have been shaped in the course of history by influences from outside and inside. These patterns describe the special characteristics of a place and thus make statements about the arrangement forms, scale, and hierarchy of the city's spatial system.

However, not all relevant characteristics, principles, or regularities in the structure of a city can be read off. There is a close connection between physical structure and urban design; less direct are the dependencies between urban layout and the economic or demographic structure of a city.

Urban structure is composed of "subunits": structure elements and regulatory elements. Structure elements are parts and links of complex settlement structures. Their type, size, and organization determine the structure of the higher-level structure. The order elements of the city can be traced back to three basic categories, from which all higher order elements are also formed:

1. Point elements
2. Linear elements
3. Area elements

The connection of the basic elements underneath and with each other to form mosaics or meshes as well as larger units is shown in the table below. These basic urban design elements can be demonstrated not only in the urban structure but also in the design and orientation structure:

Point: individual prominent physical or functional or emotional points in the urban fabric, such as

- Dominants, landmarks (e.g., church tower, high-rise building. . .)
- Focal points (e.g., important places. . .)

Line: striking linear elements that are imprinted, such as

- Route relationships (e.g., shopping mile. . .)
- Edges between homogeneous area elements (e.g., the boundary between park and development . . .)
- Distinctive linear topographic features (e.g., riverbanks . . .)
- Striking boundaries (e.g., railway line . . .)

Surface: surfaces that are imprinted as homogeneous areas, such as

- Neighbourhoods (e.g., old town neighbourhood. . .)
- Open spaces (e.g., city park. . .)

The point- and line-shaped elements can usually be defined exactly, with the area-shaped elements the demarcation is often not clearly possible.

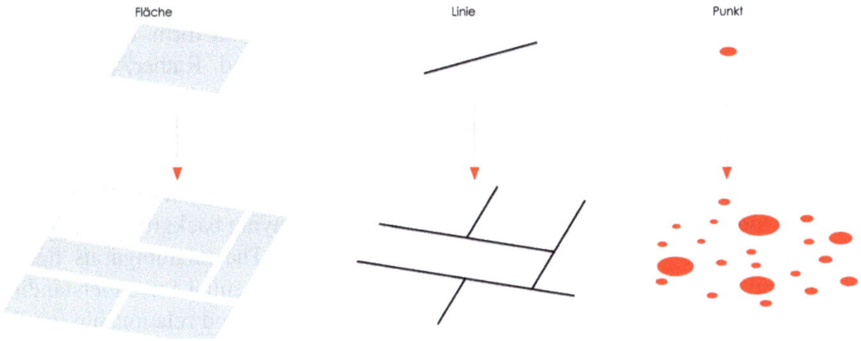

Fig. 4.18 Organizational elements

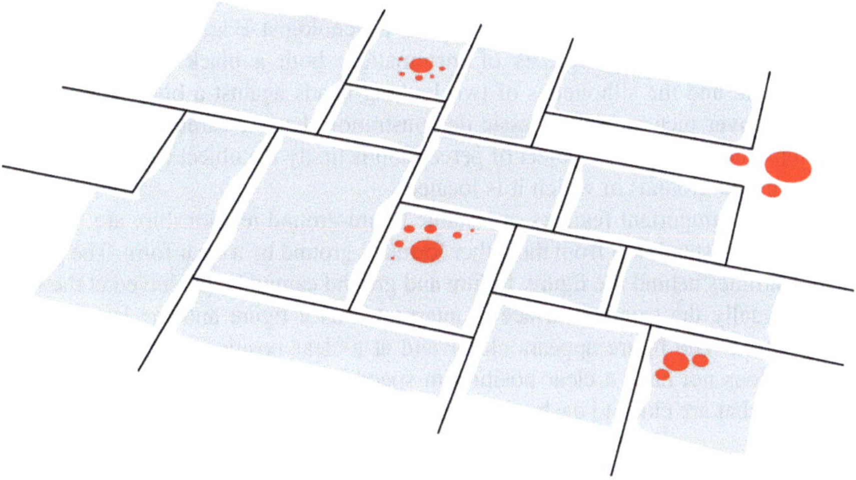

Fig. 4.19 Urban unity as a composition of elements

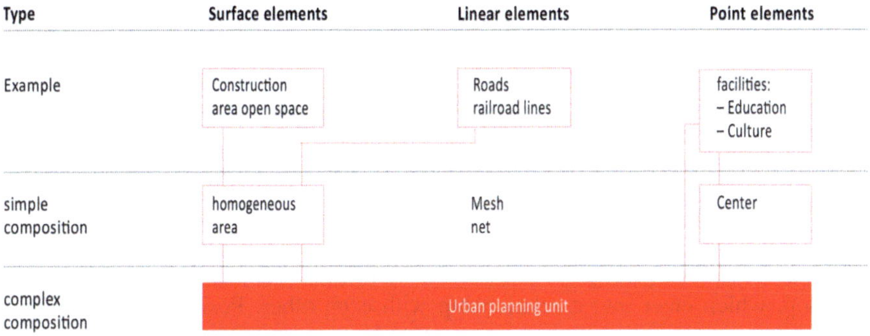

Fig. 4.20 Settlement elements

3 City Logics

3.1 Perception of the Figure-Ground Relationship

The figure-ground relationship is one of the important laws of form. It states that a figure cannot be perceived independently of its background. Rather, an observer distinguishes the essential from the inessential within a perceived wholeness. Thus a smaller area is more likely to be perceived as a figure and the larger area surrounding it as a background.

If the figure clearly stands out from the background, it is given more attention and remains more strongly in the memory than the accompanying background. One then speaks of a stable or clear figure-ground relationship. The unambiguous figure-ground relationship is of particular importance, as it is essential for understanding the content of a picture or layout. An ambiguous figure-ground relationship, on the other hand, always exists when several possibilities of perception are of equal importance. Ambiguous figure-ground relationships appear incomprehensible or boring to the viewer.

The best-known example of an ambiguous figure-ground relationship is certainly the Rubin's vase, named after the Danish psychologist Edgar John Rubin. This illustration contains two pieces of information: both a black vase on a white background and the silhouettes of two looking heads against a black background. Rubin's cover picture is the classic demonstration that the same shape can evoke different perceptions. The object of perception is firstly the object (the figure) or the situation (the ground) in which it is located.

The most important features of a stable figure-ground relationship are:

The figure stands out from the rather formless ground by a clear form. The ground also continues behind the figure. Figure and ground cannot be perceived at the same time. Usually the smaller surface is interpreted as a figure and the larger one as background. The figure appears closer and at a clear position in space, while the ground does not have a clear position in space and appears further away. Similar elements that are close to each other are perceived as one figure.

Figure-Ground Ratio in Urban Design

The figure-ground ratio is of particular relevance for urban design. According to the classical understanding of space, a space is created when two or more buildings form an interspace that enters into a relationship with each other—like the figure of a vase on the Rubin's cover picture. Accordingly, space is understood as something enclosed by bodies.

But space can also be understood as a profile of where something physical ends. The development releases a spatial profile—corresponding to the profile of two faces looking at each other on the Rubin cover. However, the inverse relationship only occurs when the ground supports a completely different category of figure, in one case the space, in the other the mass of the building.

On a site plan with a coherent built structure (mass plan Parma) the buildings act as figures and the open space as ground. From the mass, positive external spaces develop, which enter into a relationship with each other. Positive spaces allow a figure-ground envelope. In the site plan with isolated buildings on one surface (mass plan St.-Die) the exterior space acts as a residual surface. A reversal, the legibility of the exterior space as figure and the buildings as ground, is not possible in this case. Closed, coherent forms are preferably perceived as figures, they allow an inversion and are more complex in their experience of space.

Fig. 4.21 The Rubin's cover image

Fig. 4.22 Shape diagram the figure—the ground

Figure-Ground Plan

The figure-ground plan is a central medium of representation in urban planning. All other plan elements such as streets, vegetation, or water bodies are hidden in the figure-ground plan. This modification allows a clear distinction to be made between built-up (shown in black) and undeveloped areas (shown in white). The building structure is reduced to a simple ratio of figure and ground, in which a specific ratio of spatial texture to object texture can be read (see also, Rowe, Koetter 1997). With the help of such a representation the morphology of a city can be investigated. The figure-ground plan shows the essential urban supporting structure as the "physicality" of a city.

The spatial structure in the black map of a city can be compared in its entirety to a conventional gramophone record, whose grooves represent the three-dimensional imprint of a mechanical recording of sound waves. This is the layer of the physical surface of the "cityscape." In it, the built is distinguished from the undeveloped, the solid is distinguished from the space between things. One can look at each part separately. The abstract depiction of the building structure via the figure-ground plan provides the practiced reader with information about the urban spatial binding of a district, for example, the degree of coherence: to what extent do individual aspects stand for the whole, are they isolated, do they promote a spatial continuum? The age of a neighbourhood can also often be determined by its configuration, and the age of the buildings by their modular characteristics. The two-dimensional representation of the city thus becomes a nonrepresentational image of its three-dimensionality.

Fig. 4.23 Air and Water, M.C. Escher

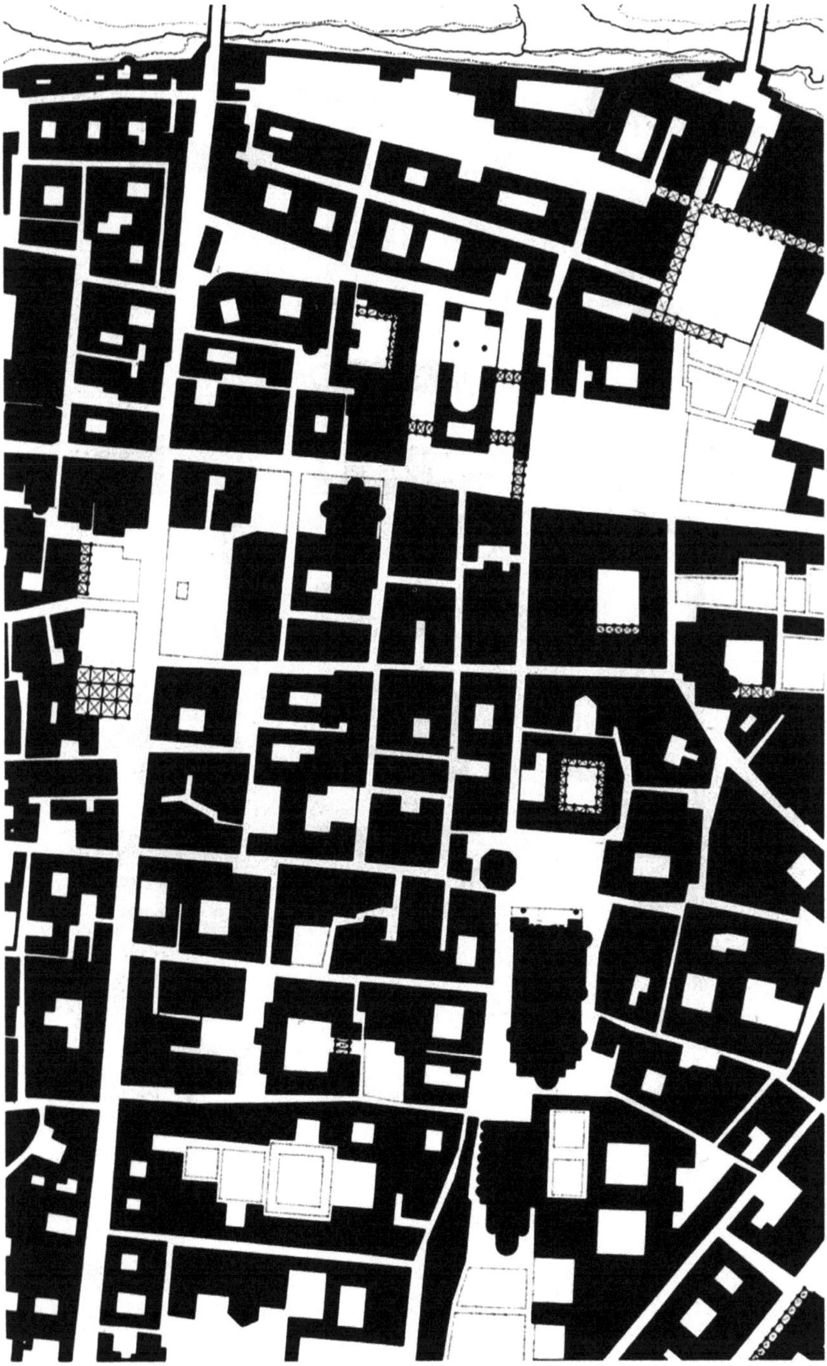

Fig. 4.24 Parma, mass plan

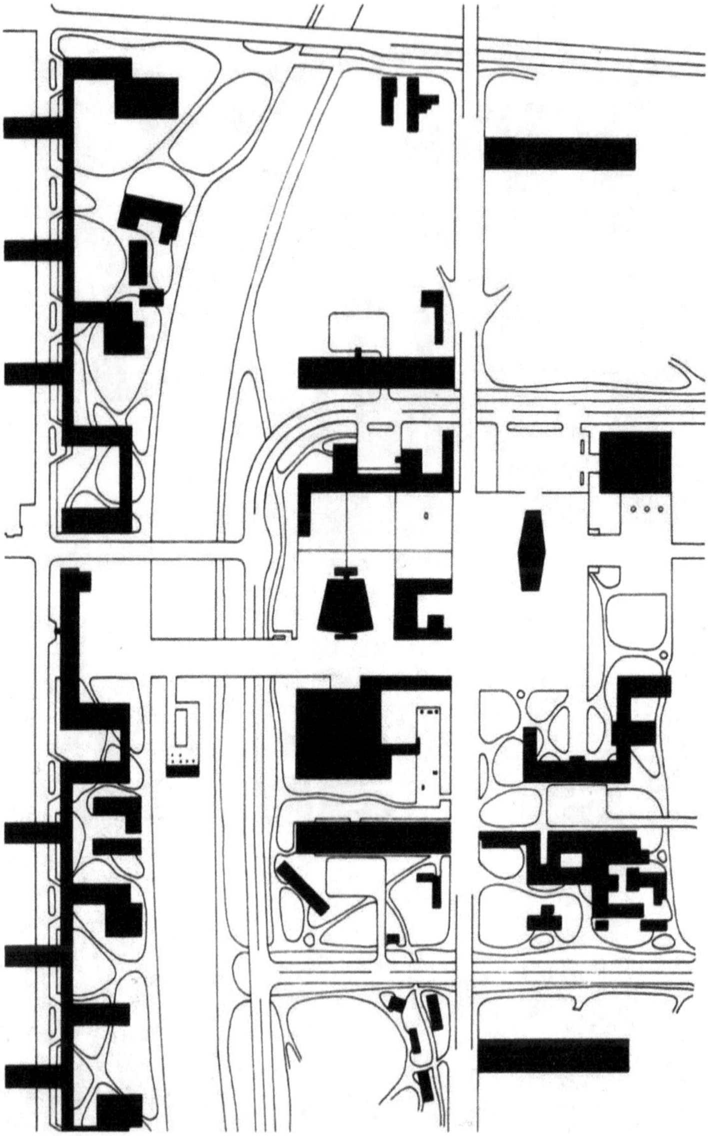

Fig. 4.25 St.-Dié, mass plan

3.2 Urban Morphology

Definition and Understanding

While the cityscape essentially refers to the visible parts of the city, urban morphology broadens the view of the city and includes the building and open space structures. The term "morphology" comes from the Greek word "morphe," which means shape, form, or appearance. The original genetic term was transferred by Goethe into the doctrine of the genesis of forms, as the doctrine of the shape, formation, and transformation of organic bodies.

Urban morphology thus refers to the principles of form and shape according to which urban ground plans were created and according to which they are constructed. The respective conditions of origin have shaped the spatial characteristics of cities in different ways.

Urban morphology is also a field of research in urban design. It is concerned with settlement and urban forms as well as with the physical forming processes within the settlement bodies. It is thus concerned with the structure of the development, the formation of plots as a basis for buildings, the building typology and the networks of development. A special focus is the historical analysis of the development and transformation of urban structures. The urban morphological investigation has long served as a basis for understanding urban structures. Its role has increased in recent decades due to the focus on the preservation and renewal of historical urban structures.

The shape of a place is determined by its geographical position, its location in a certain climate zone, or even the cultural area. Cities in comparable climate zones, for example, show similar characteristics. The structures are similarly shaped and reflect the specific morphology of the place. Through a repetition of elements such as buildings, open spaces, and topography, a texture of the location is created that specifically shapes it.

An urban morphological investigation makes the inner logic of an urban structure readable by searching for the underlying laws of buildings, surfaces, and networks. The individual elements and laws are added to form a whole, an overall structure.

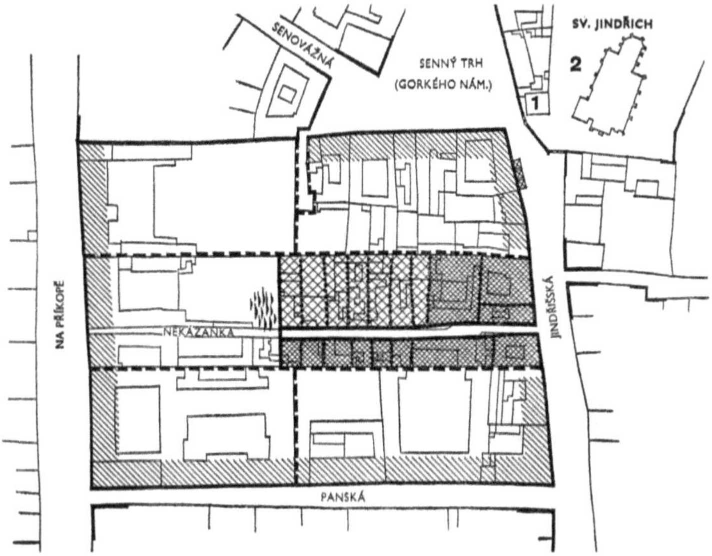

Fig. 4.26 New Town of Prague, before 1379, urban morphological investigation. Proof of the origin of a street in the central part of the Gothic block of the New Town of Prague due to subsequent parcelling

A consideration of urban morphology in the context of urban design serves to understand the laws of a place and its internal logic in order to obtain a basis for the spatial supplementation, intervention, or even further development of the structure. Urban design and architectural interventions in an existing context, in particular their scale and dimension, can be checked and adapted accordingly by analysing the urban morphology.

In an extensive study of the urban genesis and development of the historical inner city area of Prague, the urban historian Vilem Lorenc (1973) shows the formative connections and regularities of the inherent logic of historical street layout and development. On the basis of a detailed study of the foundation documents of the medieval city centre, the urban and spatial planning and development of large parts of the city centre in the last millennium could be traced in block or parcel form. An inventory of structural and spatial changes also provided a basis for the "passportization" of the buildings in the historic city centre. It included the historical documentation of the buildings' origin and those relevant to the preservation of historical monuments. The knowledge thus gained about the street layout, the layout of streets and squares, their width and visual axes or spatial-functional references serve today as a more in-depth basis for preservation and redevelopment as well as for the assessment of new construction projects in the city centre area.

Fig. 4.27 Wiesbaden, 1900

3.3 Structure of Use

The top priority of urban design and urban planning is to use the land sparingly and sustainably. Soil is a limited resource that has been increasingly appropriated by humans over centuries. This fact has increasingly affected the ecological balance.

The term "use" refers to the use of the land (including the buildings and installations erected there). A distinction is made, for example:

- populated areas,
- traffic areas,
- agricultural land,
- woodland,
- water surfaces, etc. (see also, Borchard 1974).

The populated areas are in turn differentiated according to different types of use:

- Living,
- Work (for forms of work that usually take place outside of homes and residential buildings),
- Supply (in terms of economic, social, and technical infrastructure),
- Green and open spaces (beyond building-related open spaces such as gardens),
- Transport as a generic term for facilities of all transport modes,
- Education and leisure.

The way in which a use is distributed over an area determines its use structure. The structure of use reflects the needs of the people, but also the political and economic conditions.

The functionality of an urban organism is determined by the form of usage distribution, which in turn can be influenced by social, economic, and topographic factors. The way in which uses have been distributed within an urban structure has changed considerably in the course of history. In the medieval city, the interpenetration of various uses in daily life is considered a reflection of the economic and lifestyle of the time. The industrial city, with its extreme growth and densification processes, has led to the creation of negative mixes through overcrowding of dwellings and the coexistence of industry and housing. The city of modernity with its motto of separation of functions has contributed to the disentanglement of uses and building structures.

The distribution of types of use and the associated usage structure determine the urban development quality of a neighbourhood or city.

GRANULARITY OF THE UTILIZATION STRUCTURE	SPATIAL REFERENCE	PLANNING LEVEL
coarse-grained ↑	District	Land use plan
	Neighbourhood	Urban development plan
	Neighbourhood	Urban design
	Street	
	Building	Building design
↓ fine grained	Floor Level	

Fig. 4.28 Use structure and planning levels

Fig. 4.29 Granularity of uses

Mixed use means a juxtaposition and overlapping of different functions such as living and working. The importance of mixed use for adaptable and sustainable urban structures is now generally recognised. Agenda 21 has made a decisive contribution to this.

The separation of uses, especially between living and working, leads to longer distances, which in turn results in a greater volume of traffic and a higher demand for transport routes. Resource-efficient planning requires the most intensive possible mix of uses. The granularity of the mix of uses is strongly determined by the planning levels.

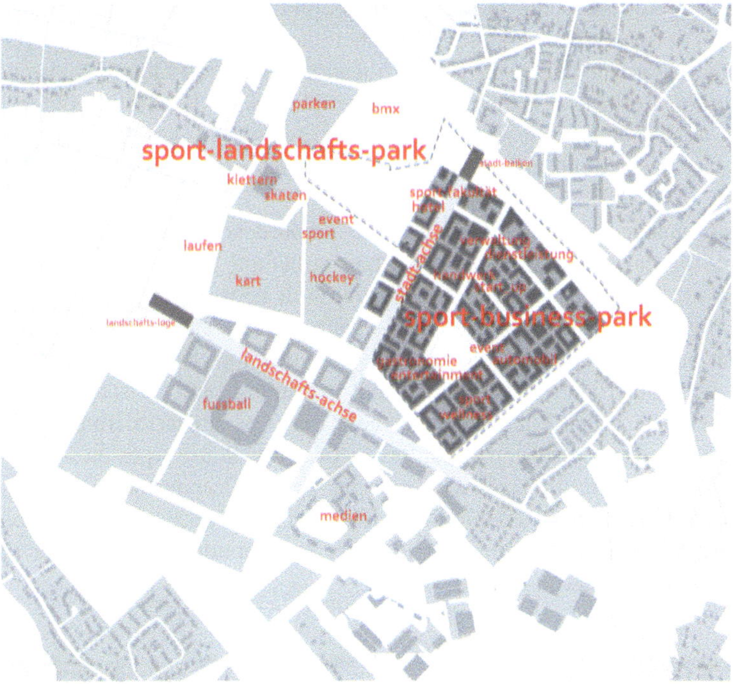

Fig. 4.30 SportPark Mönchengladbach, RHA architects and urban planners. Utilisation concept

4 Urban Structure Types

Beyond the city logics, the structure of a city is determined by urban structure types. The way in which buildings are arranged, how they form spaces through their arrangement, and how they can ultimately be used, makes them structurally determining and ordering urban design elements. The historical view of the city shows that such urban structure types have also been an expression of certain social views of the city and urban society. Nor is the paradigm of space-saving construction, which characterizes the current debates on sustainable urban development and aims at corresponding urban structural types, an invention of the present day. For this reason, there are several historical structural types that are repeatedly referred to in contemporary urban design.

Thus, the geometric-rational structure of ancient Greek and Roman cities is connected with a type of house, the courtyard house, which could be added as a basic unit to regular building blocks. This type of house allowed closed construction methods (without spaces between the buildings) and thus a particularly economical use of the existing ground. The reconstruction of the ancient Priene shows how the building block, which consisted of eight courtyard houses each, functioned as the defining structural type of the city.

The image of the medieval European city, on the other hand, is characterized by long rows of houses facing the street and their backward economic and garden areas. Scarce building land within the city fortifications required a rational division; living and working formed a spatial unit. The aerial view of Landshut's old town still clearly shows the pattern of a medieval transverse rib system. The wide street markets of the old and new town with their merchants' houses served as internal access.

Many outstanding settlements of recent urban development history follow this principle of dense construction with their structural types. The "De Kiefhoek" housing estate by the Dutch architect Oud, built between 1925 and 1930, shows how a high and at the same time high-quality density can be achieved through a compact building structure. The long rows are developed from a single module: a house with four rooms on a plot of land measuring only 70 square metres (see Fig. 4.51). The Halen housing estate (near Bern) of Atelier 5 from 1960 is also exemplary. A clever arrangement of three-storey row house types on a slope allows close links between all levels and the open space. Despite the high density, the settlement is characterised by high-quality living and open space qualities. Even these few examples show that high-quality neighbourhoods with very different urban structure types can be achieved.

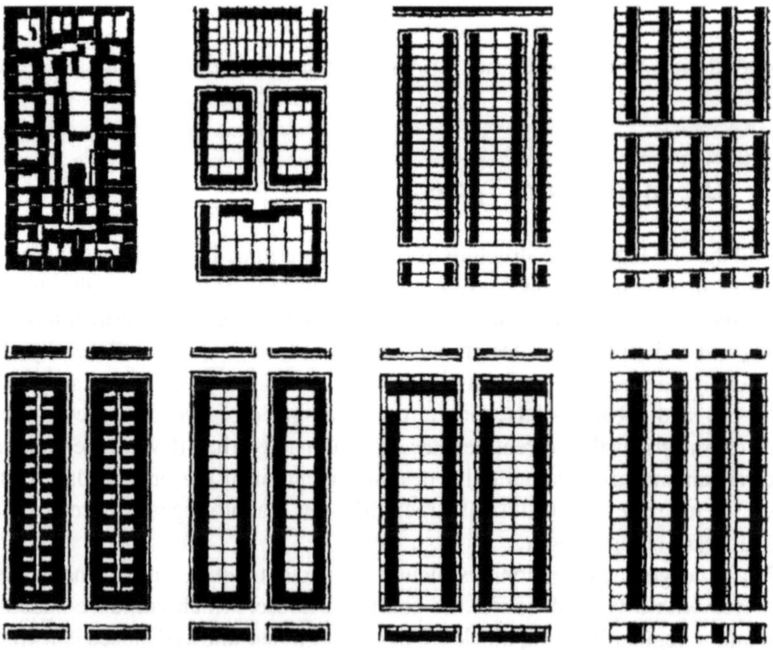

Fig. 4.31 From block to line, Ernst May, around 1926

Fig. 4.32 Reconstruction of the ancient Priene

Fig. 4.33 Old town of Landshut

Fig. 4.34 Halen housing estate, studio 5

4.1 Block

Blocks are formed by an addition of buildings and plots of land, surrounded on all sides by streets (usually four streets), with the buildings facing the street. The building structure can be a closed structure, but also a more open structure.

Fig. 4.35 Block, basic principle

Urban Significance
The block—just like the row—is one of the traditional basic urban components of the city. Over the centuries it has asserted itself as a structural typology in the fabric

of the city, sometimes more, sometimes less recognized. Within the urban fabric, the block is part of a continuous development structure and can thus be reached from all sides and added as a spatial unit.

The block is particularly suitable for mixed use, also depending on its size and the associated internal reserve and buffer areas. If necessary, the development can be extended into the interior without affecting the external appearance. This special characteristic of the block makes it a particular element of the urban fabric.

Especially in fragmented urban areas, blocks can make an important contribution to establishing or repairing urban structural connections.

Structural Principles

In addition to the differences in the closed and open nature of the development, there are various block forms with different sizes and shapes, whereby the properties are determined much more by the sizes than by the shape. With the size of the block, the scope for using the interior space increases. The form of the so-called building block, which is built on a large plot of land and intensively overbuilds the property, is an extreme expression of this urban character and is often purchased with great restrictions on the quality of use.

Spatial Development

The block is framed by streets and thus forms a clear boundary to public space. The public space creates an intensive network with the surrounding urban structure. Building lines that span the block and connected external spaces guarantee continuity in the structure and a cohesive spatial effect.

Of particular relevance for spatial development is the formation of the corner situation. On the one hand, the corner enjoys a special advantage as an overdeveloped and spatially exposed part of the block, where supply facilities (shops, restaurants, etc.) are often accommodated; on the other hand, the corner situation has many disadvantages due to the smaller plots of land to the rear, such as poor lighting, lack of expansion possibilities on the ground floor and limited privacy.

Orientation and Development

The block has a clear zoning between front and back, between the more public side facing the street and the private side facing away from the street. While the front zone is exposed to the influences of the street, the rear zone is more protected from disturbances, provided that there are no disturbing uses inside the block. Similar to the series, the block thus has a clear socio-spatial orientation between public and private areas, which in their consequence have an effect on the design requirements of the respective sides. While the façade facing the street is the front side visible to everyone and thus has to meet representative demands, the rear side facing the block interior can follow more functional and individual requirements.

The development on all sides leads to buildings that are aligned in opposite directions and thus to problems in aligning the rooms to the favourable direction of the compass. This imbalance in the orientation to the different cardinal points can be compensated to some extent by the depth of the building and the associated floor plan organisation. In many cases, the unfavourably exposed side is also completely built on, in order to compensate for the disadvantage of the location by means of an open building structure.

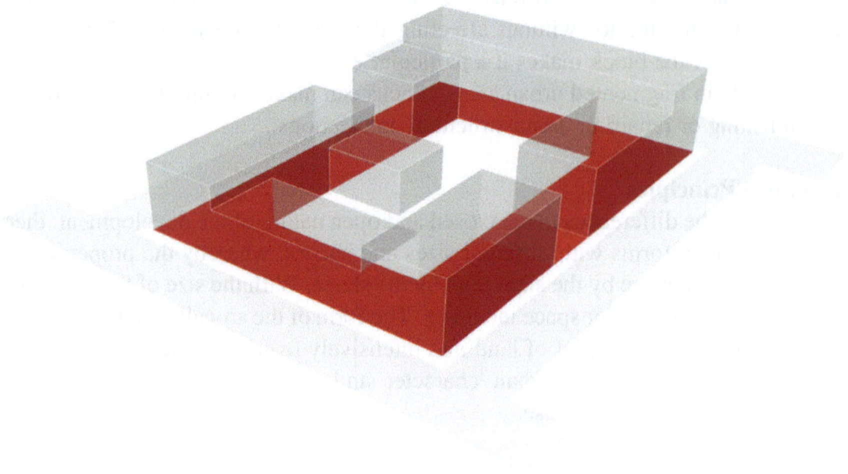

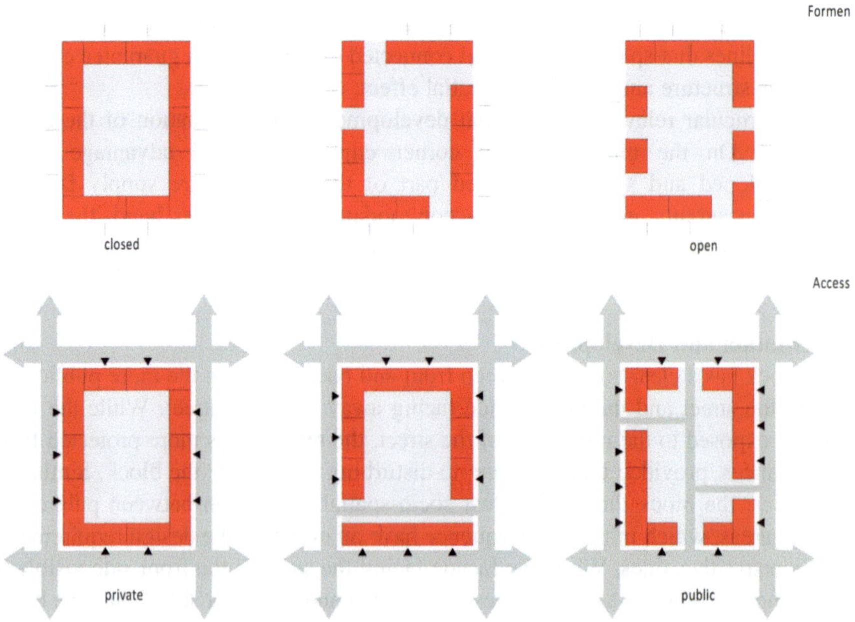

Fig. 4.36 Block, schematic sketches

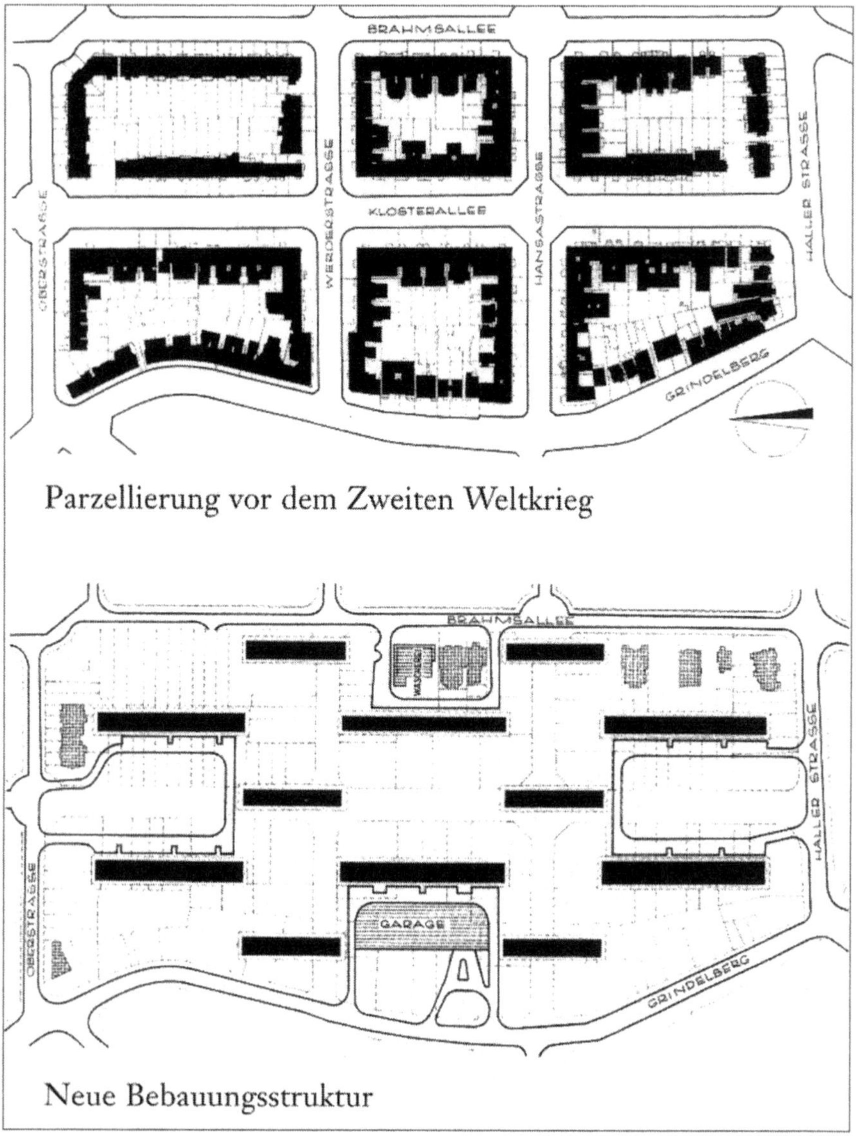

Fig. 4.37 Historic block, Hamburg, Harvesthude. Parcelling out before the Second World War

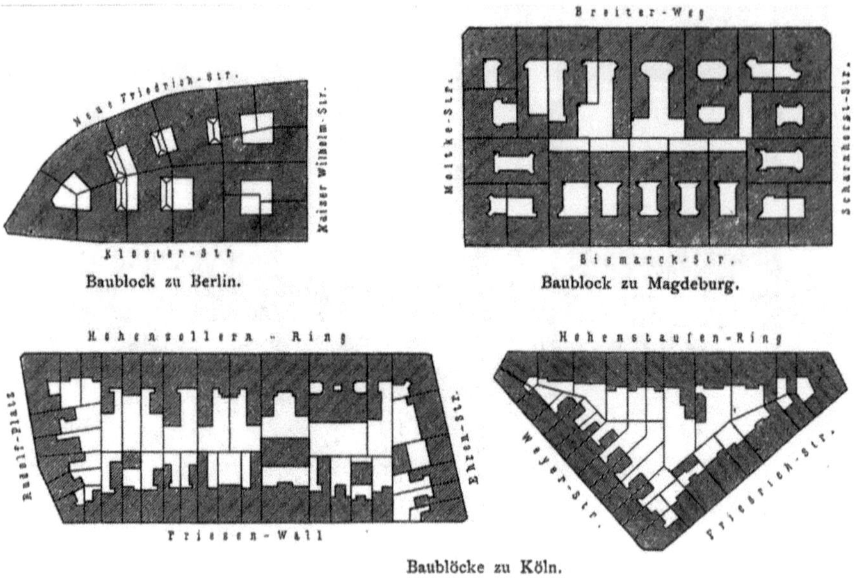

Fig. 4.38 Block historical, building blocks in Berlin, Magdeburg and Cologne, around 1890

Superblock

As an alternative to traditional blocks with a small-scale, road-accompanying development on individual building plots, the spatially, functionally, and typologically complex superblock has been gaining ground, especially since the middle of the twentieth century. Superblocks can occupy much larger areas than traditional city blocks and have their own internal structure. As a rule, they contain large-scale functional units that form a self-contained spatial unit with corresponding independent building types and achieve a high surface capacity. This type of construction is, therefore, often used for large projects in particular. The superblock is a reaction to the increasing demand for space and concentration or combination of individual functions in monofunctional and large-scale building types. In most cases, the superblock represents a compact large complex that combines a main use with supplementary functions. These can be uses such as living space, hotels, shopping, restaurants, and offices in combination with corresponding building types (high-rise, shopping mall, etc.). Characteristic features are the separation of pedestrian and car traffic and an independent internal development. Its components are traffic, parking, and open space areas in quasi-public spaces. These orient the superblock spatially, structurally, and functionally more toward the interior and only to a limited extent toward the street or its surroundings. In addition to its urban dominance through the volume or combination of independent large forms and their high-surface capacity, its strong influence on the surroundings must also be taken into account. The functional and spatial isolation from the urban fabric is to be counteracted as far as possible by means of urban design and planning by seeking capacity limits for the usable space, a spatial opening to the outside or the negotiation of open access rules for the publicly used spaces. Due to the growing criticism of urban planning and functional coarseness in late modernism, forms of superblocks are increasingly being sought in European urban design which, despite a high level of land use, adapt to the traditional spatial urban structure. One of the best-known examples in Germany is the development at Potsdamer Platz in Berlin after 1990.

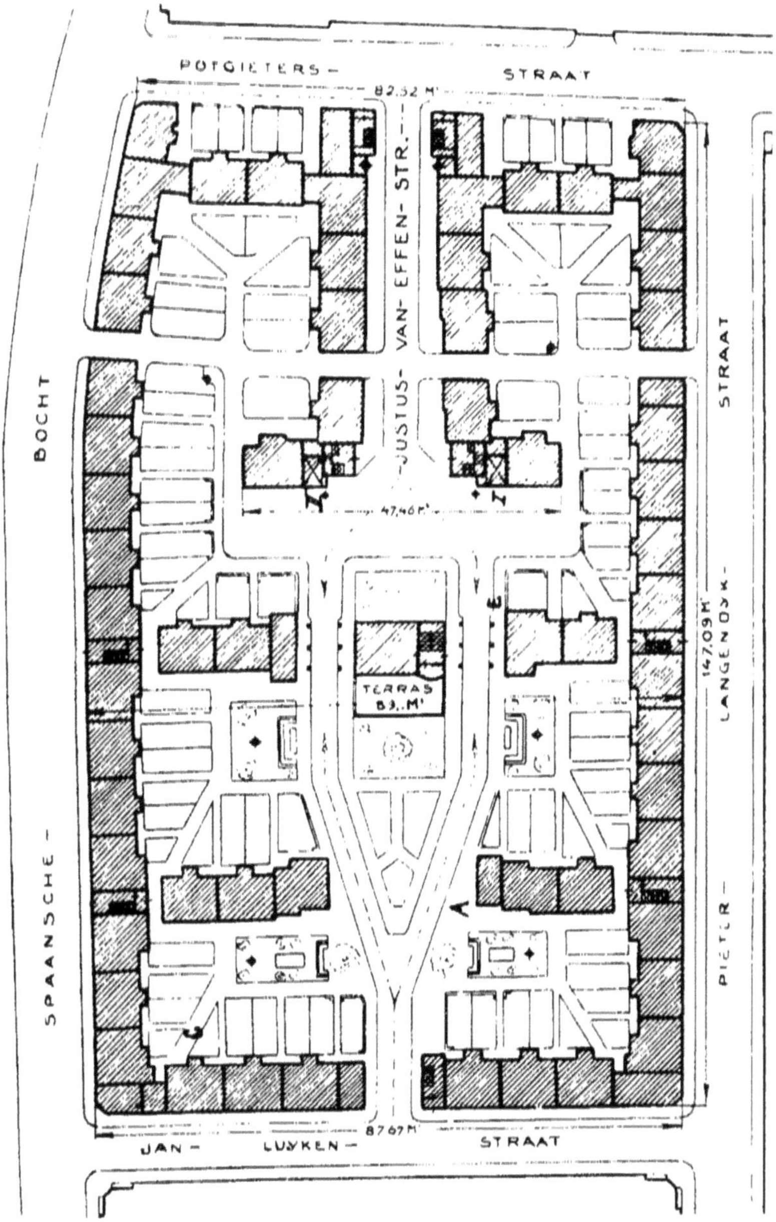

Fig. 4.39 Block historical, superblock, braces, Rotterdam 1919–1921

Fig. 4.40 Potsdamer Platz in Berlin. The Daimler-Benz Quarter forms a large functional unit with offices, a shopping centre and residential accommodation. In terms of urban development, the superblock adopts the grid of traditional block development. The Sony Centre offers a covered piazza inside the superblock, which was created by merging three blocks

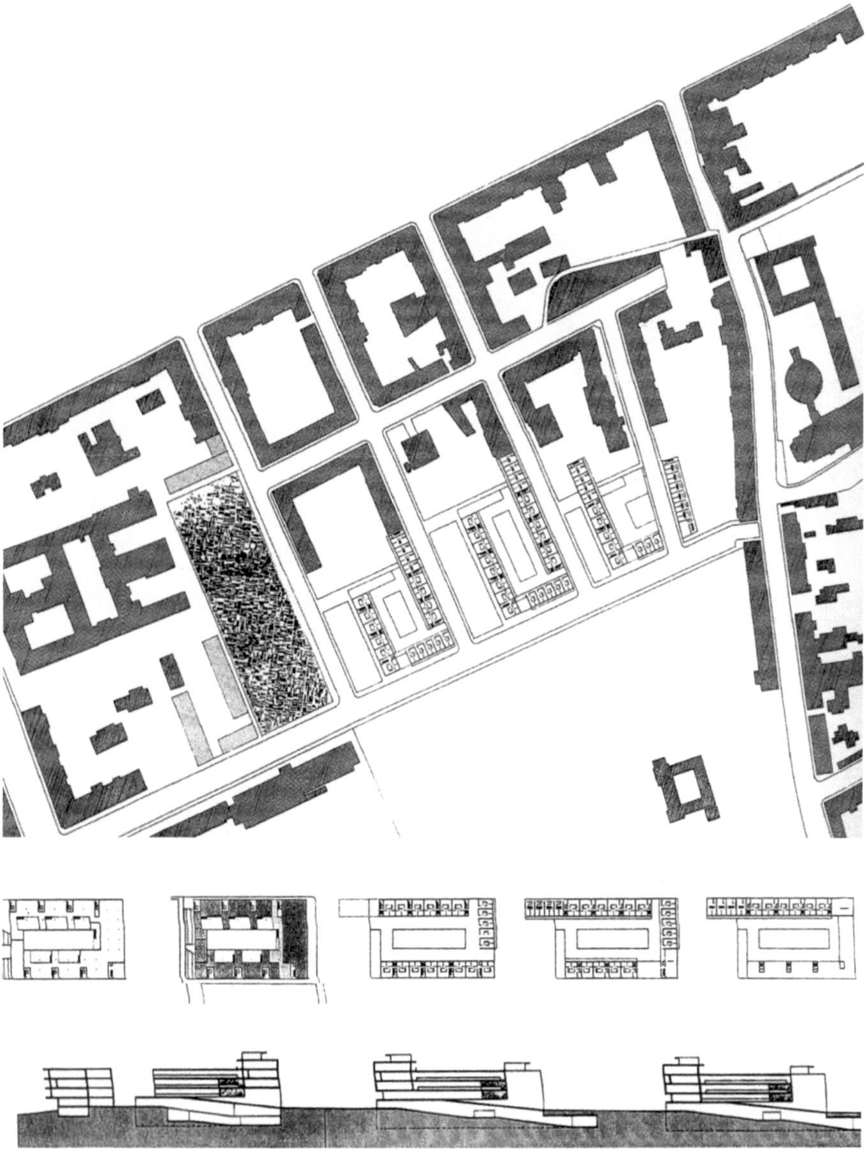

Fig. 4.41 Block modern, Graz, expansion of the historical block structures, competition Europan 4. To connect the urban structures, the exterior spaces are clearly defined. Within the housing block, private interior areas are created. A new public park will be created outside the existing building and traffic structure

Fig. 4.42 Block modern, Copenhagen, Competition Europan 9. The block as a three-dimensional puzzle, which creates an important visual axis with different directions of movement, different sunlight incidence, and ventilation. Different uses on the inside of the block and on the access side

4.2 Court

The courtyard represents an urban structural type in which buildings enclose a space, a courtyard area. This creates a protected area that is fenced off from the outside and toward which the building entrances are oriented.

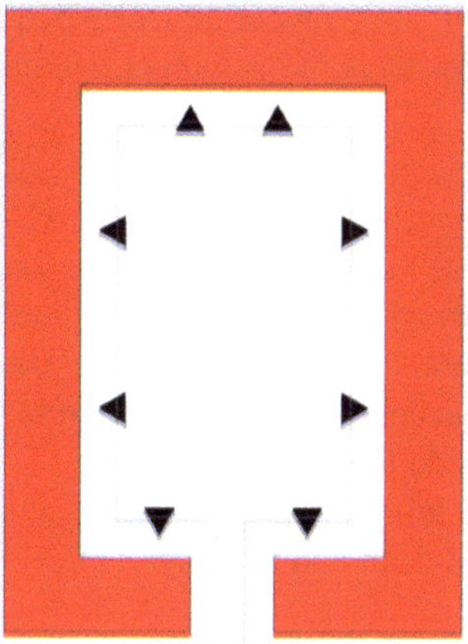

Fig. 4.43 Court, basic principle

Urban Significance

Typologically, in our cultural area, the court can be derived from traditional models such as the farm or the monastery, which is enclosed on all sides. In addition of individual buildings and parts of buildings, a space is formulated which is characterized by a certain seclusion from the context. Analogies can also be found with the courtyard house, which is a traditional building type that is widespread in Arab countries. The courtyard house is built on a single plot of land and adds up, usually without any spacing, with the neighbouring building to form a dense urban structure.

Structural Principles

The courtyard is an added element of the city. Even if this type of structure has by no means been able to attain the significance of block or row in our culture, it is nevertheless suitable for developing its own, albeit introverted, qualities in noisy or fragmentary urban structures. In several ways, the courtyard is suitable for a mixture of uses. Especially the ground floor zones are suitable for a non-disturbing commercial use (services, handicraft. . .).

Spatial Development

The term "court" is always used colloquially to describe seclusion and greater privacy. It forms an area with limited publicity for the residents or other users. The courtyard area itself is an essential element of this structural type as a reference space. The special spatial character of the courtyard, which formulates its own area between outside and inside, is well suited for communal activities of the residents, as a play area and meeting place. At the interface to public space, it assumes the function of a buffer space, which is particularly important for living.

The yard corner usually has a special function. It is—in contrast to the block—privileged throughout. It has a large proportion of land at its disposal. Courtyard and block have the same problem of orientation according to cardinal points. Due to the space-creating development, less favourably oriented buildings are also created. These location disadvantages have to be compensated by an appropriate orientation of the ground plans.

Orientation and Access

The yard can be understood as an inversion of the block. While in the block the buildings are accessed from the outside, from the street side, the buildings of the courtyard are accessed from the inside, from the courtyard area. The buildings change their orientation, because they turn their front side toward the introverted courtyard, while the back side faces the outside.

All in all, the courtyard with its development is separated from the network of public spaces. It is mostly a dead end or a private development area.

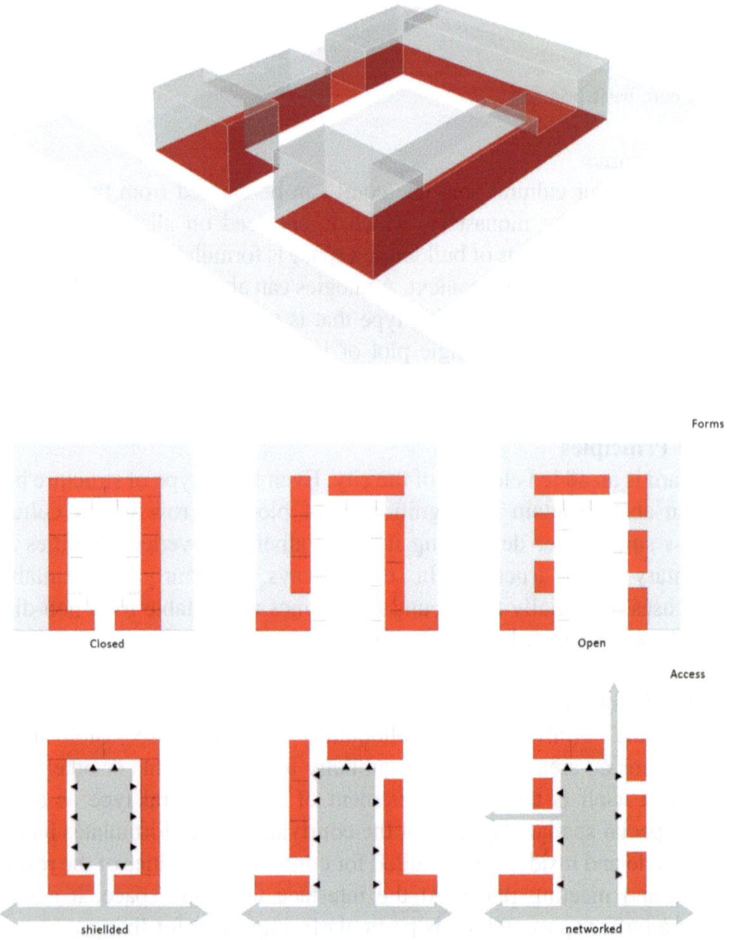

Fig. 4.44 Court, schematic sketches

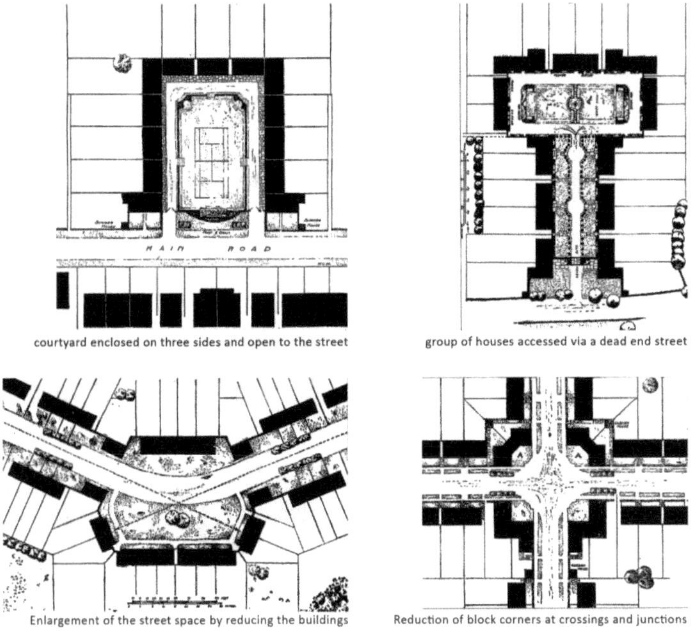

Fig. 4.45 Court historical, different historical court forms, Raymond Unwin

Fig. 4.46 Court historical, Karl-Marx-Hof, Vienna

Built between 1927 and 1930 by Otto Wagner's student and master builder Karl Ehn, the building contained 1382 apartments for around 5500 residents. One enters this city within the city via Ehrenhof and mighty gates. Only 20% of the 150,000 square meter large and 1000-meter-long area is built on, the rest is used as playground and garden area. The building contains numerous community facilities such as laundries, baths, kindergartens, a library, medical practices, and shops. The Karl-

Marx-Hof is four tram stops (about 1100 m) long, making it the longest continuous residential building in the world. The municipal building was named after the philosopher and economist Karl Marx.

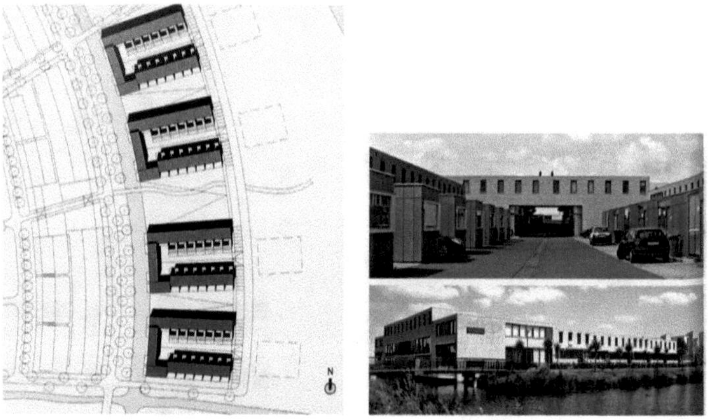

Fig. 4.47 Modern courtyard, Empel residential buildings, NL, ZEEP Architects

The four residential courtyards on the eastern edge of the settlement interpret the radial structure of the village extension in an independent way. On the basis of the urban master plan, the architects decided to condense the terraced houses into compact residential courtyards directly adjacent to public green space. The courtyards present themselves to the public street space with striking head buildings docked directly onto the surrounding canal, each of which is made up of four house units. There are no private gardens, but each apartment has a large balcony that floats above the water.

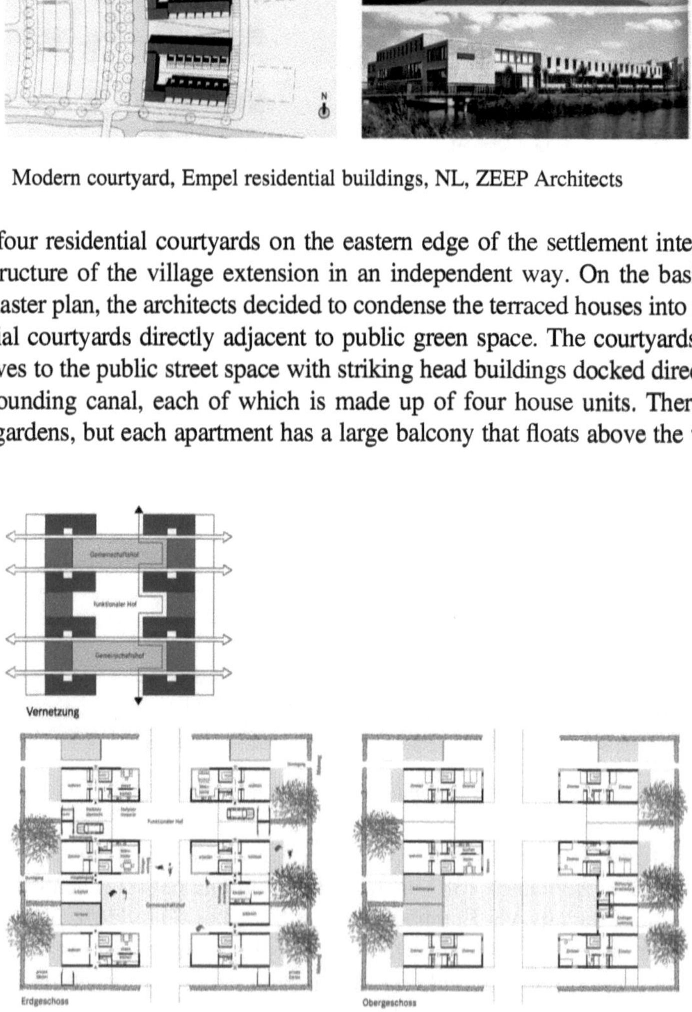

Fig. 4.48 Hof Münster, RHA architects and urban planners, competition entry

The orientation of living is possible to the courtyard as well as to the garden. A functional courtyard with a parking space and storage room alternates with a communal courtyard where the apartment entrances are located.

4.3 Rows

Rows are created by linear addition of buildings and plots along a street. The buildings can be lined up in an open or closed construction and take on the most varied spatial patterns.

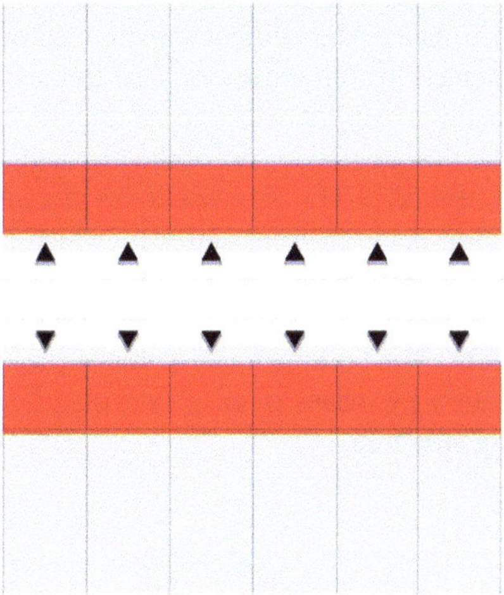

Fig. 4.49 Rows, basic principle

Urban Significance
A row is one of the oldest building blocks of the city and, because of its multiple uses, is one of the basic elements from which most cities and villages are built. Today it can be found both on urban city streets and in less densely populated residential areas. As a structural typology, the series can be arranged along access roads and streets as well as used as an independent arrangement pattern. Depending on the spatial context, different uses can be found, from heavily mixed-use urban building blocks to mono-use building blocks. In principle, the ground floors are particularly suitable for a wide range of uses (shops, gastronomy, services, etc.) in combination with the street as a development space.

Structural Principles
The row is created by adding buildings and plots along a road that can be straight, bent or curved. A distinction can be made between single-sided and double-sided rows, whereby each side of the street can follow its own geometry. Due to the flexibility of the design, the row can react to different topographic situations. As a rule, the row can be easily integrated into an urban network determined by other

structural typologies (such as solitaires, blocks, or courtyards). It is particularly suitable for supplementing and repairing gaps and fragmentary structures.

Spatial Development

The clear orientation toward the street creates clearly defined urban spaces. The spatial effect is determined by the unity or openness of the building structure. The transition area between street and building, that is, the transition from public to private space, is of particular importance for the urban spatial effect. The design of this area in the form of front zones, front gardens, parking spaces, etc., is decisive for the quality of the design of an urban space.

Orientation and Development

Characteristic of the series is that the buildings and the entrances or accesses to the buildings are always oriented toward the street. Due to the clear orientation toward the street, it clearly distinguishes between front and rear sides. The additively arranged buildings define the street-side part of the plot and at the same time ensure a great openness in the rear area. This clear orientation and zoning creates a clear socio-spatial differentiation between front and rear, which plays an important role especially for uses such as living. As a consequence, this clear orientation also means that the orientation according to the direction of the compass is determined by the arrangement of the streets. The buildings can compensate—at least partially—for disadvantages in orientation by means of an appropriate floor plan organisation or the shape of the building structure.

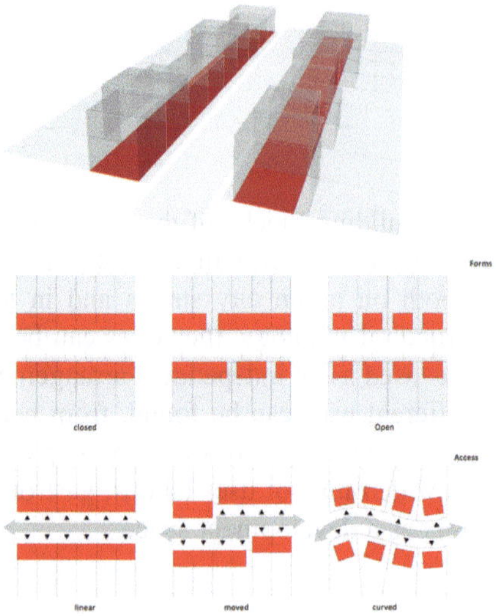

Fig. 4.50 Rows, schematic sketches

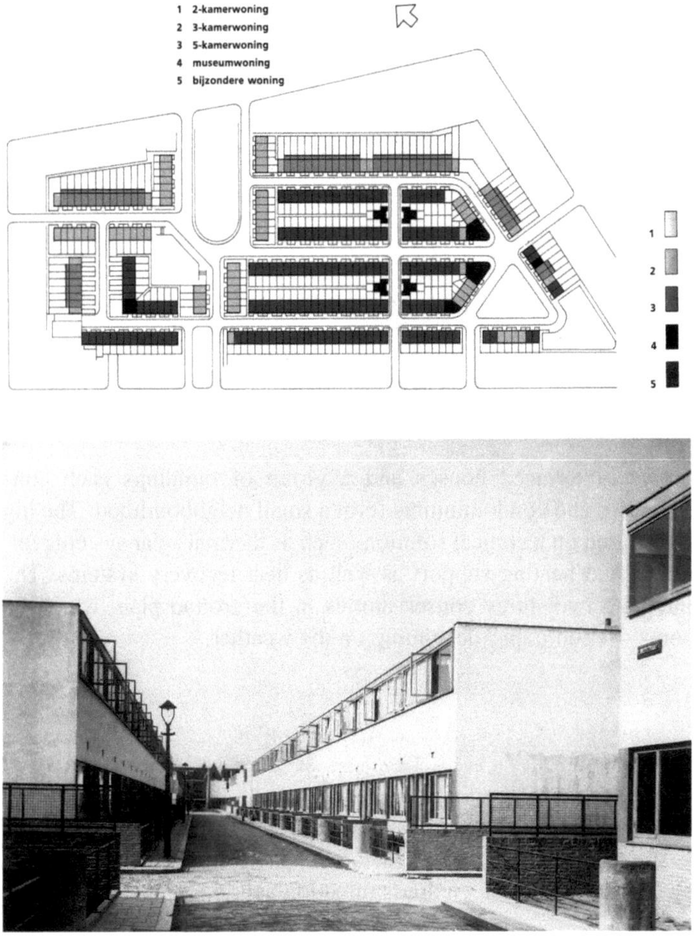

Fig. 4.51 Historic row, "De Kiefhoek" housing estate, Rotterdam, NL,—Oud, site plan

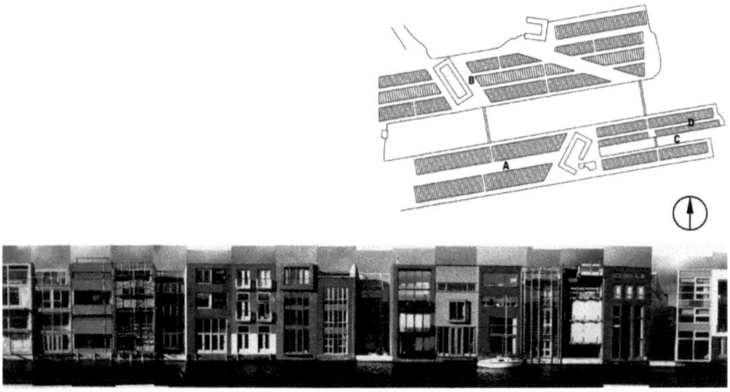

Fig. 4.52 Modern series, Borneo and Sporenburg, Amsterdam, NL. 100 row houses per hectare

Fig. 4.53 Modern series, eco houses Weihoek, Roosendaal, NL, DAT Architecten, Architecten Werkgroep, Tilburg

Two rows of terraced houses and a group of buildings each consisting of courtyard houses and condominiums form a small neighbourhood. The high-energy efficiency is based on technical solutions such as thermal solar systems for domestic water heating and heating support as well as heat recovery systems. The terraced houses integrate two-storey conservatories in the ground plan, which function as buffer rooms or "sun traps" depending on the weather.

4.4 Line

Lines are linearly arranged buildings and plots. They deliberately detach themselves from the street space and follow their own laws such as optimal organization, favourable orientation to the compass direction and topography.

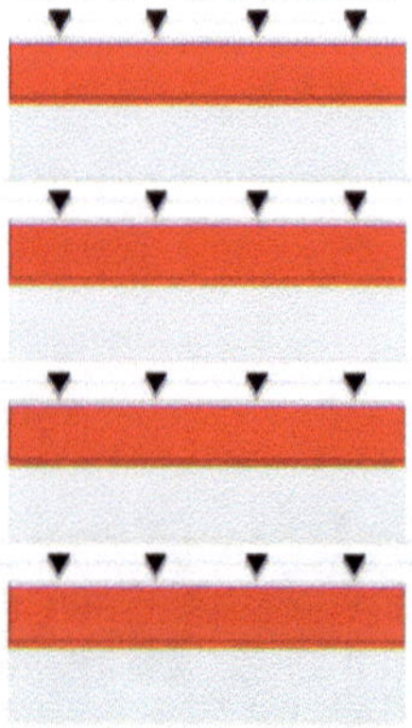

Fig. 4.54 Line, basic principle

Urban Significance

In Germany, terraced housing emerged in the late 1920s as a reaction to the poor housing conditions in the densely populated blocks in the inner cities. In the development of this structural typology, the desire to improve the housing conditions, especially the lighting of the apartments, especially for the lower income groups, was at the forefront.

In the modern age as well as in the phase of reconstruction after the Second World War, terraced housing has played an important role. The current discussion about a more energy-oriented urban development has led to a real renaissance of the line. The north-south orientation of the buildings offers advantages for active and passive solar energy use.

Both the line concepts of the more distant as well as the more recent history of urban development show that by adding up the same lines there can be no or only a slight spatial reaction to specific contextual conditions. The urban structure type of the line stands for a functional and spatial segregation, which makes it difficult to meet the demands of urbanity. For this reason, the line is often referred to as an unurban or even less urban building structure.

Structural Principles

The line is created—similar to the row—from an addition of buildings and plots, which however follow their own laws and are organized independently of the superordinate development system.

The line is the result of a rationalization of the living space floor plans, the serial production as well as the so-called "living hygiene," because the line can be rotated in the orientation of the living spaces so that the desired orientation toward the sun is achieved. The orientation toward the sun (east-west or north-south lines) creates complete autonomy from the surrounding context and from the street layout. Rationalization also includes the serial repetition of the building itself and the renunciation of special urban conditions of a location. Due to its neutrality toward the location, the line has an "unurban" quality. This spatial impression is reinforced by the monofunctional use of lines and the lack of contact between the building use and the public space.

Spatial Development

Lines deliberately turn away from the city system and form independent, self-referential organizational units. Depending on the type of building, different spatial effects can arise. While an addition of single-family houses can create a differentiated effect of the individual house, linearly organized multistorey residential buildings tend to create the impression of closed large structures or large building volumes. The spatial effect is decisively determined by the meeting of different open spaces, private front areas, and rear gardens. The rows do not have to be arranged parallel to the development system; a differentiated spatial effect can be created by jumping and shifting.

Orientation and Development

A special feature of the line is that the gable sides or line headers are oriented toward the main indexing. The line buildings themselves are oriented toward their own secondary development, which is often dedicated as a private path or designed as a cul-de-sac. Due to the one-sided development, the front and back sides of the buildings meet and are usually separated from each other by spatial elements (basement substitute rooms, walls. . .).

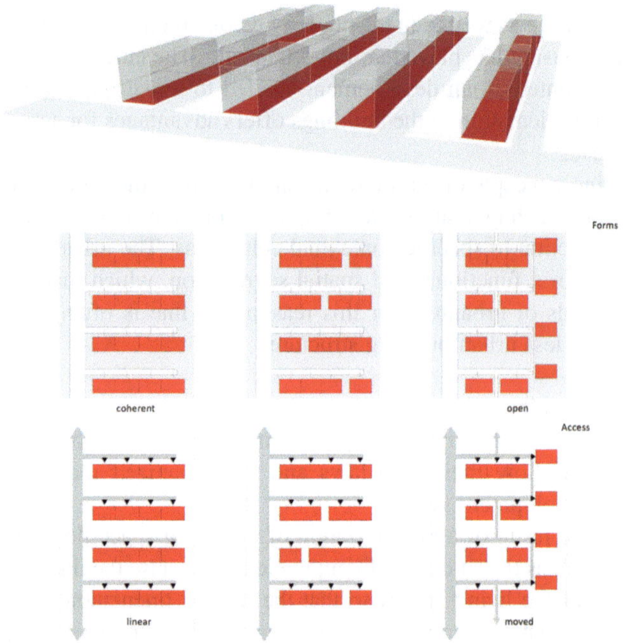

Fig. 4.55 Line, schematic sketches

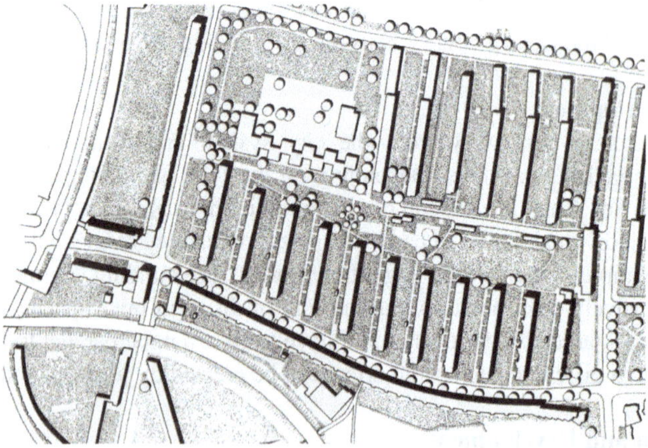

Fig. 4.56 Line historical, Siemens City, Berlin, housing estate in Berlin-Charlottenburg-North, Hans Scharoun

The large housing estate Siemens City was built between 1929 and 1931 under the direction of architect Hans Scharoun in Charlottenburg North as an eastern extension of Siemens City in Spandau. In two construction phases, almost 1400 apartments were built between Heckerdamm and Mäckeritzstrasse in accordance with designs by Walter Gropius, Otto Bartning, Hugo Häring, Fred Forbat, and Paul Rudolf Henning among others. Almost 80% of the units have one and a half and two and a half rooms.

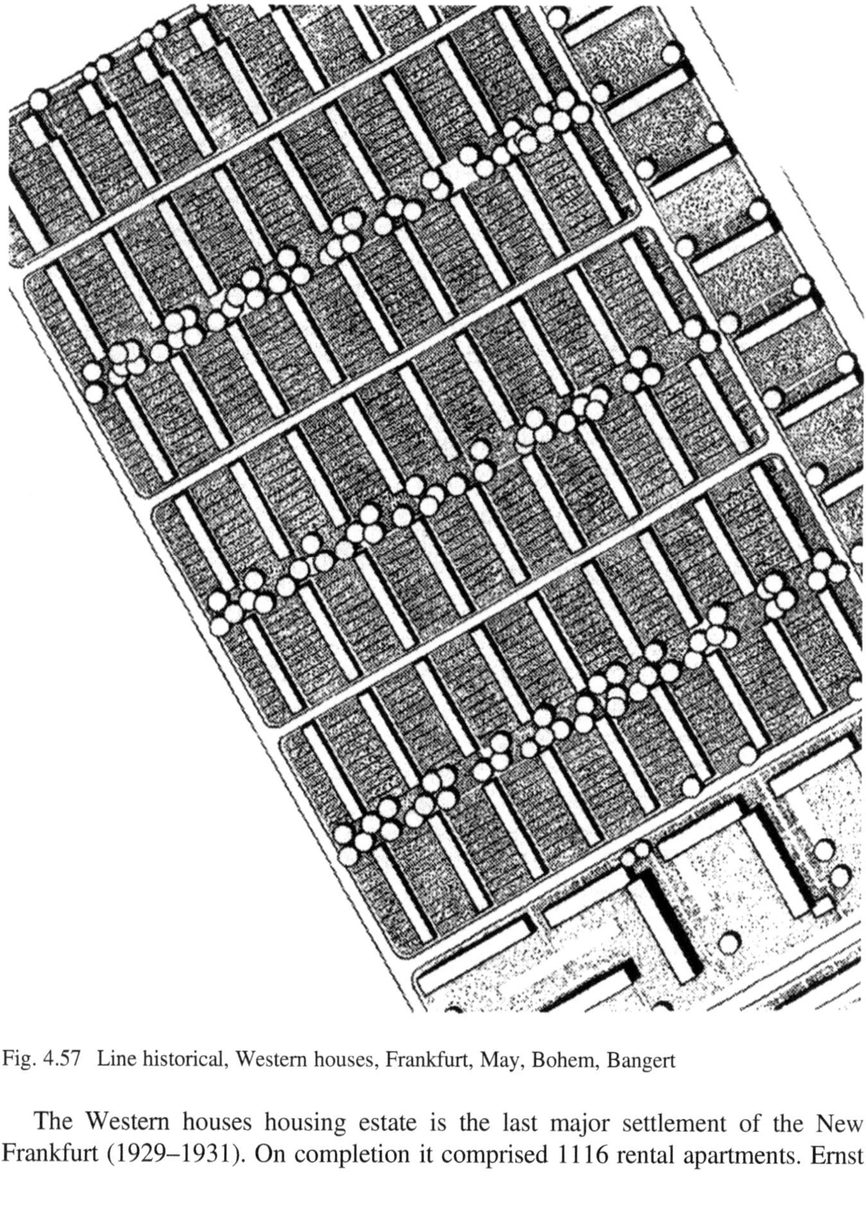

Fig. 4.57 Line historical, Western houses, Frankfurt, May, Bohem, Bangert

The Western houses housing estate is the last major settlement of the New Frankfurt (1929–1931). On completion it comprised 1116 rental apartments. Ernst

May was responsible for the management of the housing construction program and the preparation of the general development plan as a city building councillor. He was supported by Wolfgang Bangert, Eugen Blanck, Herbert Boehm, and Emil Kaufmann, among others.

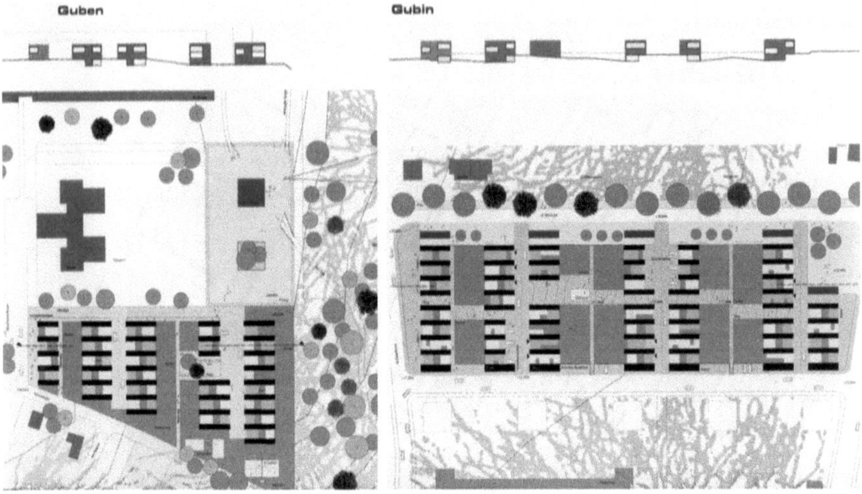

Fig. 4.58 Line modern, Guben-Gubin, Competition Europan 5

The design aim was to upgrade both districts through an overarching green concept and the creation of new relationships between them. Derived from its specific location in the landscape, each district is given its own open space theme, which will be formulated in the new residential quarters.

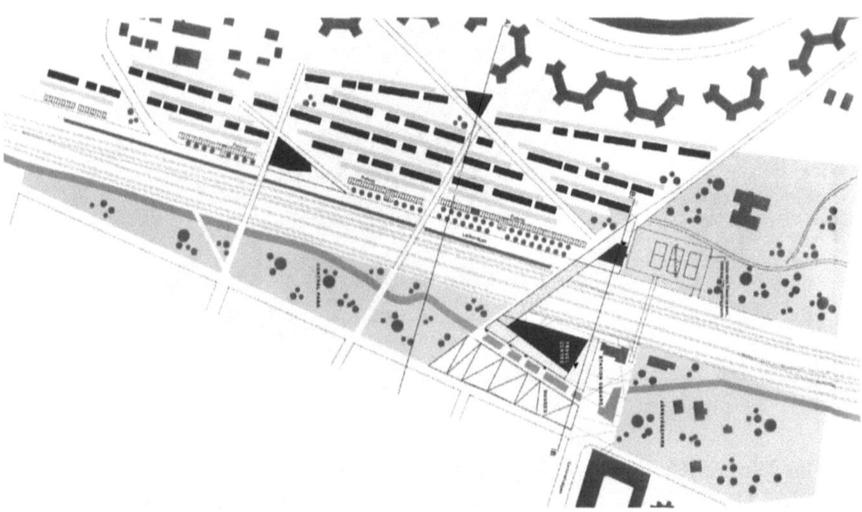

Fig. 4.59 Line modern, Upplands Väsby, Sweden, Competition Europan 9

The settlement area is intersected by a railway line. New connections will be created by mixed-use buildings, a pedestrian ramp, and an underpass. Lines run parallel to the route of the railway line; these allow for different types of housing. All apartments and gardens face west.

4.5 Solitaire

Solitaires are buildings that either stand isolated in the landscape, such as individual farmsteads, or buildings that cannot or should not be connected to others because of their underlying concept or size.

Fig. 4.60 Solitaire, basic principle

Urban Significance
Solitaire has been more or less pronounced in all phases of the history of urban development. The type of free-standing building can be found both in the rural cultural landscape as a farmhouse and in the form of stately buildings such as castles and palaces, but also in the form of public buildings such as the town hall, church, and theatre.

In the Renaissance phase the type of villa was particularly evident. In the nineteenth century, solitaires were often used as end points of road axes. During the industrialization, the production facility, the factory, was removed from the structure of the city as a solitaire. Modernity has given the solitaires a special significance as an addition of autonomous structures surrounded by generous green spaces.

In recent urban development history, high-rise buildings in particular, but also free-standing towers with office or residential use, have taken on the role of solitaires. Today, this form of solitaire is discussed in locations where there is great pressure for the economic exploitation of a site or where spatially effective "dominants" are sought.

The variety of uses shown shows that the solitaire is suitable for a wide range of different uses. In most cases, however, it is a specific use per building. Mixed uses often occur as additions, for example, as a service use in the basement floors of high-rise buildings.

Structural Principles

The form of solitaires is often determined by the inner function and the desired external effect. Therefore, the solitaire appears in different geometries: as tower, disk, cube, cylinder and other organic forms. Even when the solitaires are used in combination with other solitaires, the spatial relationship between them is minimal.

However, solitaires also appear where urban structures break off or dissolve at the edges and individual buildings still exist. Solitaires are also used as conscious settings to set accents in the cultural landscape and thus to create landmarks and identification.

Spatial Development

The solitaire stands out from the rest of the development as a spatial accent. As a rule, it occupies a larger spatial area, which enables it to show little consideration for the neighbouring buildings. The solitaire is not only autonomous in terms of the size of the building but also in terms of architecture and floor plan organization. The outer form of a striking urban landmark is often not in harmony with the qualities of use. The relationship between external appearance and internal functions often does not appear to be sufficiently resolved. Due to the distance to the surrounding buildings, the solitaire is visible from all sides. There are not the classic front and back sides, as we find them especially with the block and row, but the solitaire triggers a design demand to all sides of the building.

Orientation and Development

Both in terms of orientation and development, the solitaire claims a certain autonomy from the context. The differently used solitaires thus also show a variety of development possibilities.

Due to the exposed situation, the building and its use offer a wide range of possibilities for orientation and alignment to the respective cardinal points. The solitaires themselves, in particular the high-rise buildings, can cause a negative impact on their surroundings through shading. Recent discussions about the influence of high-rise buildings on the "hierarchy of the city" (see Chapter VII.3) in Cologne and Munich show that high-rise buildings not only have functional consequences but also decisively impair the status of traditional buildings in the historical context of the city.

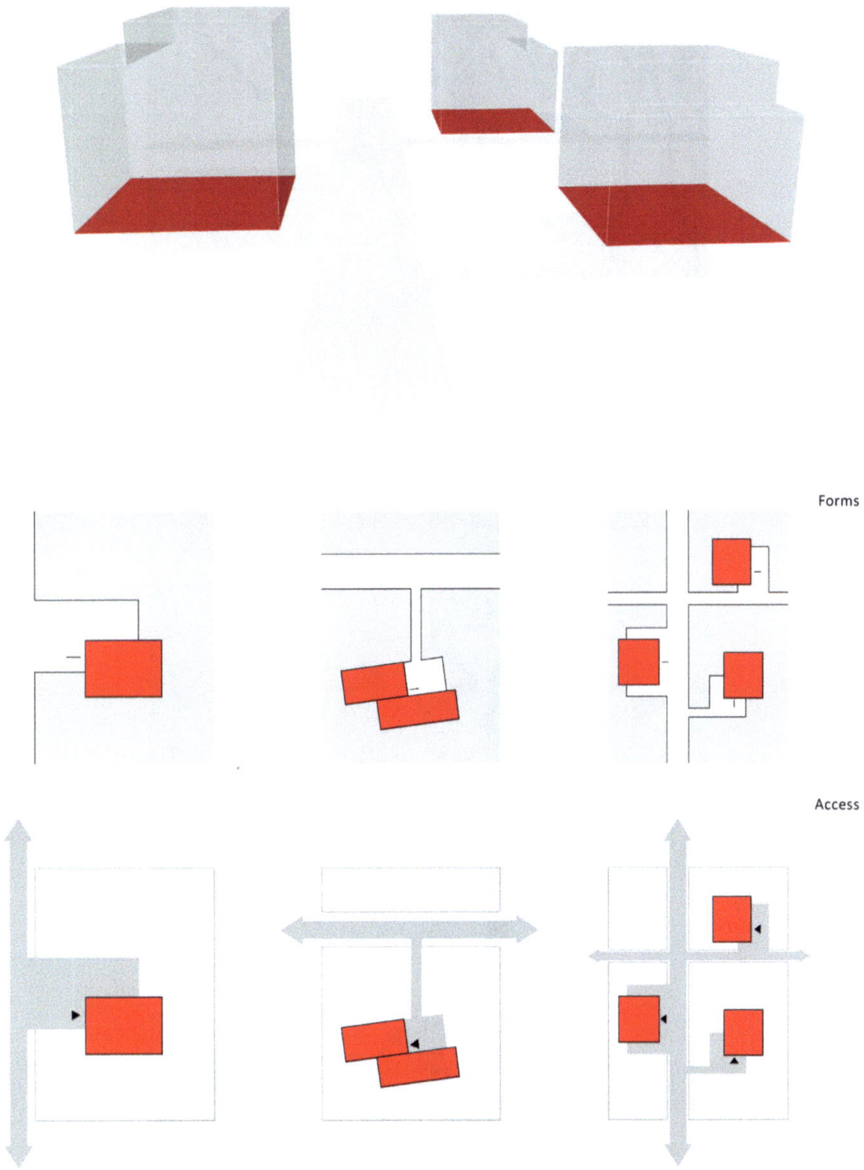

Fig. 4.61 Solitaire, schematic sketches

Fig. 4.62 Solitaire historical, administration building Düsseldorf, 1955–1960, Helmut Hentrich and Hubert Petschnigg with Fritz Eller, Erich Moser, Robert Walter, and Josef Rüping. The "Three-panel house" juts out of the green area under which the parking garage is hidden. The emphatically slender shape characterises the silhouette of the city of Düsseldorf

Fig. 4.63 Solitaire historical, Villa Badoer, Fratta Polesine, Italy, Andrea Palladio, 1556–1563. Partial view. The villa and its estate were built directly on the banks of the Scortico. It is oriented equally to all four cardinal points

Fig. 4.64 Solitaire modern, urban villas Star road Kassel Kassel Penkhues Architects, Kassel

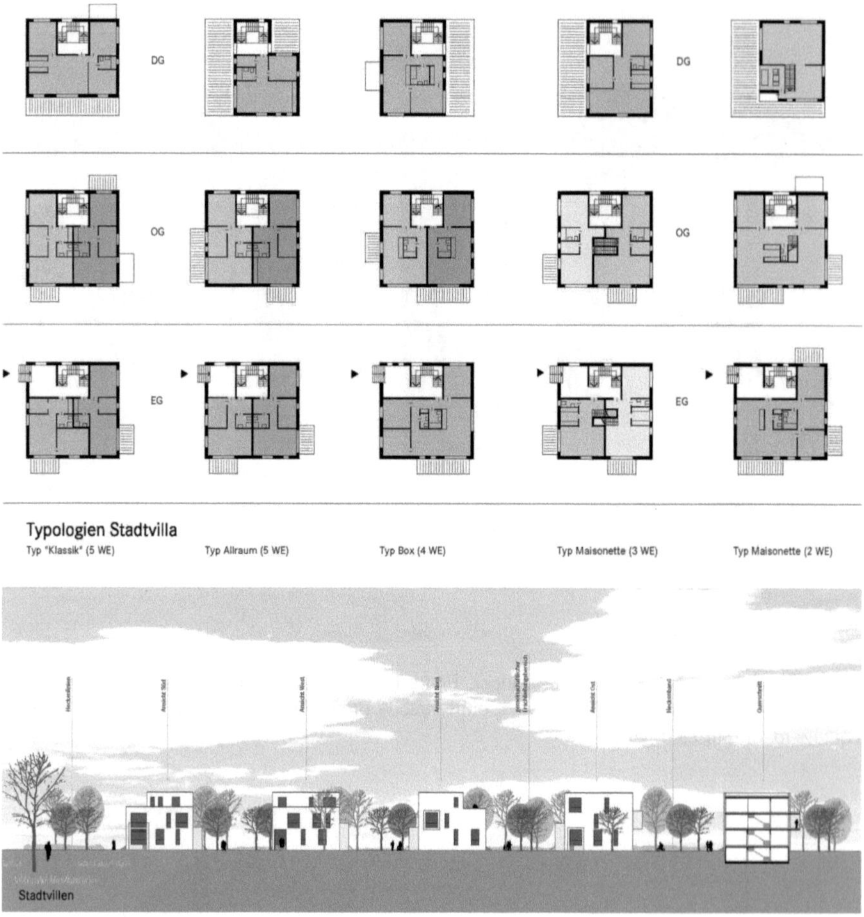

Fig. 4.65 Solitaire modern, City villas Remscheiddtvillen Remscheid, RHA Architects + Urban Planners. City villa type with different floor plan design possibilities

4.6 Clusters

A cluster is an urban structural typology in which an arrangement of buildings follows its own compositional logic. The English term cluster means grape, bundle or swarm.

Fig. 4.66 Cluster, basic principle

Urban Significance

Clusters already appeared during the industrialization of the last century as workers' housing estates, in the garden city movement as larger housing groups and, since the 1960s, increasingly as suburban settlements. Such projects are often used to communicate certain urban design models (such as space-saving construction or similar). In the case of housing projects, cluster development often expresses the idea of communal living together.

Overall, cluster structures are occupied by one use, mixed uses occur less frequently due to the limited size and the strong spatial relationship of the individual building blocks.

Structural Principles

This cluster construction is often referred to as a group construction, as a group of objects is combined to form an organizational unit. Each element of the group refers to the other, so that there is a mutual interaction of the individual components of the cluster. Clusters can occur in open and closed construction, resulting in different spatial impressions of density and openness.

Spatial Development

An essential feature is, therefore, the spatial separation from the building context, often caused by a deviating form or geometry of the arrangement. Because of their separation from their surroundings, clusters are independent parts of the city. They, therefore, facilitate recognition and thus identification by the residents. As a consequence, however, this separation of context can mean that the connection to the surrounding urban structure is reduced and, in extreme cases, clusters can develop as islands in the structure of the city.

Clusters are, therefore, organised according to their own spatial laws. They can follow geometrical laws or—caused by the topography or other spatial conditions—be more organically modelled.

Orientation and Development

Clusters follow their own internal organisation and development principle, which is distinct from the superordinate network. They are often arranged around a common area, a courtyard, a square, a green space, and make a spatial reference to it. What is important here is not so much a specific relationship to public space—as in the categories dealt with so far—but rather a composition according to its own laws.

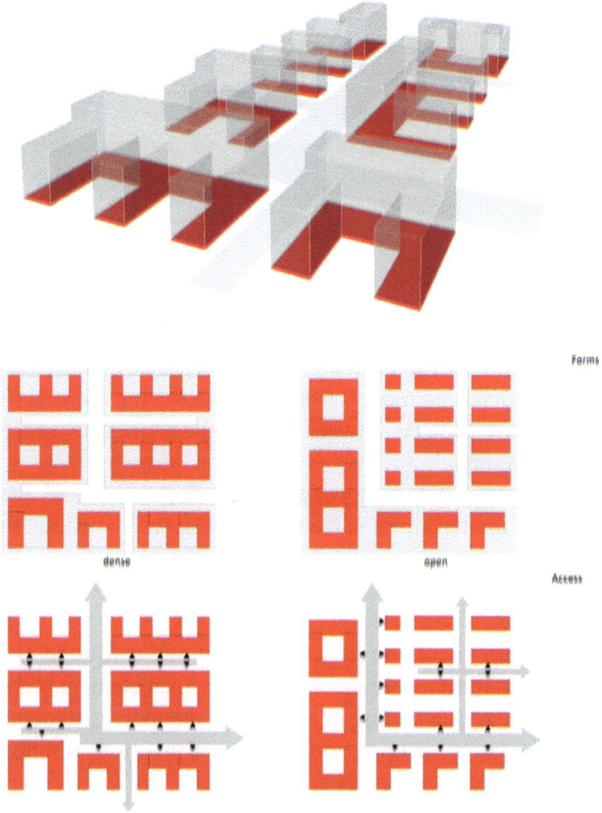

Fig. 4.67 Clusters, schematic diagrams

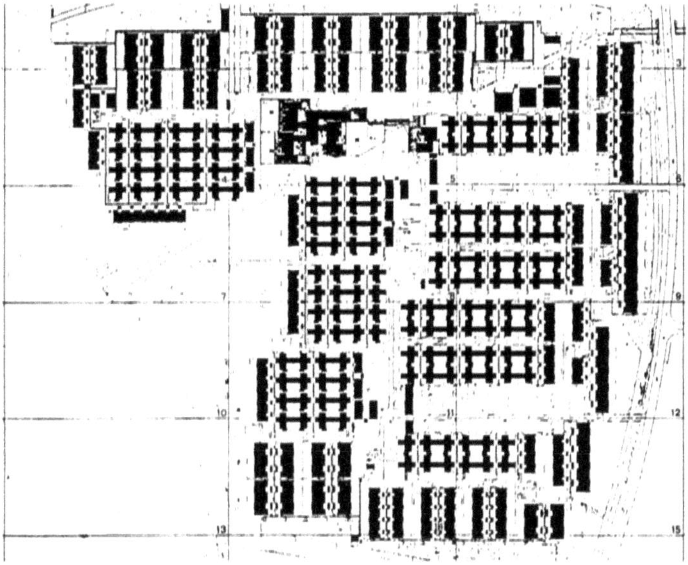

Fig. 4.68 Cluster historical, settlement Galgebakken Albertslund, Denmark, 1968. The settlement consists of a carpet-like, low-rise, high-density development. The main idea is an intimate low-density housing estate with small private open spaces and extensive community facilities

Fig. 4.69 Historical cluster, design stucco house goods, Atelier 5, 1987. The competition design shows a type of compacted construction. The design was not realized, but its principles were taken up in the project Ried-Niederwangen, 1991

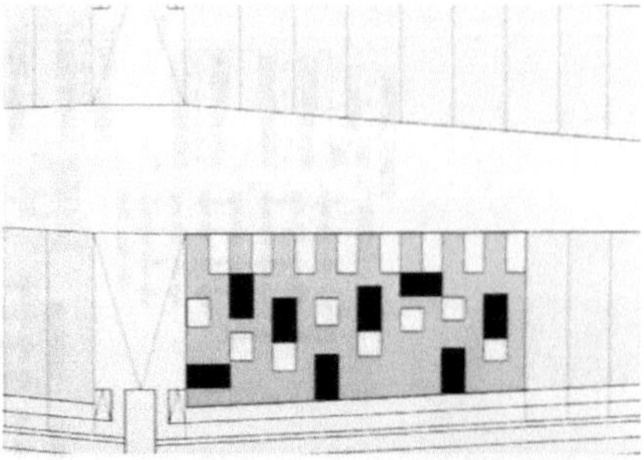

Fig. 4.70 Cluster modern, villas on Klein Rieteiland, Amsterdam Ijburg, OeverZaaijer architecture and urbanism, Amsterdam, NL, 2007. Combination of the design principles of a patio house with those of a detached house. Interplay between introverted and extroverted rooms

5 Tradition and Change in Urban Structures

The twenty-first century is considered the urban age—the century in which, for the first time, the majority of humanity lives in cities, a total of about three billion people, one billion in the north and two in the south. In the industrialized countries almost every second city dweller is a "city dweller," in Latin America every third, in India and China every tenth (see also, Ribbeck 2005). The global wave of urbanization first hit Europe and then North America, half a century later Latin America, and is now overrunning Africa and Asia.

In the discussion about urban age, the focus is usually on the rapidly growing mega-metropolises: Mumbai, Cairo, Sao Paulo, Shanghai, Istanbul, or the traditional global cities such as New York, London, Tokyo, Moscow. These multimillion cities are only one aspect of the global urbanization process, because the majority of city dwellers do not live in the megacities, but in small, medium-sized, or large cities below the five million mark, which are more often organized into large agglomerations.

The phenomena of globalisation are not only felt in industrial metropolises, they also put pressure on metropolises in the south of the world. Building structures are being reshaped by changing demands on living and working. In the southern metropolises in particular, the difficult balancing act between meeting the demands of the elite on the one hand and the basic needs of the masses on the other hand is becoming apparent. Parts of the city are being transformed into "gated districts" accessible only to certain population groups, or into insecure ghettos in which the people living there remain among themselves.

The megacities in Asia, Africa, or Latin America are "population monsters" in which people keep themselves afloat without large resources. A transfer of North American energy and land standards would lead to ecological collapse.

The dynamics of spatial and social development have always produced new models for urban development. Thus, Europe has shaped the nineteenth century, North America the twentieth century and it is evident that the standards for urban development of the twenty-first century are being set in the Far East. Global urbanization has not only brought about enormous quantitative upheavals but has also provoked qualitative leaps. Many of the cities that are being built around the world can no longer be measured by traditional urban design values. They have their own qualities: contrasts and breaks, dynamics and change or simply improvisation. For this reason, it is understandable that the spectrum of building forms and urban design patterns is more diverse today than ever before. The crucial question is how to assess and evaluate these international developments, how to deal with social polarisation and local building cultures and what quality of life is created.

Parceled Block

The subdivided block offers space for different building sizes on several plots. The front sides of the buildings face the public street space. Despite distances between the buildings, there is no relationship between the outer, public area and the privately used inner area of the block. This type of block is used in multifamily housing or to accommodate mixed uses in close proximity.

Closed perimeter block/block edge development

The closed block follows the street on all sides and presents itself as one building on the public side. The interior of the closed block is private and available for the residents. The density that the block creates is emphasized by several floors.

Open perimeter block/open block edge development

Like the closed block, the open block perimeter development is characterized by a higher number of storeys and is also oriented toward the public street space. The inner area of the block is accessible from the street and, therefore, has a semipublic character. This type of block was used especially in the twentieth century.

Mat block

The area block makes maximum use of the plot and thus creates a very high density. This density is supported by a high storey density. Examples of this block typology are the Berlin Block or the block development in Barcelona.

Mat fabric

Intensively used space structures can be found especially in European city centres. They are used to accommodate different uses on a maximally utilized area. Only streets and public spaces represent free spaces within the intensive development.

Field

The structure of the field is determined by buildings that are arranged on public or semipublic open spaces and are not oriented toward public street space. This typology is typical of the way urban planning was handled in the first half of the twentieth century and was applied in varying degrees of density and height.

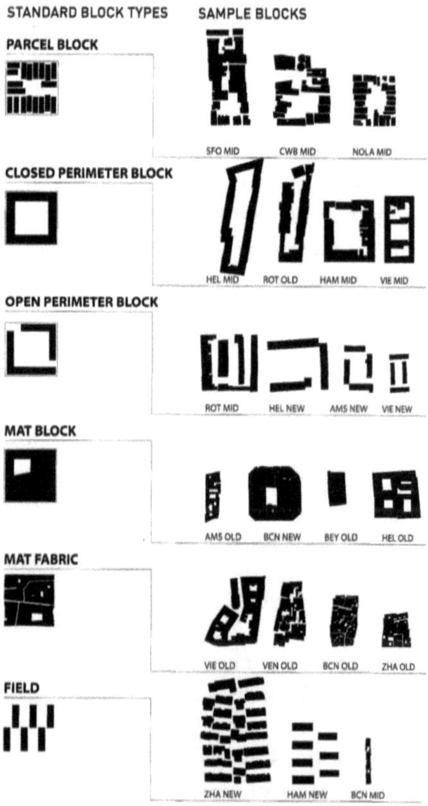

Fig. 4.71 Block types. AMS—Amsterdam, BCN—Barcelona, BEY—Beirut, CWB—Curitiba, HAM—Hamburg, HEL—Helsinki, NOLA—New Orleans, RED—Rotterdam, SFO—San Francisco, VIE—Vienna, ZHA—Zhanjiang

5.1 The European City

The quality of the European city is based on a long history, which includes Roman origins and the Middle Ages, Renaissance and Baroque, Wilhelminian and modern. If its characteristics are named, then the urban compactness and homogeneity, the small-scale parcelling and mixture of uses, as well as the urban density and the relevance of public space are mentioned. In the international context, the European city has an average density and height.

With respect for its historical heritage and warned by exaggerated modernist models, the European city has been slow to adopt new urban and building typologies.

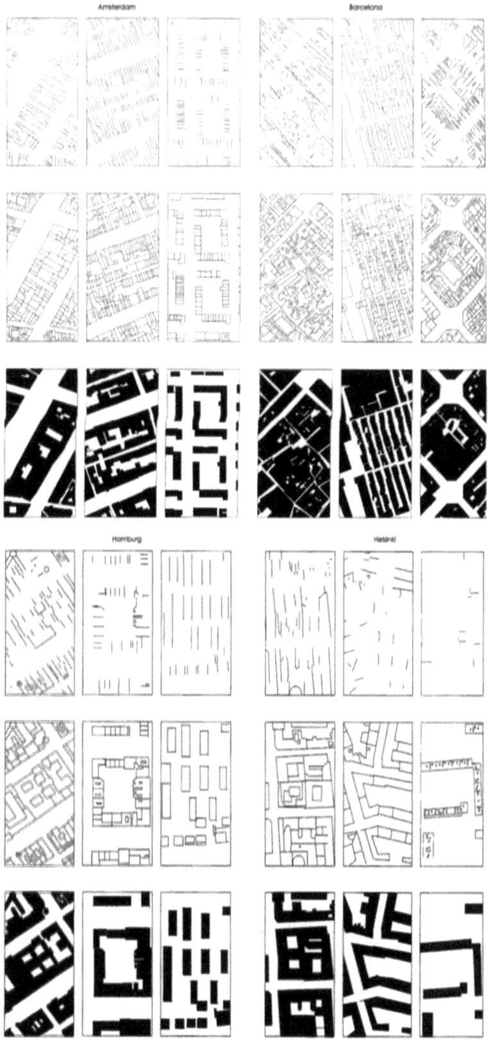

Fig. 4.72 Block types in the European cities of Amsterdam, Barcelona, Hamburg, and Helsinki

5.2 The US City

The traditional features of the US city are the mostly chessboard-like, orthogonal street grids. The incipient suburbanization with an area growth beyond the city limits has been accompanied by an emptying of inner city areas. Despite this spatial expansion, the number of inhabitants and jobs remains roughly the same. This is followed by the emergence of "edge cities," which, as peripheral cities, are competing with the "central business district," with its predominantly business high-rise buildings. The North American city is thus characterized by its vertical "central business districts" and the seemingly endless "suburbs." New developments and strategies in the US urban design such as "Smart Growth" and the "New Urbanism" attempt to create new qualities in the sprawling structures of the "suburbia," drawing on historical European structures.

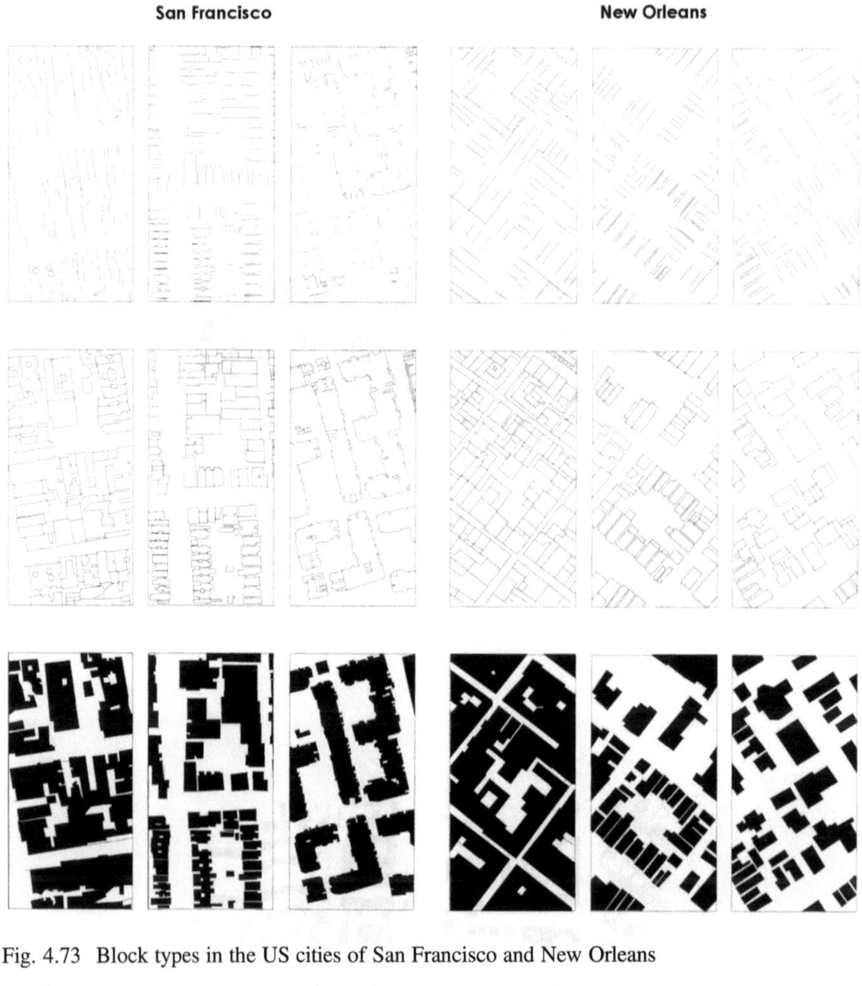

Fig. 4.73 Block types in the US cities of San Francisco and New Orleans

Fig. 4.74 The US Suburbia. Suburbia in California (left) and around Boston and Washington (below)

5.3 The Asian City

In contrast to the European or US-American city with its specific significance, the Asian city encompasses urbanization developments on a vast and geographically diversified continent, the shaping of which by different cultures led to different forms of cities. In the densely populated metropolises, the urban development structures are characterized by an alternation of small-scale local and traditional structures as well as the overall urban or globally oriented ultramodern "urban high-performance engines" of extreme height and density. These mostly cope with the enormous capacity and functional demands of use in superblocks. Public space, often limited to streets, plays an important role in cities despite its scarcity. In the Middle East, historically valuable urban structures are often threatened by lack of protection and wars. Urban planning in China is often accused of relying on an unbroken belief in progress and destroying cultural identity by removing traditional urban structures. In Japan, too, there is a freer approach to urban design, despite pronounced traditions. The new cities of East Asia are not shaped by historically grown citizenship and public participation, but rather by "public-private partnerships" and "joint ventures," whereby investors invent new urban and building typologies with great freedom.

Fig. 4.75 Block types in the Asian cities of Beirut (left) and Zhenjiang

Fig. 4.76 Images of the Asian city. Scheme of a superblock, alternation of small-scale structures inside and large-scale structures at the edge of the block. Tokyo from above, 2008, Photo: Jürgen Krusche. Shanghai, 2010. Shanghai, Redesigned public space at the federal government, 2010

Literature

1. Understanding the Place and the Code

El Khafif, Mona 2009: Inszenierter Urbanismus. Stadtraum für Kunst, Kultur und Konsum im Zeitalter der Erlebnisgesellschaft. Saarbrücken: Vdm Verlag Dr. Müller.

2. Typologies and Layers of the City

Wiegand, Jürgen 1973: Funktionsmischung. Niederteufen.
Peterek, Michael; Bürklin, Thorsten 2007: Basics Stadtbausteine. Basel, Boston, Berlin: Birkhäuser.

3. City Logics

Borchard, Klaus 1974: Orientierungswerte für die städtebauliche Planung. Hrsg: Institut für Städtebau und Wohnungswesen der Deutschen Akademie für Städtebau und Landesplanung, München.
Rowe, Collin, Koetter, Fred 1997: Collage City. Basel, Boston, Berlin: Birkhäuser.

4. Urban Structure Types

Curdes, Gerhard 1995: Stadtstrukturelles Entwerfen, Stuttgart, Berlin, Köln: Kohlhammer.
Jonas, Carsten 2009: Die Stadt und ihr Grundriss. Tübingen, Berlin: Wasmuth.

5. Tradition and Change in Urban Structures

Berman, lla, El Khafif, Mona (2008): URBANbuild Local Global, San Francisco.
Ribbeck, Eckhart 2005: Die Welt wird Stadt. Stadtbilder aus Asien, Afrika. Lateinamerika. Berlin: jovis.

Urban Components of the City 5

1 Urban Component Open Space

Ulrike Beuter

Many cities have now grown into an extensive urban landscape. The dissolution of the boundaries between city and countryside has led to a patchwork-like image of the built on the one hand and landscape on the other. At the same time, landscape should not be confused with nature. Rather, it is part of the diverse cultural landscape in which the old city centres are only accents. Against this background, cities and landscape are no longer the traditional antagonists, but merely different expressions of the cultural imprint of our environment. The idea of making cities with landscape is not at all far-fetched. On the contrary: it is obvious.

Landscape can easily form an equal part of the urban fabric in the city. The function of the landscape varies according to the respective situation. In the inner city context, landscape can support and strengthen urbanity as an urban development element. Landscape and open space have often become the hallmark of cities. This means that a differentiated range of high-quality, well-maintained, and easily usable open spaces is increasingly becoming the hallmark of a city. A well-networked open space system is thus increasingly developing into an important location factor, for example, in the settlement of high-quality jobs, and is thus becoming an important economic factor for a city in addition to its ecological significance. The bond between young families and a city is also determined not least by the qualified supply of open spaces.

In contrast, in view of their tight budgets, cities, and municipalities are hardly in a position to adequately maintain or modernise their open space stock or even develop new open spaces. Thus the question of improving the quality of life through high-quality open spaces becomes existential and calls for action. In addition, the gap between growth regions and regions in the process of shrinking is widening. In boom regions, landscape is increasingly being consumed, that is, the physical quality of the

C. Reicher, *Urban Design*, https://doi.org/10.1007/978-3-658-34370-5_5

region suffers. In "shrinking cities," land becomes available that can be returned to the eco- and open space use system.

Historical Development

In the past, the garden was created as a protected place surrounded by walls, fences or hedges, defying the wild, untamed nature or being excluded from the open, cultivated landscape. As a kitchen garden it served to supply food and to cultivate medicinal plants.

In the Middle Ages, private gardens were usually located in front of the compact stone city and functioned as private kitchen gardens. They were, despite the gardening, mostly places of rest and relaxation. Within the walls of the city, squares, alleys, paths, and courtyards represented the stone "free-space" system, which regulated supply and disposal, had a connecting function, represented economic (market) and assembly spaces (folk and church festivals) and reflected political power (court, parade ground).

In the Renaissance, the embodiment of the divine in nature, beauty, and harmony was again recalled. The feudal gardens of the Italian Renaissance, with their complex geometry and symmetrical harmony, bear witness to these ideas.

Fig. 5.1 Breakfast in the country, Manet, 1863 (left), album cover rock group Bow Wow Wow (right), 1981

The people moved out of the narrow city to its gates for certain festivals, went into the countryside to celebrate.

Larger, park-like gardens were reserved for the ruling class, the nobility, or clergy until the eighteenth century and were not open to the public, such as the park of Versaille (Baroque, Absolutism) or the English Gardens of the Romantic period.

With the increasing self-confidence of the merchants and the free citizens, as well as due to prospering trade, in the eighteenth and nineteenth centuries, ornamental squares, citizen and public gardens were created for the pleasure of the citizens. The first public green spaces created by the citizens in Germany were the ramparts in Bremen, which still characterize the image of the city centre today.

In the course of industrialisation and the associated densification of large urban workers' neighbourhoods of the Wilhelminian period, Volkspark was created at the beginning of the twentieth century, which was intended to consciously contribute to the hygienic and health improvement for the population living in unhealthy conditions as usable spaces. Sports areas, gymnastics meadows, paddling pools, light, and air baths were part of the health programme.

Parallel to this development, the small-scale factory settlement structure was taken over from England, primarily in the phase of the coal and steel industry, and later the garden city idea of Ebenezer Howard, in order to bind the constantly fluctuating skilled workers and with them the know-how acquired to the production site. The large utility garden areas were vital free spaces for the employees. At the same time, they were retreats or supplementary spaces for the very narrow, overcrowded apartments.

After the Second World War, rental gardens were also a widespread, vital strategy for supplying the population in multistorey housing construction or extensive burial grounds on the periphery of cities. The postwar reconstruction was also based on the Athens Charter (1933), which, as a reaction to the inhumane living conditions in the dense urban structures, called for a disentanglement of the usage structures, divided by generous green spaces, and propagated a clear greening of the residential areas.

In the prosperous 1960s and 1970s, this principle of separation of functions was further developed and culminated in the traffic-oriented city. The initially tempting approach of a safe network of footpaths independent of the road network increasingly turned into the opposite, with street spaces developing into purely functional, insurmountable barriers in the middle of cities. Important open spaces and streets that had shaped the cityscape were lost, and districts were cut off from one another. Many cities lost their historical reference to the river, which was separated from the urban space by streets. Today, many cities are concerned with the deconstruction of these problematic traffic axes into appropriate street spaces that are suitable for everyday use and integrated into the urban design.

The ecology movement in particular opposed functionalized urban development with its demand for the improvement of the living environment. It advocated a reconquest of the urban space by its inhabitants, on the one hand, and on the other hand, the consideration of ecological concerns in order to safeguard natural resources and to increase the ecological diversity in the city.

A look at the historical development of urban open space shows that the content and characteristics of open spaces always reflected the current sociopolitical and cultural attitude of the respective epoch.

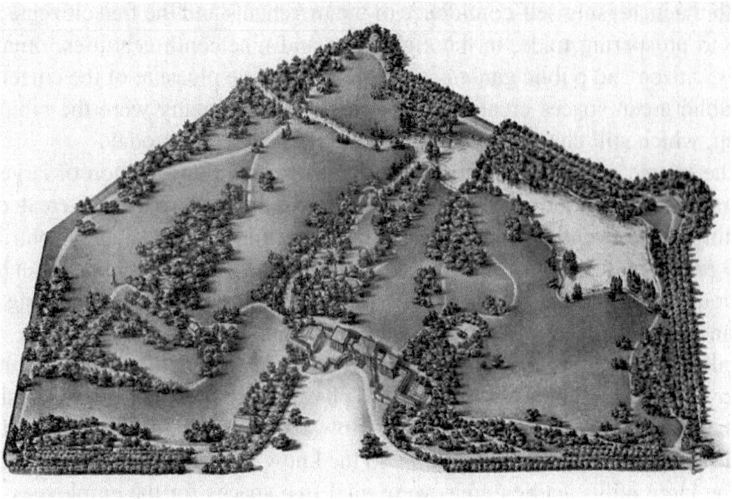

Fig. 5.2 Engraving of the English landscape park of Stowe

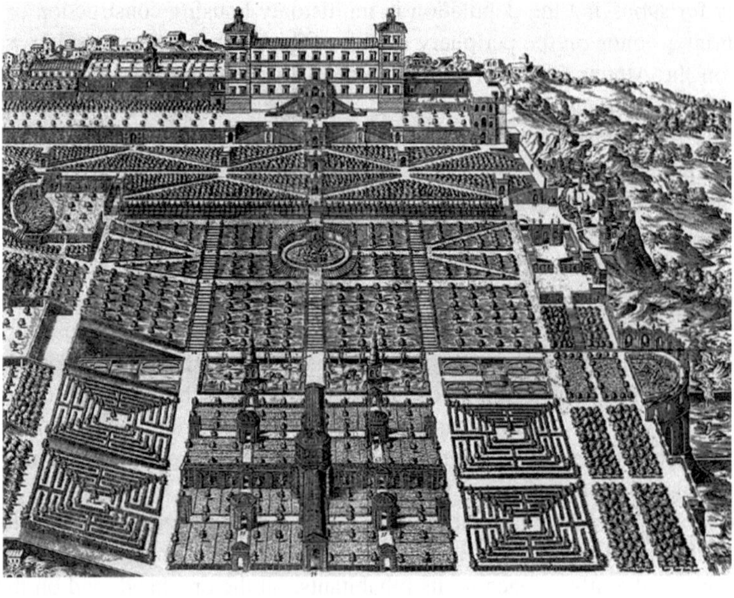

Fig. 5.3 Engraving Renaissance Park of Villa D'Este, Italy

Today's Challenges

The current topics of urban development also have an impact on the open space system and design. Demographic shrinking processes lead to more usable space for the remaining residents. This creates good conditions for a better quality of life and a positive ecological balance. Nevertheless, an increasing segregation of open space use is taking place, particularly in attractive neighbourhoods, which is due to social displacement processes.

The aging of the population poses new challenges: reduced mobility of this growing population stratum must also be taken into account in open space and public areas. Barrier-free access, recreation, and exercise areas suitable for the elderly, as well as walker-friendly design of surfaces are also a current topic outdoors to enable elderly people to live a life as independently and self-determined as possible.

Fig. 5.4 Ideal green system of the city of Münster

Climate change is affecting fauna and flora to a large extent. An improvement of the local climate can be achieved by sufficient vegetation: A fully grown deciduous tree consumes about 10,000 I CO_2 and produces 9000 I oxygen per day. At the same time, high vegetation acts as a temperature buffer, as the ground is protected from extreme heating or cooling by shading.

Requirements for the Planning and Design of Open Spaces

In view of tight public budgets, it is important to structure open spaces in such a way that they achieve the triad of high aesthetic value, high usage, and equally high ecological quality with the least possible maintenance. The quality of the open spaces and thus the satisfaction of the residents with their living environment depend not least on the state of maintenance of the open spaces. With suitable concepts, care can become a decisive design factor through targeted and effective measures.

With the increased cultural diversity of our society, the demands and usage habits of open spaces have also become more diverse. This fact must be taken into account by open spaces. They should be open for use, that is, they should be interpretable in a variety of ways, they should meet the increasing urge to move, that is, they should not only offer available, robust areas but also a differentiated network of paths with different worlds of experience and distances for young and old.

The possibility of appropriating free space for individual use is of great importance for the quality of life. This is especially true for family-friendly housing, if families are to remain in the city and not migrate to the rural periphery. Gardens are, therefore, not only in demand in single-family home neighbourhoods but also as tenant gardens in multistorey residential buildings.

The following types of open space can be distinguished in a hierarchical order, ranging from large-scale landscape areas to small-scale residential gardens, which together form a networked open space system:

- Outdoor landscape (agricultural area, forests, water bodies),
- Settlement edge,
- Brownfields, interstices, "no man's land,"
- Green links,
- Open spaces of the industrial estates,
- Green infrastructure (cemeteries, sports areas, allotment gardens),
- Parks,
- Street spaces,
- Squares,
- Neighbourhood green areas,
- Residential open spaces, private gardens,
- Playgrounds, retreats/dysfunctional rooms.

Landscape Space in the Outdoor Area

Both for the urban nucleus surrounded by landscape and for the dense urban-landscape patchwork of metropolitan areas, landscape represents the basis of the available (urban) space. The landscape, space of primary production, that is, biomass production, represents the natural resource in the ecosystem and is thus the basis of human life. On the one hand, the landscape space in outdoor areas is characterised by the natural conditions such as topography, geology, soil, water balance, climate, fauna, and flora. On the other hand, landscape is characterised by the anthropogenic forms of use such as gardening, agriculture, and forestry. These forms of use are dependent on social and cultural conditions and shape the different landscape images.

The interaction between landscape and urban space is particularly visible in the outer areas of large cities with strong pressure of use that emanates from the settlement area. From the city's point of view, the landscape is regarded as a recreational area. In order to avoid conflicts of use, it is important to create good path connections with a circular character, which open up the area not only economically but also "touristically." In conurbations with extremely high pressure of use, concepts must be developed for the peripheral zones which link agriculture with the demands of the inhabitants.

Settlement Edge

In dealing with the peripheral zones to the landscape, a clear demarcation between urban and landscape space must be formulated in the area of the settlement margins. Different approaches are conceivable here: On the one hand, a clear structural demarcation from the landscape can form a readable settlement edge. The interlocking between urban space and landscape with finger-like green spaces leading into the urban space is another possibility to form transitions between urban and landscape, as well as buffer zones in front of settlement edges. Bordering dense settlement structures on the landscape, transition zones are to be created which take up open space usage claims from the settlement area and which at the same time have an effect on the landscape area with a high ecological quality. They can include compensatory measures of the impact regulation and necessary facilities for rainwater treatment (e.g., troughs/lgols). The design of these areas must take account of the adjacent landscape type. The landscape is not to be excluded, but visually included.

Brownfields, Interstices, No Man's Land

Structural change has left large areas of wasteland behind. For many cities it is not possible to develop or use these areas in terms of urban design or green planning. They are therefore successively being conquered by vegetation. However, there is an opportunity in this. They can help to improve the local climate and promote biotope networking. In addition, they represent important retreat and roaming areas for different user groups. In regions undergoing structural change, these (often rail) lines are being expanded to create new routes through the city out into the countryside in order to link urban areas.

Fig. 5.5 Landscape space outdoors. Large-scale agriculture Rhineland, Roman road

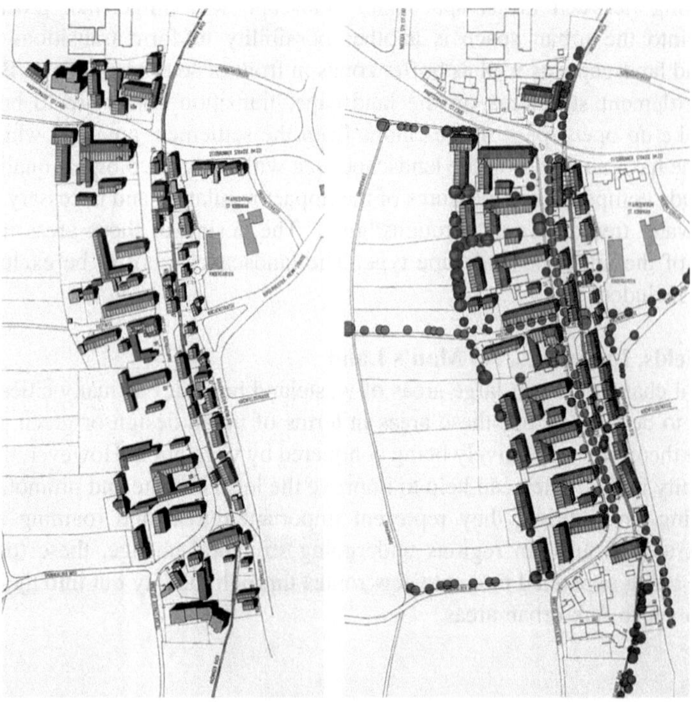

Fig. 5.6 Settlement edge. Constructional extension of the edge of the old town centre of Unterhaching. The yards form a clear edge to the open field with the trees (right)

Fig. 5.7 Schurenbach waste tip

Fig. 5.8 Duisburg North Landscape Park

Fig. 5.9 Tenant gardens Gladbeck

Fig. 5.10 Communal living spaces

Green Links

Green links are path connections away from the roads. These can be green gaps between the neighbourhoods, but also run along the abovementioned infrastructure bands. This increases the permeability and thus the variety of experiences in the city.

Green Space Infrastructure (Cemeteries, Sports Areas, Allotment Gardens)
Cemeteries are usually large green spaces with a high proportion of woody plants. They are contemplative recreation areas and are usually designed according to traditional patterns. New approaches with gravestones or crosses in wide meadows, as they were already developed in Sweden or Switzerland in the 1960s, can hardly be found in Germany. Burial in so-called peace forests comes close to the desire for a natural design of the final resting place.

Sports areas are assigned to schools or larger parks, due to the larger catchment area and the necessary distance to residential areas. A division of sports areas by rows of trees is desirable in terms of shading and spatial design quality, but from the point of view of care, it is usually rejected by those responsible. Informal sport is becoming increasingly important today. For this reason, the integration of open sports areas in parks is an important aspect, even if high-quality competition areas for club sports must remain fenced in for reasons of care.

Allotment garden sites are regulated by the Federal Allotment Garden Act (Bundeskleingartengesetz—BKleingG) and are reserved for noncommercial horticultural use. The plots of the allotment gardens should not exceed 400 m^2. Arbours may not cover more than 24 m^2 of ground area. The allotment gardens are important supplementary living spaces if there are no tenant gardens available in the living area. The allotments should be easily accessible from the living areas even without a car. Since the statutes of allotment garden associations cannot fully satisfy special needs for garden use, there are experimental approaches to alternative forms of gardening: multicultural gardens, ecologically oriented gardens, gardens with small animal husbandry, which can also vary in terms of plot size. Public access should always be guaranteed.

The green space infrastructure, that is, the designation and establishment of cemeteries, sports areas, and allotment gardens, is determined by the urban land use planning.

Parks
Parks are the representative green spaces of a city and at the same time important recreational areas. They are no longer just a backdrop for a Sunday stroll, but are part of everyday life, for relaxing during the lunch break, for volleyball in the evening, for picnic meetings at the weekend.

The classic image of a park since the English Landscape Park—meadow with trees, water, and topography—has been established as a common image for centuries. All three elements are important building blocks of an open space: the tree is space-creating because it brings the third dimension into the open space. It not only provides shade but also oxygen and at the same time consumes CO_2. It slows down the wind and mitigates heating by the sun during the day and cooling at night.

The meadow is the "user interface" on which the activities can be mapped. It can be a shorn carpet of lawn, but also a waist-high flower meadow, or both, when shaves are cut into the meadow, thus creating new spaces again and again.

The topography is the prerequisite for the spatial imprinting of the surface. According to the respective function, the surface can be flat and smooth for playing. It can become a hilly landscape in whose valleys visitors can retreat or be atmospherically shaped by water.

Parks today are varied, interpreted, and created in different ways:

- The conceptual park
  The park will be developed on the basis of an overarching idealistic concept, which creates unusual connections. Abstract structures are superimposed in a deconstructivist manner and form an unusual space for the modern type of amusement park.
- The park as interpretation of landscape
  This park version idealizes and exaggerates the type of natural space or the development of landscape.
- The park as interpretation of cultural landscape
  This type takes up agricultural land use as a park theme and, with this irritation, prompts visitors to reflect on the traditional park image and to look at the agricultural landscape with different eyes.
- The park as a "pattern"
  Close-meshed patterns are developed into park structures that cover space almost uniformly, creating a maze-like impression through addition and allowing for small-scale appropriation.
- The linear park
  Abandoned or "capped" infrastructure bands are developed into linear parks that open up surprising perspectives into the city and offer new open spaces in districts that are insufficiently supplied with green space. Burdensome road barriers are covered or abandoned, thus overcoming the separation.
- Neighbourhood green spaces
  They have the task of structuring the neighbourhoods, providing play and recreational facilities close to the homes, fulfilling ecological compensation functions and, if possible, regulating the use of rainwater.
- Residential open spaces
  Residential open spaces—especially in the form of home gardens—are important components of urban quality of life. They should be of a high variety of uses and design quality to make living in the city attractive again.

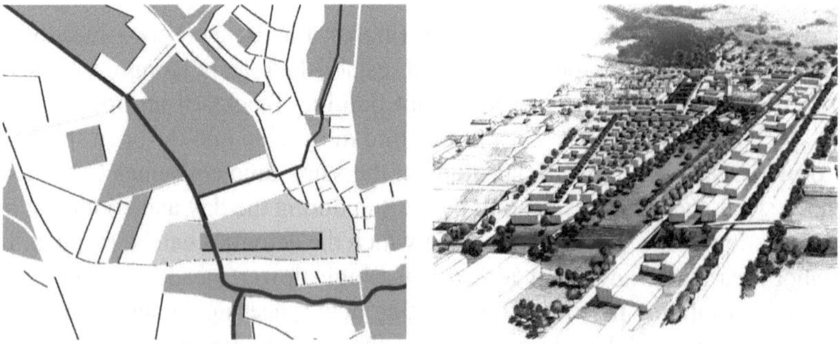

Fig. 5.11 Düsseldorf Gerresheim, RHA architects and urban planners. Development of a brownfield site in Düsseldorf Gerresheim

The following design criteria shall be taken into account: On the development side, attention must be paid to a differentiation of the degree of public access. A buffer zone between the public access area and the private entrance is desirable, which is both mediating and at the same time usable and not only decorative. On the residential side, the transition between living space and garden/yard space must also be given special attention in multistorey residential construction in order to avoid disturbances. It must be checked whether tenant gardens are possible and in demand at the respective location. Gardens directly assigned to the ground floor apartments should be made possible in any case. A garden path at the rear of the building provides access to the communal play and recreation rooms and, if necessary, to tenant gardens surrounded by hedges on the upper floors. This garden path should be linked to the other superordinate path system. This applies accordingly to the gardens of the various types of single-family homes.

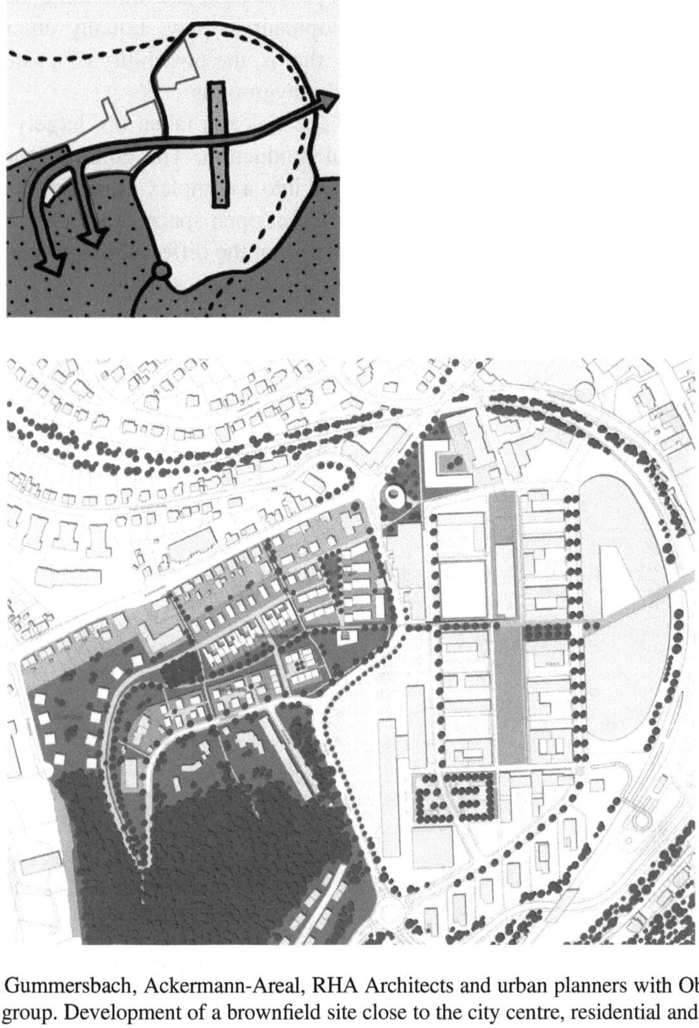

Fig. 5.12 Gummersbach, Ackermann-Areal, RHA Architects and urban planners with Oberhausen planning group. Development of a brownfield site close to the city centre, residential and commercial, open spaces as central elements

Playrooms, Retreats, and Dysfunctional Rooms

In the city today, scope for appropriation is not a matter of course; in many cases it must be designated as such. Dysfunctional spaces that can be changed in the game are becoming less and less available. This means that the playability of the entire facility must first be taken into account when designing the living environment. The everyday world should be reinterpretable for children's play, in order to give children's imagination as much reason to play as possible.

When creating the designated play areas, the design should be integrated into the overall facility and the arbitrary stringing together of functionally one-sided play equipment should be avoided. In addition to the dysfunctional rooms, the apartments should offer play areas for small children that are as visible as possible and retreats for school children and young people that contrast with the regulated, orderly living environment. The balance between adventure and safety, between uncertainty and security, between order and chaos, between provocation and imitation of adults is of decisive importance for the child's development in play. Equally crucial is the independent expansion of spatial horizons, that is, the possibility of a safe route to the respective age-appropriate facilities and playgrounds.

Through successive settlements, landscape space is taken up, largely removed from the ecological cycle and agricultural production. The cultural landscape is overlaid with settlement areas and developed into a complex urban space, which is structured by a differentiated system of different open space categories. This open space system is to be developed in such a way that the different usage requirements of all population groups can be met within adequate accessibility.

One-sided functional open space structures are no longer up to date, as they do not sufficiently meet the complex requirements of the urban network.

Fig. 5.13 Berlin, Graceland

Fig. 5.14 Gevelsberg, Ennepe Park

2 Urban Component Public Space

Public space and the life that takes place in it shape the identity of our cities. Streets and squares are figurehead and backbone of urban development, whether in the neighbourhood, at city level or in a regional context. Public spaces are of essential importance for cities; they formulate the framework, are structuring elements of urban space and thus make it readable.

Public space refers to the areas in our cities and settlements that assume public functions. These functions are, essentially, development and residence. The term "public space" is comparatively young in our language usage and has only been used in this form since 1950. Previously, we had rather talked about "public streets and squares" or even "public facilities."

The terms "public space" and "urban space" are closely linked. Urban space is used to describe all cavities in settlement structures formed by buildings. In this definition the term coincides with the so-called negative space, that is, the areas left free by buildings and facilities. This three-dimensional hollow form is formed by the negative imprint of buildings. Consequently, this urban space is created through the interaction of several dimensions: the physical dimension, the use, the materiality as well as specific conditions and factors that characterize the respective location.

The public space with its network of streets and its sequence of squares is an expression of the quality of life in a city, but at the same time an indicator of the respective society. The character of the public space is an expression of a certain culture and way of life, which in turn is influenced by climate, tradition, current social models and last but not least by technical standards.

Historical Development

The origins of public space, especially the European Square, are closely linked to the emergence of urban society. In Europe, the term "square" is as old as the cities themselves. In the search for the oldest squares, one comes across the cities of Lato and Gurnia on the Greek island of Crete, where squares with staircases served as places to advise citizens. In Greek antiquity, the squares were called "agora" and also served as a meeting place for the political decision-making of the people and as a market for the exchange and trade of goods. The squares played a central role in the city as places of political, economic and cultural centre. The Romans took over these structures and developed them further.

The public space as we know it today is an invention of the nineteenth century. Only in the course of industrialization and the accompanying urbanization were public parks and squares created. A previously unknown densification and differentiated distribution of uses in existing cities began. Planners gave cities a new appearance by adapting to rationalities of mobility and economy and acting according to new urban design principles—the grid, the axis and linear streets, taken over from the Romans—are among the expressions of this.

The possibility of mass-producing cars triggered radical changes in urban design in the 1920s and found its expression in paths, streets, and squares that were developed or converted according to the rules of individual mobility. This mobility made it possible to separate the functions of living and working spatially more clearly. As greening of the city simultaneously became the central concern of modern planning, the car-friendly streets, squares, and parks shaped public spaces of cities.

Fig. 5.15 Hattingen, upper market with weekly market, around 1905

In the structured and loosened up city, which was decisive for the reconstruction phase after the Second World War, public space tended to dissolve. The centres were expanded to make them suitable for motor traffic, streets intersected the neighbourhoods and public places often degenerated into traffic junctions and islands. When negative consequences in inner cities also became apparent for trade, further functional separations occurred. The first pedestrian zones were set up in the centres.

Change in the Proportions of Street Spaces

If you look at streets in different historical phases, you can see different regularities. Medieval streets were narrow, often a proportion of height (H) to width (B) of 2:1 could be observed. Renaissance and Baroque cities had 2 to 3-story buildings and a ratio of building height to street width of 1:1 to 1:2. In the nineteenth century, a width-to-height ratio of 1:1 was established in many cities. The permissible building height corresponded to the street width. Today, we find very different spatial characteristics of street spaces depending on the functional requirements and the location in the urban structure.

Fig. 5.16 Hattingen

Consideration of the Form of the Square

The squares of the Middle Ages seem closed, because mostly only narrow alleys lead out. The irregularity of the squares leads to dynamic spatial impressions and constantly changing perspectives. The squares usually approach the ideal proportion of 3:1 in relation to the surrounding buildings, which gives the impression of a closed inner space. Just as the walls influence the character of a room, so the facades of houses influence the effect of squares. Even after centuries, the retention of narrow plots of land still led to varied yet similarly constructed facades. Churches, town halls, and other special buildings were incorporated into the square and street walls. The entire square area was passable and usable.

Over the centuries there have been different ideas about the nature of the ideal place: its dimensions were large enough to accommodate the Assembly, but again not too large so that the speaker could be clearly heard everywhere. Theoretically, the central square was to result from the order of the city. According to the four cardinal points, the two main streets were oriented, which were to cross exactly in the middle of the street. This intersection was in turn extended into the square. But what was supposed to be according to the rules was often difficult to realize in many places due to the topography. One has to mention the Campo in Siena that has the distinctive shape of a shell. At the upper edge there is the market fountain, at the opposite side, the Palazzo Pubblico creates a worthy stage background. Spatially, one has on this plaza at each point of view the impression to be in the centre of the city space.

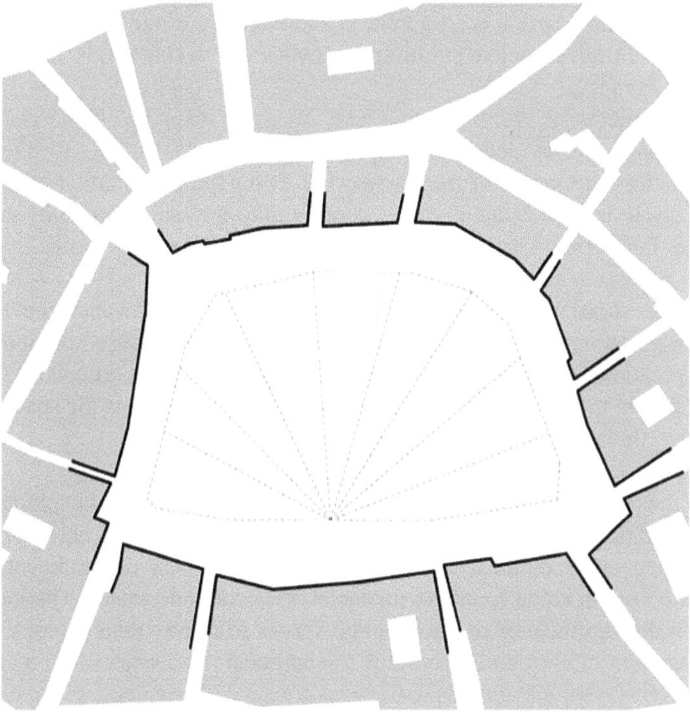

Fig. 5.17 Campo in Siena, map

The erosion of medieval city rights by absolutism was accompanied by a transformation of public space. While urban space had previously been an extended usable space for citizens and a place of communal decision-making, it now became a stage for the representation of absolutist power.

In the nineteenth century, with the boulevard and the large-scale urban spaces, the transformation from pedestrian city to traffic city took place. Specialized squares such as traffic squares, jewellery squares, building forecourts, and parade grounds developed.

The end of mixed use and closed development, as propagated by modernism, inevitably led to an end of the traditional square. Only in the remains of the older structures did the former qualities remain.

Today's Rquirements, Designs, and Planning

The concept of "public space" requires a systematic differentiation of its meaning and function in order to do justice to its complexity. Its use takes place in different ways: as a traffic space, as a consumption space, as a communication space, as a recreation space, etc. This already names an essential characteristic of public spaces—their multifunctionality (as in the Middle Ages). Six essential aspects can be assigned to the multiple meanings of public space: aesthetic, cultural, social, ecological, political, and economic.

- Aesthetic aspects
  The aesthetic quality of public space has a decisive influence on the cityscape. As the city's calling card, it is a figurehead and indicator of the attractiveness of any city. A beautiful, attractive public space invites you to stay and is the prerequisite for well-being.
- Cultural aspects
  Public spaces are of elementary importance for the city culture, for the external presentation, the image of the city as well as for the presentation of the people who stay in it. Good public spaces contribute to the tangible quality of a city's culture. Furthermore, many cultural events depend on public spaces.
- Social aspects
  Numerous social processes take place in public space: On the one hand, processes of appropriation, communication, socialisation, play and sports activities, leisure and recreation take place. In this respect, the public space is particularly valued by children. On the other hand, however, the public space is also the space of a city in which threat, insecurity, and exclusion are experienced.
- Ecological aspects
  With the establishment of the goals of sustainable urban development, the consideration of ecological aspects gained in importance in the early 1990s. From then on, ecological issues were treated in urban development, at least officially, on an equal footing with social and economic issues. The main focus was on the significance of these public spaces from an urban ecology or urban climate point of view and their value as additional open spaces for city dwellers.

- Political aspects
  The connection between politics and public space has existed since the late Middle Ages. In city descriptions, for example, the shape and condition of streets and squares are taken to indicate "good government." Moreover, in the course of the outsourcing of sovereign tasks, the public space is considered one of the last core businesses of direct political responsibility. On the other hand, however, public space is also used as a platform for uprising and demonstration of citizenship against politics.
- Economic aspects
  In preindustrial times, the public space in form of marketplaces, city squares, and streets was a reloading point for all kinds of goods. But crafts and trade were gradually "domesticated" and shifted to department stores and production halls. The public space remained of economic importance above all as a transport space. For a long time, only the "hard" location factors (road, rail, etc.) were of importance for company settlements, but with the structural change from an industrial to a service society, "soft" location factors are increasingly coming into play. Culture, nature, recreational, and leisure facilities, the image of a city, etc., are increasingly in demand by employees and thus become an essential location factor for companies. Public space is able to influence these factors, just as it has a growing influence on the location value of real estate and entire city districts. Neglected or heavily trafficked areas can lead to a migration of use, whereas the upgrading of public spaces can have the opposite effect and contribute to an increase in location values.

Fig. 5.18 Cologne, Domplatte

Knowledge of these individual aspects is relevant for urban design, as is knowledge of their interaction.

Current Developments and Challenges of Public Space

The negative developments that have occurred in recent years and decades, such as loss of function, competition for use, privatisation, commercialisation, neglect, and vandalism, keep planners busy. Some of the approaches to solving the problems are still in their infancy, but the trend "from the marketplace to the car park. . ." has in many places been reversed into a development ". . .from the car park to the marketplace."

Two problematic tendencies are shown as examples:

- Privatization

  Associated with the privatisation of public space are primarily shopping malls, shopping arcades, or other building complexes that are privately owned, but are made accessible to the public—albeit potentially selectively. Privatisation in the sense of a change of ownership from public to private cannot be confirmed by planning practice.

  On the other hand, there is an undisputed tendency to assign urban development to project developers or even individual large users. The project developers are not limited to their actual property but are also responsible for road construction measures, the design of green spaces, or the creation of parking spaces. In the course of this, there is an increased privatisation, that is, a shift of activities (especially in the area of consumption and leisure) to privately created and managed seemingly public spaces. This form of privatisation has many negative consequences: it forces selective, social exclusion, thus reinforcing the tendencies already observed in cities. The city is becoming more and more equipped for the adult customer with purchasing power. Old people, children, the poor and sometimes also women are marginalised. As a result, the city loses a key element of urbanity, namely its colorful public, and thus loses quality overall. However, this should be seen as a general development in urban planning and not as a specific phenomenon with regard to public spaces.
- Commercialization

  The trend toward the commercialisation of public spaces is closely related to their privatisation. The result of this development is a broad spectrum of uses with different demands on public space: gastronomic offers, festivals, events, exhibitions, sports, and cultural events, etc., are increasingly found in public space. Tourist, pub, and promenade miles are emerging, some of them artificially created, others historically grown and upgraded. All of these uses are, to varying degrees, geared to giving public space the qualities of diversity, metropolitan atmosphere, flair, and intensity of use. Since this commercialisation often takes place in these new public spaces or other highly frequented spaces in the inner cities, a dangerous gap is created between the commercialized spaces and the less significant locations, which are quickly affected by vacancies and emptying on the ground floors.

Fig. 5.19 Duisburg pedestrian zone, open for business on Sundays

The commercialisation of central public spaces is already "everyday business" and will increase rather than decrease in the future as a result of the ongoing "competition for attractiveness" of cities. The challenge is to face this development from an overall urban perspective in order to avoid spatial selectivity between the centre, districts, and the outskirts of the city and to manage these diverse measures in an urban-friendly manner.

Design Elements

- Spatial typification of public space

The different categories of urban space and development systems can be viewed from different points of view:

- the design and spatial qualities or
- the technical requirements (functionality)

Both levels are important for urban design, because they must be considered in context and combined into a single unit in urban design.

Different categories can be distinguished in the spatial expression of urban space:

- Enclosed space—fluid space
  We refer to urban spaces as enclosed space, which are formed by continuous vertical lateral boundaries. Such boundaries can be formed by buildings, by hedges or by dense tree plantings. Clear spatial boundaries create a clear distinction between "front" and "back" as well as between "private" and "public." In fluid space, on the other hand, clear spatial edges are missing. Positive and negative space are approximately equal. Often the interspaces also dominate, as in scattered settlements or even industrial estates. Optical guidance and orientation are provided by paths and streets. Without these linear orientation aids, it is often difficult to use them. This distinction means for the design process that an idea has to be developed of what kind of space is to be created and with which elements this goal is to be achieved.
- Linear space—square space
  By linear spaces we mean above all the streets, which primarily present our cities. The streets are lined up with uses, they are dynamic and very complex structures with a wealth of functions, including aesthetic ones.

  The aesthetic image of the street is often a reflection of social reality. As public spaces, the streets are accessible to everyone, they connect the different worlds of a city—no matter how contrary they may be—in a common public network of movement.

  For urban design, the spatial effect is important in relation to the management and design of street spaces.

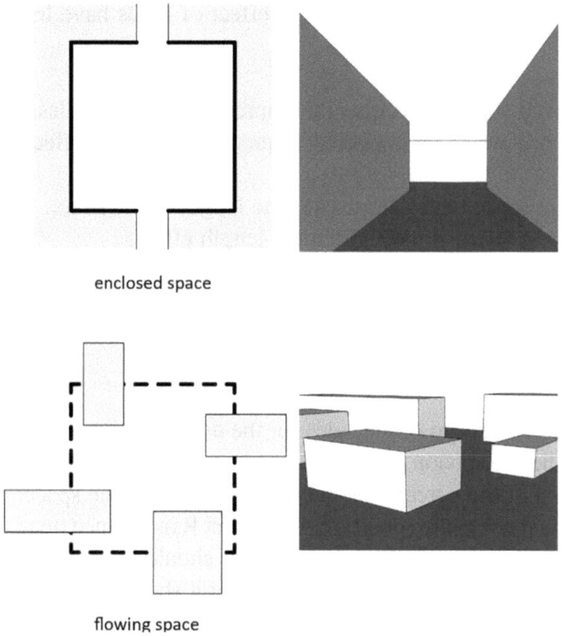

Fig. 5.20 Enclosed space, fluid space

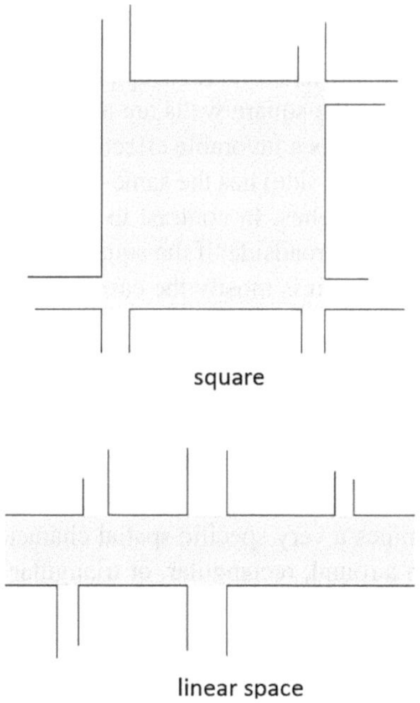

Fig. 5.21 Linear space, square space

Investigations of the length and depth effect of roads have led to the following result:

- Convex curved space walls give the impression of an "endless" street space.
- Concave space walls in staggered sequence create the effect of limited room sections.
- Long straight space edges emphasize the length of the street.
- Curved or bent space walls shorten the length effect.

The different arrangement of the space walls leads to different spatial effects, which are relevant for design.

Spatial Effect

Vitruv proposes a 3:2 ratio of the sides for the dimensioning of a space and its ideal proportions. The dimension of the square is an essential element in achieving a favorable ratio. For the perception and spatial quality of the space, the ratio of wall surfaces to floor space is also decisive. In the first Renaissance treatise, Leon Battista Alberti (1404–1472) recommends that a square should be twice as long as it is wide and that the height of the buildings surrounding it should be between one third and one sixth of the width of the square, in order to achieve a pleasing harmony of proportions.

Lower buildings generally make a square appear larger, higher ones a square smaller. To be able to call an area a square, its width should also be three times the height of the building. In addition, the choice of the location of the buildings on the base area is important for the overall impression, whether on the broad side in the centre (width square) or on the narrow side (deep place). The location and quality of the architectural elements of the square walls are also decisive for the open space.

Thus, the deep place only has a favorable effect if the dominating building in the background (i.e., on the narrow side) has the same dimensions, as is usually the case with the main facades of churches. In contrast to the broadside square, where the dominating building is on the broadside: if the square is in front of a building with a predominantly broad effect, as it is mostly the case for city halls, the square should also get a similar broadness.

Forms

What dominance is given to the actual shape of the square, what spatial effect does the deviation from the orthogonal ground plan, the trapezium, the polygon, the circle, the ellipse have? The ground plan, which defines the square in two dimensions, thus determines a very specific spatial character. It makes a big difference whether you are in a round, rectangular, or triangular space.

While the triangle defines the formal expression of the idea of the path, the circle is regarded as a manifestation of the central; the ellipse, however, as a combination of both forms. The Roman amphitheatre was created from the circular form—as an ellipse, for better arrangement as a stage to accommodate as many spectators as possible.

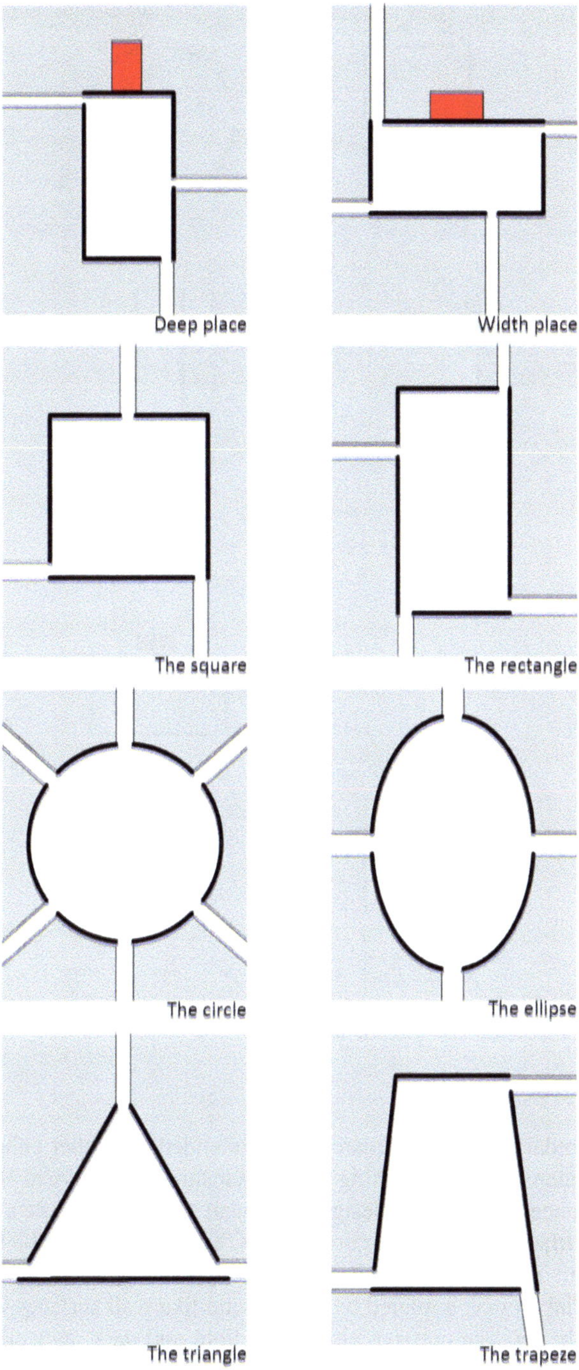

Fig. 5.22 Square forms

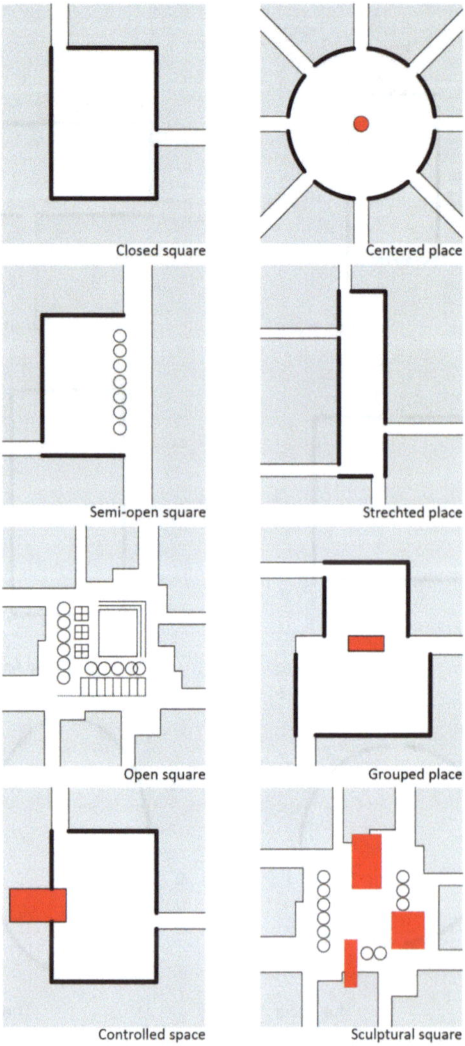

Fig. 5.23 Square types

The trapezoidal shape of a square is also intended to further enhance the stage effect—with open sides and flanking space enclosures. A theatrical atmosphere can be further enhanced by a special design of the wall surfaces to create a suction effect, with curved surfaces, irregular alignment lines, even with wave-like boundaries.

The contours of the square walls also form the end of the ceiling of the entire room. The aerial view is bordered by a silhouette-like wall surface—and the sky is perceived as the most impressive element of light and dark, as well as constantly changing play of colors and movements.

Towers and even ridge turrets are used to create the effect of the silhouette as an important characteristic, so that the building, as it rises up into the blue vault, simultaneously captures it, binds it to itself and returns to the building until it is built (Knirsch 2004, p. 12).

Functions

The shape and use of space always depend on its location and size. While a large square in the city centre is used as a central marketplace for events, a smaller square in a residential area is more contemplative. Therefore, squares can be classified into different categories, all of which have different uses and sizes.

There are places in front of churches or public buildings known as cultural squares. Public life takes place here in the form of various events and appropriations. Due to the mostly central location of these squares in the urban fabric, shops, public buildings, representative office and commercial areas, restaurants and museums are settling around the square, which benefit from life on the square.

Another type of square is the station forecourt, which is usually located between the station and the city centre. Station forecourts are first and foremost "a traffic lock, a transformer that brings pedestrians, cyclists, bus, and public transport travellers and finally car travellers onto the train and back. This requires bus and tram stations, tram stops, bicycle stands, multiple access roads, short-term parking, taxi stands and permanent parking spaces" (Aminde 1994, p. 115). The size of station squares must be determined according to the type of station, especially if other shops, public facilities, and hotels must be taken into account.

In residential areas and districts, neighbourhood squares are very popular because they serve as meeting places for residents and are reserved exclusively for them due to their intimate character. Neighbourhood squares tend to be smaller, according to the buildings surrounding them. Neighbourhood squares play a multifunctional role within the settlement. They can be used as mixed areas, recreation rooms, playgrounds, bicycle rides, and last but not least as a festival place for a district or neighbourhood.

Design

In the design of public spaces, the integration into the architectural context plays an important role. The location within the urban fabric and the functions associated with it determine the requirements for the design of public space. Due to its use, the central city square has to meet different requirements and criteria than a neighbourhood or neighbourhood square. The material and design of streets, paths, and squares should complement and enhance the effect of the space-enclosing walls.

Furniture and Materials

The design of public space includes the conscious creative use of materials and furnishing elements. Local floor coverings and their traditional, material-appropriate laying gave our old villages and towns a characteristic and unmistakable appearance. Even today, this site-specific use of materials can help to integrate streets, paths, and parking areas into their surroundings and stimulate identification with the place where people live. Covering materials can be deliberately used in harmony with the materials of the adjacent facades, but can also form an exciting contrast to them. The same applies to the materials of the furnishing elements. It is desirable to create a creative connection between the different elements.

Atmospheric Design

The use of light in the design of public spaces has gained in importance in recent years. Until the eighteenth century, cities tended to remain unlit, at best there was a waking light at the city gates or a lamplight falling from a window. For about 200 years, there has been a literal escalation of the light media, which changed the nocturnal cityscape more radically than any mechanization. Lighting concepts in public space are to be seen in dialogue with the lighting of important objects. They should support the overall design idea and not counteract it.

Possible Uses

The public space is to be seen in interaction with the adjoining uses, especially with the ground floor uses. A harmony of design and usage requirements is what makes public space a lively place. The motto here is: a space for appropriation is more important than superficial design.

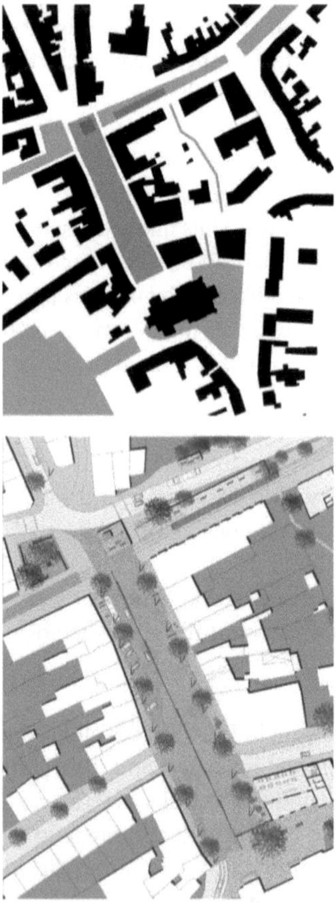

Fig. 5.24 Würselen, RHA architects and urban planners. Design of the public space

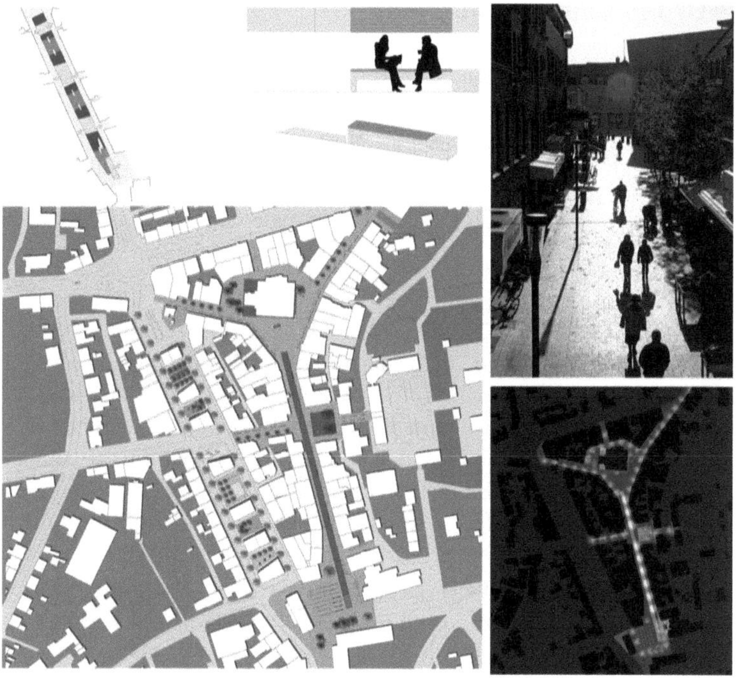

Fig. 5.25 Ahaus, RHA architects and urban planners. Design of the public space, mission statement and zoning (top left), overall concept (bottom), atmosphere in the pedestrian zone and lighting concept (right)

Ecological Concerns: Minimising Sealing

A major goal is to keep soil sealing to a minimum. In many cases, this concern cannot be realized in central, inner city situations that require an urban solution from a functional and design point of view.

In the case of neighbourhood and neighbourhood pitches, permeable surfaces (such as gravel lawns, open-pored paving, etc.) are preferable, if the traffic load permits this. If possible, the water from paved areas of the site should be let through into the ground in adjacent seepage systems in order to support the natural water balance of the ground and its functions. Planting along roads, access roads, and car parks also has various design functions. Used correctly, it helps to create and structure spaces, emphasise directions, screen off areas and enliven surfaces. In addition to their design functions, plants also have microclimatic and urban hygiene functions. Through their evaporation, they reduce the air temperature on hot days, provide shade and filter dust, which contributes significantly to the quality of use.

The consideration of various design aspects of public space—from the overall urban design concept to the details—makes it clear that a functioning concept can only be created if all levels are combined to form a convincing overall picture.

3 Urban Component Living

An essential task of urban design is the formation of the city as a place to live and the design of a liveable city. Residential buildings not only serve to meet individual housing needs, they are also the essential substance from which the city is formed as a community and social space. Urban design has always been obliged to provide living space and to find answers to basic conditions such as population growth and, today, shrinkage.

Historical Development

From the Stone Age to the Middle Ages, living was predominantly realized in the simplest forms of housing, with hardly any functional separation within the buildings. Often humans and animals lived together in a very small space. The living together of several generations inevitably led to very close social contacts. With the age of industrialisation, rural exodus, that is, a massive move from the countryside to the city, began. The construction methods within urban areas became more and more dense, and as a result, the rented apartment, which was closed off from the outside, was created, in which the rooms were accessed via a corridor. The economic circumstances of the inhabitants determined the size and equipment of their apartments, families—especially from the working class—lived together in a very small space, but minimum standards for apartment sizes were set.

Fig. 5.26 London in 1872, engraving by Gustave Doré

Since industrialisation, especially from the beginning of the twentieth century, shaping the Urban component living has become more differentiated. At the beginning of the twentieth century, cities grew less and less by building their centres, but by exploding into increasingly fragmented and loosened peripheral settlements.

After the Second World War, a reorientation in urban planning and urban development took place. The rejection of the big city was promoted in different ways. Increasing motorization led to the expansion of cities. Housing estates were built on the outskirts of the city and the urban areas were increasingly functionally separated. The inner cities profiled themselves as trade and service centres and less and less as production or residential areas.

The suburbanisation processes intensified from the 1960/1970s onward with the construction of large housing estates in the environs of prosperous cities. Traffic-oriented planning promoted the change of use in the cities.

While accessibility has improved for service and commercial uses, often in the city centre, housing has been pushed out of the city to greenfield sites with cheaper land and rents, resulting in what Wolfgang Pehnt has called “endless scattered settlements of single-family homes” that “preclude a profitable means of mass transportation.” This development has not only had unpleasant consequences in terms of settlement structure, but has also massively increased the volume of traffic and contributed to the spatial dispersion of supply facilities and infrastructure.

From the mid-1970s onward, the signs of a “renaissance” of urban living began to appear not only in Germany but throughout Europe, and a large number of new types of house and apartment developed.

Demographic change, the increasing individualisation of society, shrinking processes, and rising regional differences influence the housing markets, require new tasks for the housing industry in the future and demand new strategies.

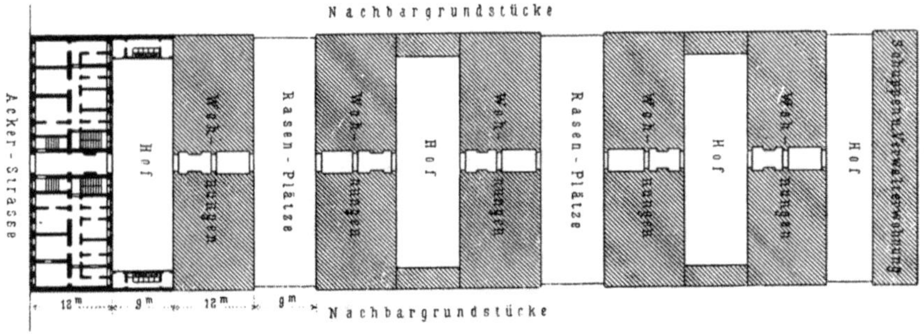

Fig. 5.27 Property in Beriin, Ackerstraße 132/133 (built 1873/1874)

Requirements, Designs and Planning

Good housing construction is more than just the individual house: transport connections, urban development, accessibility of infrastructural facilities, the development situation, and the shaping and design of intermediate areas have a decisive influence on the quality of living and social interaction of the residents. This requires an orientation toward different demands and resident groups. The individualisation of society demands not only typical residential floor plans but also more flexibility in new and existing buildings. Due to the differentiation of household and family constellations and the resulting demands on living, new floor plans must be implemented, and existing floor plans must be adapted to changing requirements.

Planning Handling

The Urban component living makes use of very different design elements and urban structure types, which are oriented toward the needs of the residents, the functions of different types of housing and external conditions.

Fig. 5.28 Werkbund Exhibition "The Apartment," Stuttgart, 1927. 21 individual buildings, 17 architects, including Scharoun, Hilbersheimer, Le Corbusier

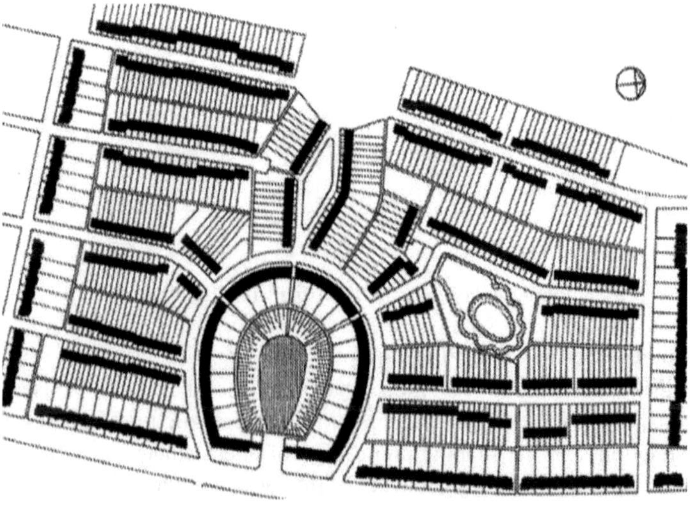

Fig. 5.29 Horseshoes Colony (Hufeisensiedlung), Neukölln, Berlin, 1925–1933, Bruno Taut

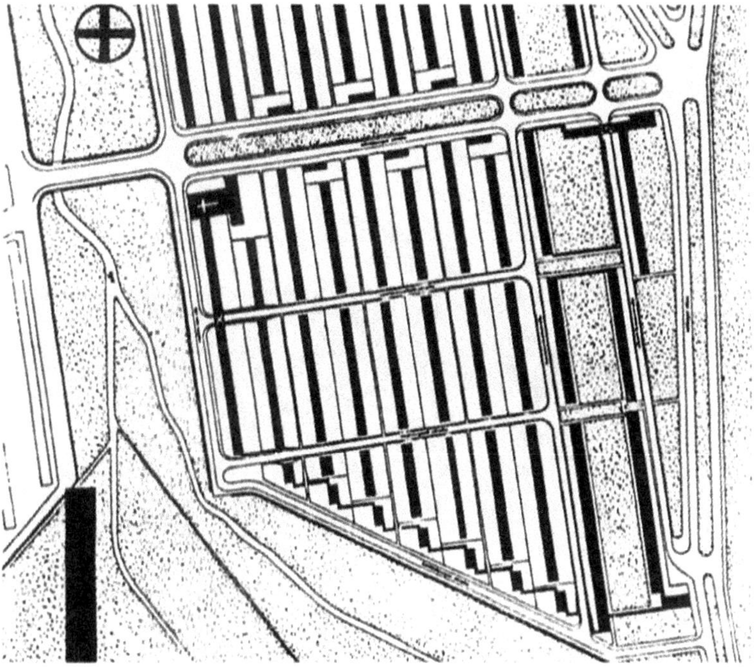

Fig. 5.30 Dammerstock, Karlsruhe, 1929–1931, Walter Gropius

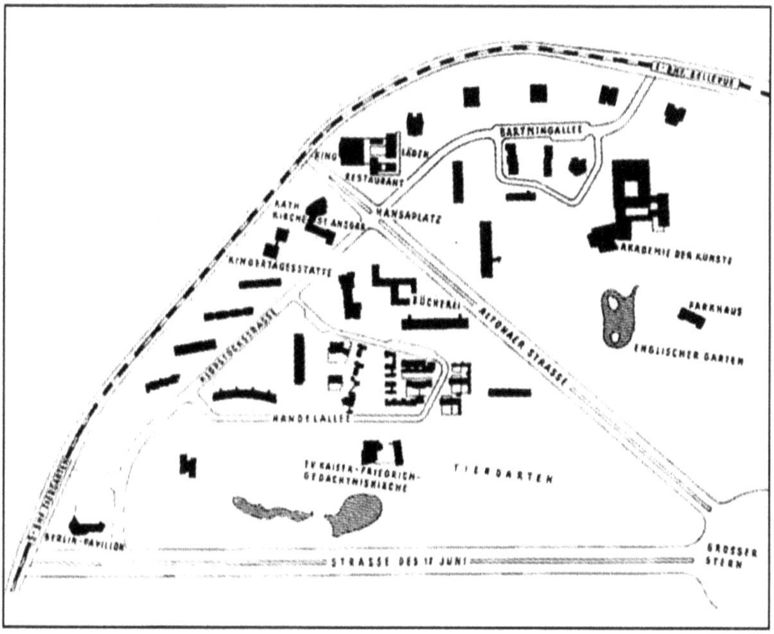

Fig. 5.31 Hansaviertel Berlin, 1957, Gropius, Baumgarten and Niemeyer

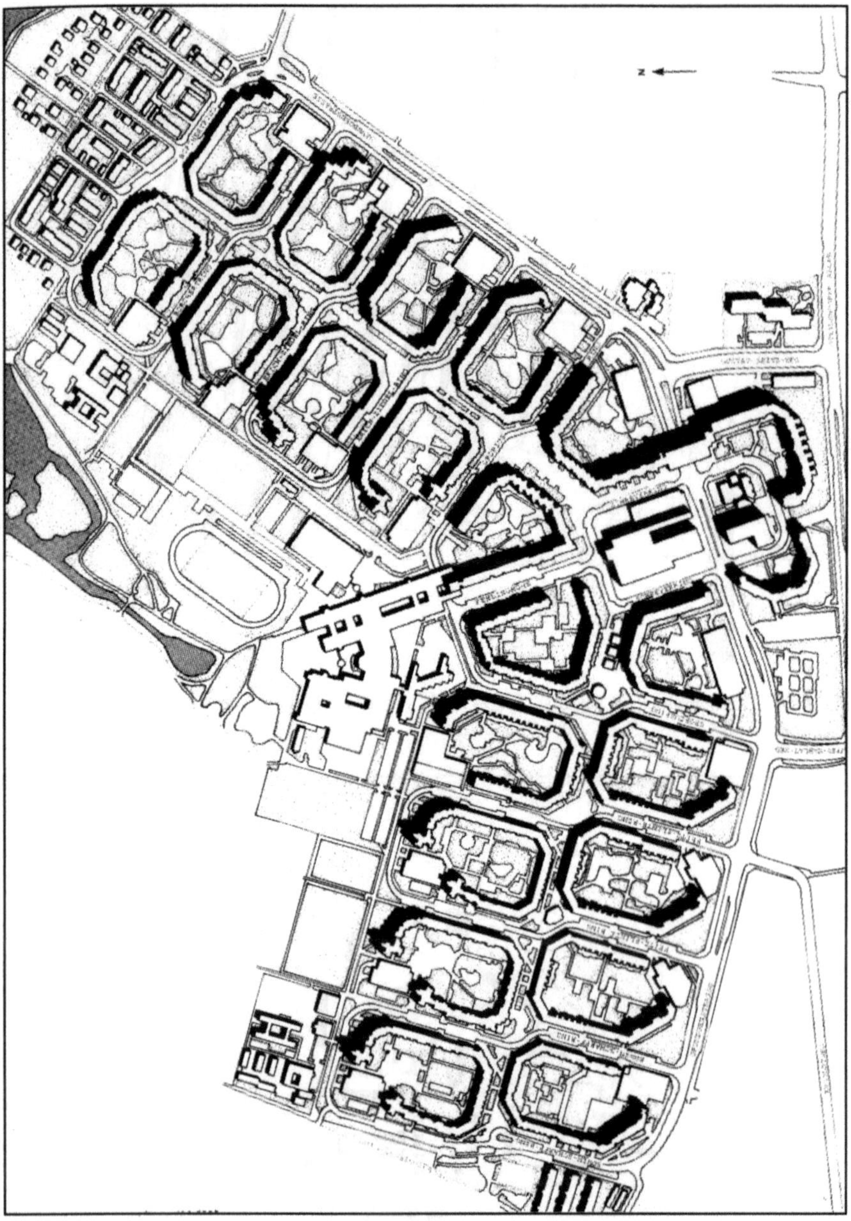

Fig. 5.32 Hamburg-Steilshoop, 1961, H. P. Burmester, G. Ostermann/Candilis-Josic, Woods

Usage Structure

The mix of living, working, services and supply should be taken into account in urban design. The mix of uses can be achieved within buildings, for example, by locating retail or services in the representative ground floor zones and apartments on the upper floors. At the neighbourhood level, too, the mix of uses can be achieved by locating utilities, services close to the dwellings, etc.

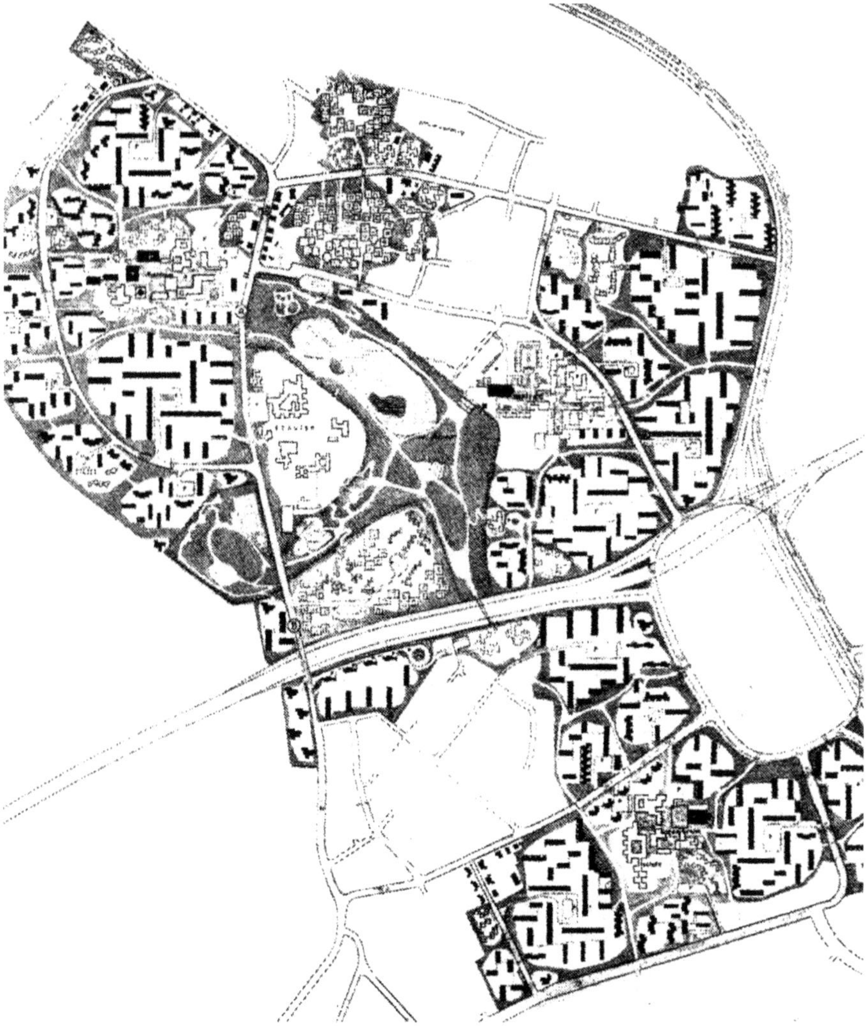

Fig. 5.33 Northwest City of Frankfurt a.M., 1962–1968 (Planning)

Fig. 5.34 Urban Sprawl, Sun City near Phoenix, Arizona, USA

Building Structure and Residential Area Types

The structure of a residential area, the arrangement of its buildings into streets, squares, and green spaces, is closely linked to the system of development.

The selection of certain types of structures is based on local conditions such as plot size and layout, topographical conditions, existing connecting routes, development and the financial capacity of the client.

The typification of residential areas is based on the building structures, a distinction is made between

- **Multifamily house**: multistorey residential construction, designed as timber-framed buildings, point houses and high-rise residential buildings and

- **Single family house**: Detached single house, semi-detached house, terraced house, chain house.

These types of housing often go hand in hand with different forms of ownership, multi-storey housing as rented housing, single-family houses as condominiums.

The tasks for urban development planning can be different, mostly mixed concepts with internal zoning are required, that is, the two forms of housing and ownership should be realized in a certain mixing ratio in the individual construction phases. In this way, a social mixing of the residents in a neighbourhood is also pursued. The basic principles and types are presented separately below.

Single Family House

House types

The **detached single-family house** has the necessary minimum distance to all property boundaries and thus to the neighbouring buildings. Single-family houses are more expensive than other types of housing due to their higher material, space, and energy consumption.

A **semidetached house** refers to two independent single-family houses built next to each other, separated by a fire wall. The private garden is located at the rear of the building.

Row houses are groups of houses with at least three independent residential buildings built next to each other. Middle terraced houses must be built up to the two lateral property boundaries, end terraced houses must comply with the respective distance regulations on the detached side of the house. They have a smaller building width and are, therefore, a space-saving building type. The private gardens are located behind the buildings.

Garden courtyard houses or angle houses are usually laid out as rows of L-shaped floor plans. The advantage of this construction method is the high density and the privacy due to the inner garden courtyard.

In the case of the **atrium house**, the inner courtyard is surrounded on four sides by living spaces, which makes very good use of the property. Flat sloping roofs and a low level (usually single storey) are necessary in order not to shade the private area inside. Atrium houses are often not built as individual buildings, but as connected rows or clusters.

Building Positioning and Space Formation

In the strict arrangement of single-family houses along a residential street, it is important to break up the monotonous structure, for example, by grouping several houses around a common access yard. The position of the building allows different open spaces with areas in front and behind the house.

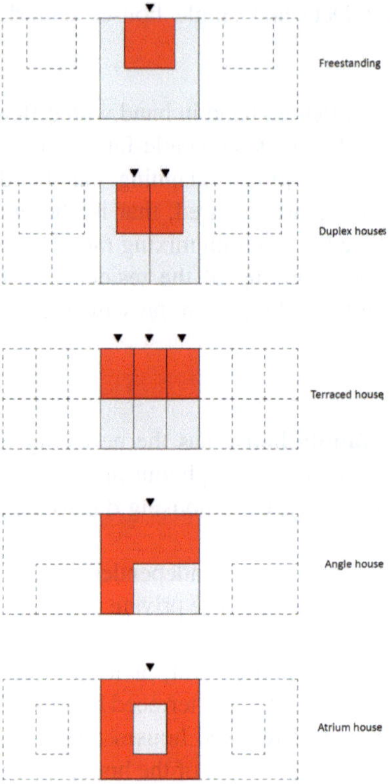

Fig. 5.35 Single-family house types

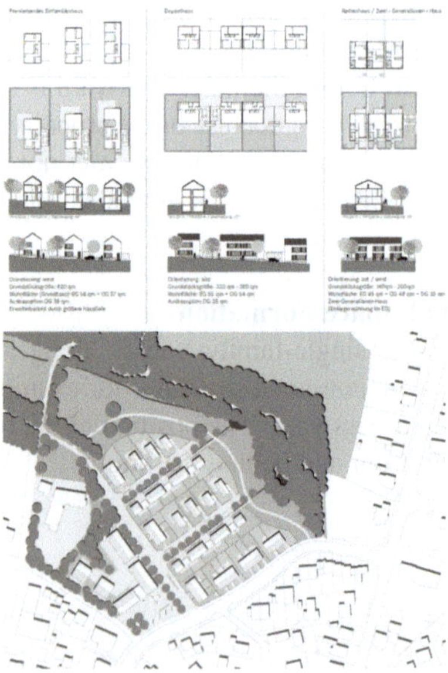

Fig. 5.36 Living Wefelshof, Bielefeld, RHA Architects and urban planners

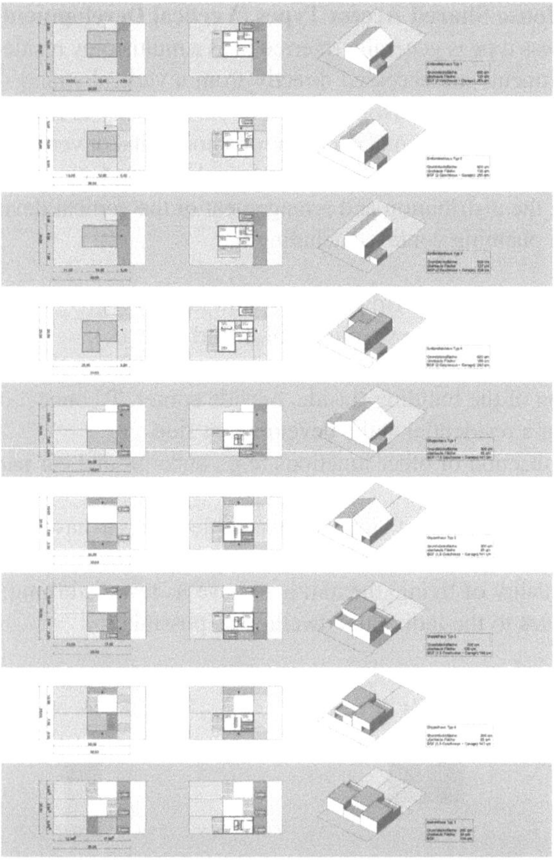

Fig. 5.37 Single-family house development Knollroad, Osnabrück, ASTOC Architects and Planners, Cologne—Lützow 7 Garden and landscape architects, Berlin. The future residential development Knollroad is located in a "green finger" that extends along a floodplain and thus takes over important ecological and recreational functions. Different types of owner-occupied housing can be realized on the plots

Multifamily House Shared Access Types (Vertical Development)

The shared access type is generally referred to as a multistorey residential building in which the apartments are accessed directly from a vertical strand—a staircase and possibly an elevator. The horizontal distribution is usually in front of the building.

The basis for the fundamental decision for a horizontal or vertical development of a residential building is already created with the urban design. Conversely, preferences for the distribution and arrangement of the vertical developments result from the urban planning context, including the:

- Form of the development (row, block, point house . . .),
- Orientation (direction, view, sources of disturbance . . .),
- Connection to a road network (development . . .),
- Constellation of the building (inside, outside corner, T-connection . . .),
- Definition of a residential and a development side
- possible connection of other functions (e.g., underground car park).

From an economic point of view, the main objective is to have as few staircases as possible in order to provide access to as many residential units as possible. With regard to the quality of living, the aim is to have as few disturbing, direct, but also identifiable routes to the individual dwelling as possible.

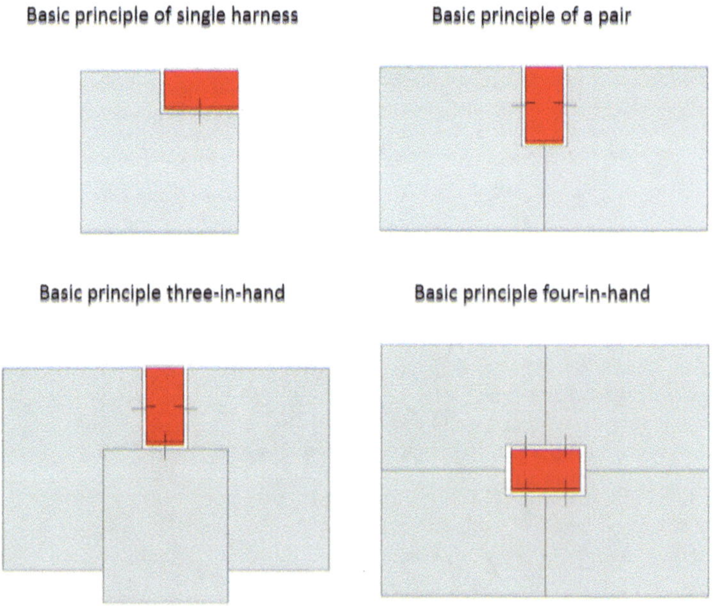

Fig. 5.38 Shared access types

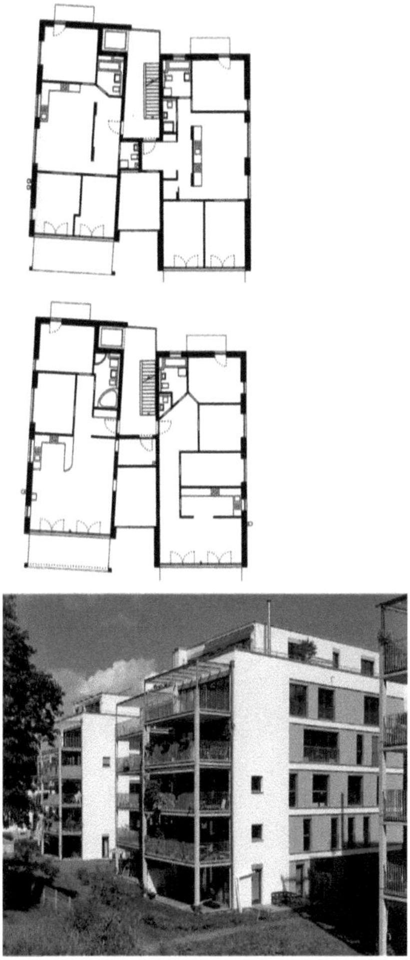

Fig. 5.39 Tensioners, mirror houses Freiburg, Amann I Burdenski I Munkel Architects and general planners, Freiburg

(Arbour) Gangway Types (Horizontal Development)

A basic distinction is made between staircase and pergola access. A pergola access should offer a quality of stay which makes it usable as an extension of the living space. Pergola accesses can be useful if it is necessary to connect an elevator. An aisle type is generally described as a multistorey residential building in which the residential units are lined up along a horizontal traffic axis and connected to each other. Vertical access is via stairs and, if necessary, elevators, which can be positioned as required. Depending on the location of the corridor, a distinction is made between single or double-legged types—outside or inside corridor type.

Preferences for horizontal development systems may already arise at the urban planning stage, which must be taken into account:

- Orientation (direction, sound sources, view . . .)
- Outdoor space situation (public, semipublic space)
- Building constellation (block, line . . .)
- Connection to a road network

With the possibility of apartment access at almost any point, an external entrance provides the basic prerequisite for a flexible division of various apartment sizes. The designation of "barrier-free apartments," also for upper floors, in combination with an elevator as well as an outside entrance is a particularly economical and communication-friendly solution. This solution is particularly suitable for a high proportion of smaller apartments.

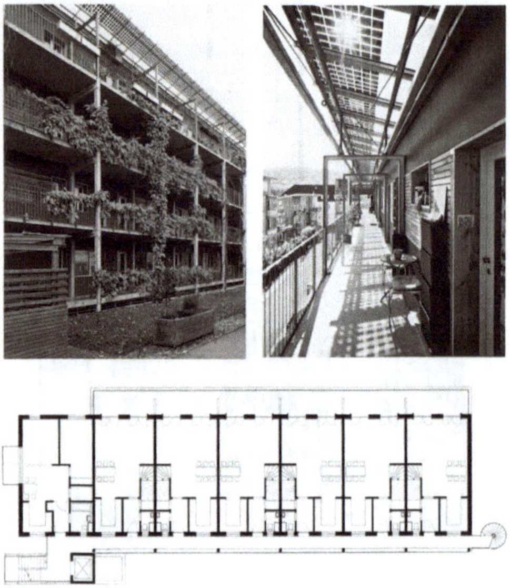

Fig. 5.40 (Pergola) passage types examples

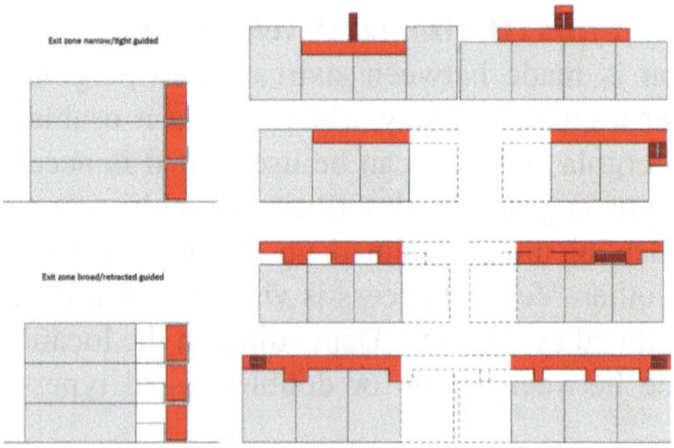

Fig. 5.41 Variations of (arbour) gait types

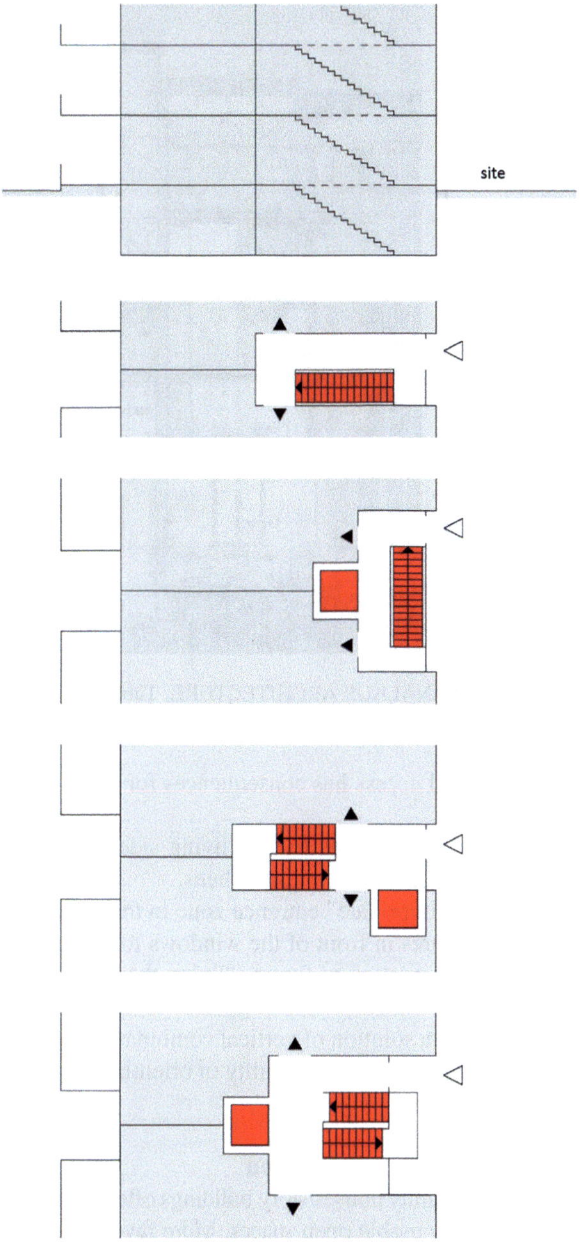

Fig. 5.42 Relationship of the building to the site

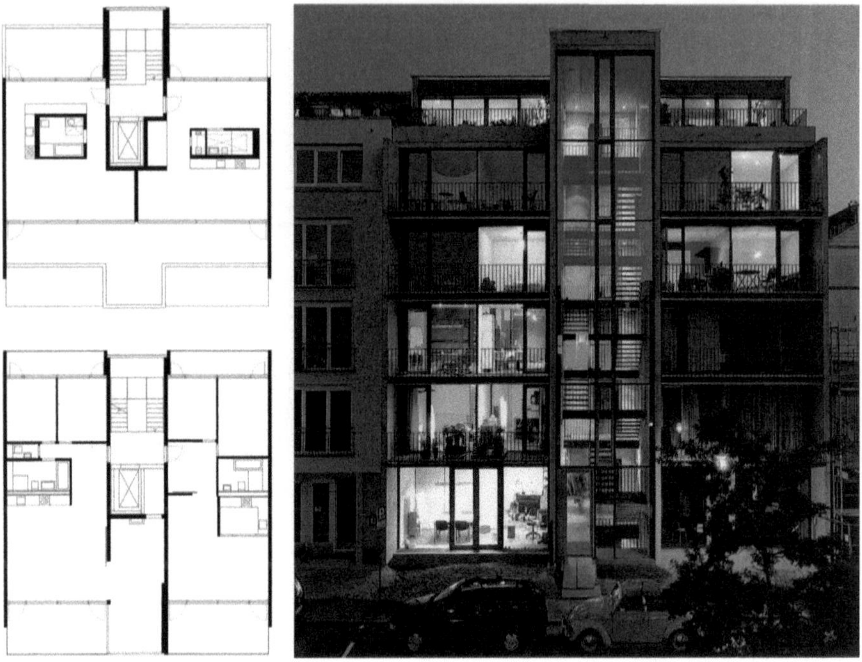

Fig. 5.43 Townhouse, NOENENALBUS ARCHITECTURE, Tübingen. Development/floor plan organisation (left), view (right)

Development of external access has consequences for inner and outer zoning:

- One-sided orientation of all rooms to the "living side"; strict zoning of living/ bedrooms and shielding sanitary rooms/kitchens,
- Creation of a usable "semiprivate" entrance zone in the entrance area,
- Distance-forming measures in front of the windows to the outside passage,
- Option for two-sided orientation of living ("living through").

The high living value of a solution of vertical combination (maisonette) results—as with the terraced house—from the possibility of orienting the second floor to both sides without restriction.

Building Positioning and Space Formation

Housing estates with multifamily houses/story buildings often have well developed, but poorly structured and poorly usable open spaces. More favourable for the use of open spaces are buildings that are related to each other in their development and location.

In multi-storey residential construction, a division is possible both in the building itself and in the living environment. On average, the buildings are accessed via a common street and the street space is spatially structured by trees and structural means, for example walls and pergolas. If the street only serves the residents, its width can be significantly reduced in favour of small residential gardens in front of the ground floor apartments. Behind the houses, residential gardens can be laid out for the ground floor apartments as well as neighbourhood and tenant gardens for the upper floors; they lead over into the communal and public areas.

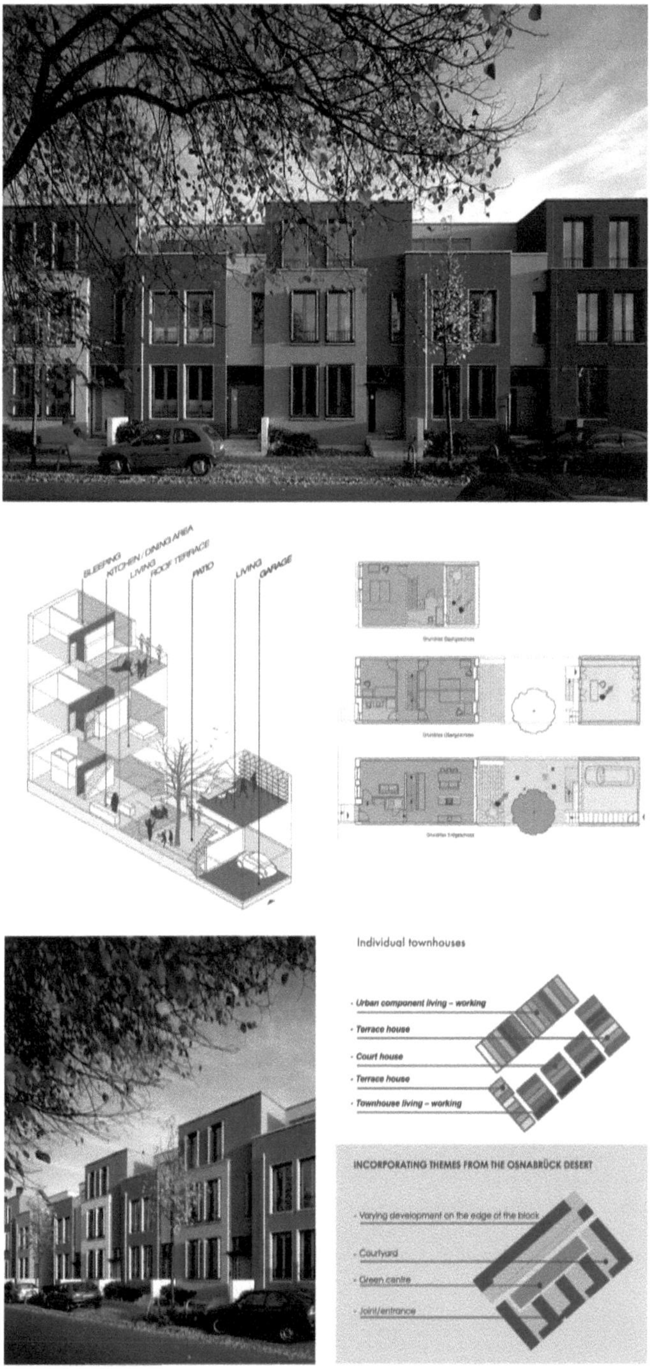

Fig. 5.44 Osnabrück Desert Urban District, Carsten Lorenzen APS, Copenhagen, DK

Individual townhouses
RECORDING OF THEMES OF THE OSNABRÜCK DESERT:

Fig. 5.45 Fischers Höfe, Hamburg, Carsten Lorenzen, Copenhagen, DK

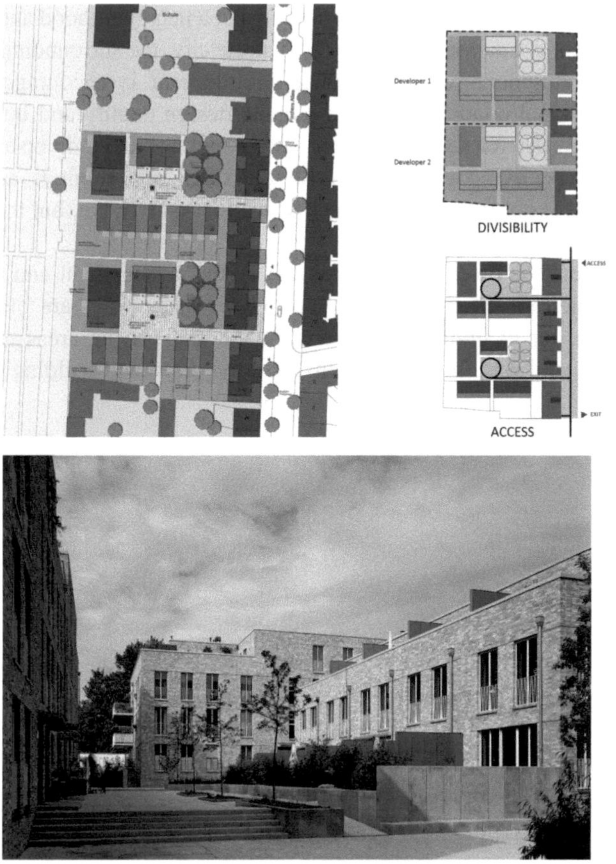

Fig. 5.45 (continued)

Open Space/Residential Environment

Housing is not limited to your own four walls; the design of the living environment plays a major role in the satisfaction of living. The living environment and the built neighbourhood are to be understood holistically as living space and as a spatial and social network of relationships. The demands placed on the living environment are manifold. It should offer space for individualisation, space for community activities and for retreat, contributes to the health of living and is important for the small-scale climate from an ecological point of view. A good living environment enables communication, encounters, and movement.

In the immediate vicinity of the house and apartment is the private open space as garden, courtyard, balcony, loggia, or roof garden. These private open spaces also characterize the communal or public areas of the living environment. The connecting links between private and public space are the house entrances and front areas as well as the semipublic residential streets, paths and squares. Together with green links and green corridors, they link the individual subareas of a neighbourhood and lead into the extended residential environment—the district, the city, the surrounding landscape.

A positive attitude toward the living environment is encouraged by certain structural and social conditions. These include design quality and human scale as well as opportunities for self-determined design and use and a diverse, near-natural environment. The living environment should offer opportunities and space for development. It is not the size of a green space that is decisive, but a differentiated open space and its design.

The open space is essentially determined by the arrangement and design of the buildings. Therefore, the development and green space plans are of great importance. In them, for example, public, communal, and private areas are defined in addition to the land area, building depths, building lines and building boundaries.

Fig. 5.46 Tartu, Estonia, Competition Europan 9. Integration of new buildings into existing structures. The main focus is on the nature reserve in the centre of the area, where an ecological upgrading takes place through an optimized connection to the regional network

Zoning and Use

An essential prerequisite for the individual use of the living environment is its sensible division into private, communal, semipublic and public areas. Like the skin of an onion, the living environment is ideally divided from private to public space into various successive and merging areas. The decisive factor here is the transition between private and public space, that is, the communal and semipublic areas. If these are missing or if the transitions between private and public are too hard, the quality of the living environment is reduced.

Terrace, balcony, and loggia, the traditional open-air seats at the house, allow you to participate in the life of the living environment, but still retain most of your privacy. During the planning stage, requirements such as protection against views, wind and weather, a sunny, and as quiet a location as possible, as well as sufficient dimensions and room depth must be taken into account. Balconies and loggias are the most common form of outdoor seating in multi-storey residential buildings.

Conservatories and glass annexes enable—almost independent of the weather—a feeling of "living outdoors." However, they can not only extend the living space, but can also—if properly planned—act as a climate buffer in the transition area between home and living environment.

The front area of the house is a part of the living environment that is experienced and used daily. It serves both as a means of access and as an area of residence and contact between the private and public spheres. Depending on the arrangement and design, vestibules offer space for different uses, for example, for shelters and storage facilities, space for business, work, repairs, and gardening activities, as well as an opportunity to stay and play.

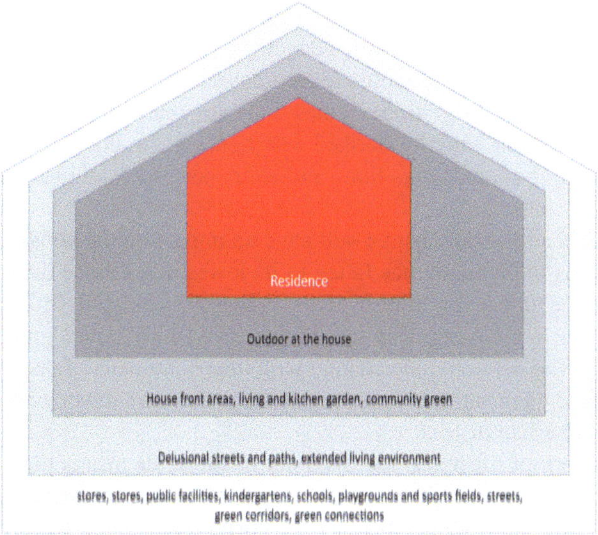

Fig. 5.47 Principle of zoning (onion principle)

Residential streets and roads are part of the immediate residential environment and the local transport network. In addition to the development of residential areas, they serve as a place to stay, meet, and play. In order to avoid uncertainties in behavior, residence and development functions must be recognisable to everyone. Pedestrians and cyclists are to be given preferential treatment. In order to minimise disturbances caused by motor vehicle traffic, more and more streets in residential areas are being extended or rebuilt with traffic calming measures.

Trends in Housing: Setting the Course for the Future

Housing is part of the basic social equipment of every human being, even if at the moment the discussion about the future of housing focuses strongly on housing as an economic good. The sale of large housing stocks to private investors, the privatisation of municipal housing companies and, last but not least, the construction boom in the area of housing for the elderly, which promises a return on investment, have almost forgotten the importance of housing as an important social good.

Nevertheless, the home and its surroundings are still an unrestrictedly important "good," even if this has been subject to serious change processes in recent decades and major challenges lie ahead.

With regard to suburbanisation, we have seen a turnaround. For some years now, there has been agreement that "the migration tendencies in almost all European metropolises are showing clear signs of slowing down" (FAZ October 31, 2003). A not so insignificant contribution to this is being made by rising energy prices. The trend of people returning to the cities in search of accommodation has now also reached the large and medium-sized cities of the Ruhr Area. The path from the housing estate back to the urban residential neighbourhood is the first step toward a city that reunites urbanity and individuality as an openly accessible, communally used system.

In addition to this generally positive development, we must note that in times of a relaxed housing market, households leave unattractive residential neighbourhoods and settlement areas. Migration studies indicate that a relevant proportion of emigrated households initially looked for housing in the cities or in certain neighbourhoods, but did not find an offer corresponding to their housing preferences.

Demographic and social change will have an impact on the housing market. The number of births in Germany has fallen to its lowest level since the Second World War. At the same time, every fourth person is over 60 years old. In 20 years' time—assuming we do not reverse this trend—we could have more grandparents than grandchildren. The demographic change with the phenomena of decline, increasing ageing and heterogenisation of the population also has direct consequences for housing and the urban district.

The patchwork of lifestyles, the diversity of lifestyles means that pluralism and individuality will become more prevalent in the housing sector in the future. The housing forms of tomorrow will become a colorful reflection of the diverse, individual lifestyles. An increasingly differentiated society with diverse and different housing needs requires an adequate range of new and experimental forms of housing. Flexibility is also gaining in importance in the residential sector. New communication technologies are bringing living and working closer together.

It will be difficult to realise the differentiating demands in the existing housing stock. As a consequence, this means that despite relatively high vacancy rates in certain urban neighbourhoods, new residential construction is required at the same time, with the focus on meeting changing qualitative demands. The demand for flexibly usable, still densely populated living space in central urban locations is growing, also due to the increasing importance of exogenous factors such as climate change and the resulting imperatives for ecologically responsible, energy-saving living in urban density.

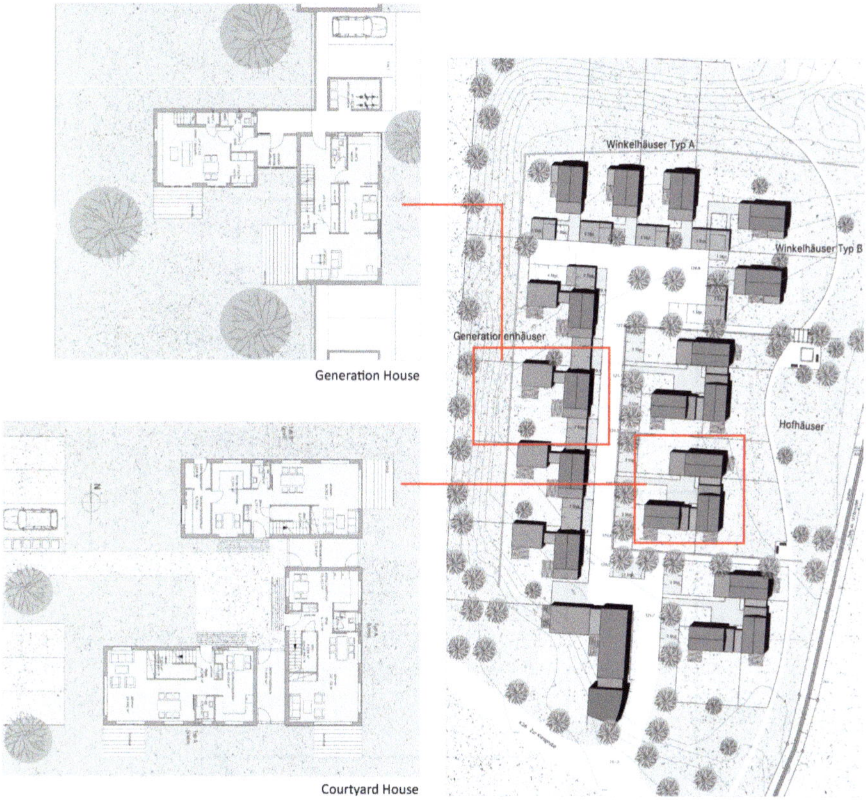

Fig. 5.48 Multi-generation housing project, RHA architects and urban planners. Multi-generation housing project, Schweich Generation houses, courtyard houses

Experimental Forms of Living

The trend toward diverse housing structures, which is accompanied by a change in the age structure of the population, calls for innovative and experimental ways of living. Flexibility, neutrality of use and the special flair of an apartment or house are increasingly in demand on the housing market:

- The “growing house” offers possibilities for long-term expansion.
- Multigenerational living creates a coexistence of different groups of residents through the housing mix and common areas.
- The Boardinghaus and Hometel can satisfy the demand for temporary accommodation.
- Floating homes are developed at water locations to meet the desire for a special atmosphere.

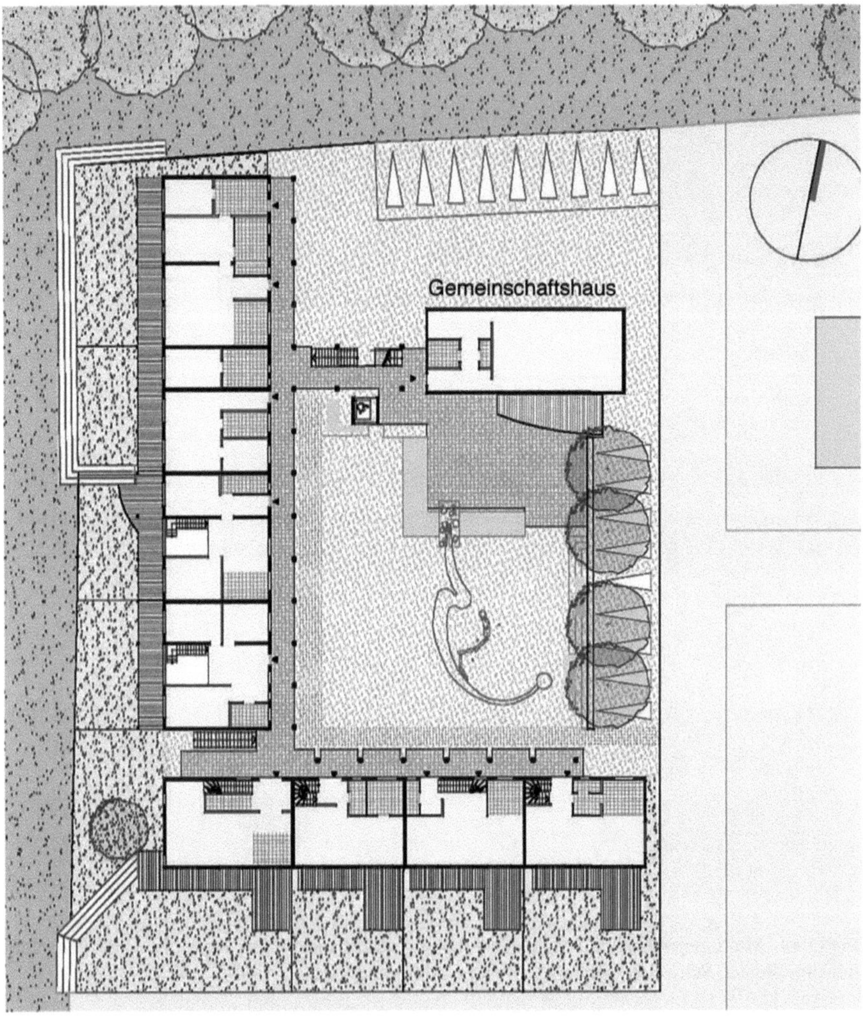

Fig. 5.49 “WohnreWIR” at Tremoniapark, Dortmund, Post—Welters Architects and Urban Planners GmbH, Dortmund. On the site of the disused coal mine “Tremonia” on the edge of the city centre, an ecologically oriented housing project for all phases of life has been developed on behalf of WohnreWIR GbR. 21 residential units between 55 and 150 sq. m, all of them barrier-free, are grouped around a common patio

Fig. 5.50 Floating home, IBA Lake, 2010

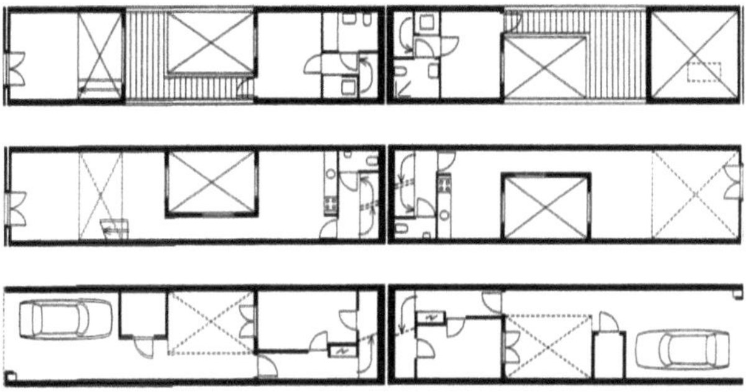

Fig. 5.51 Borneo Sporenburg, Claus en Kaan Architects, Amsterdam, NL. Floor plans ground floor, first floor, second floor

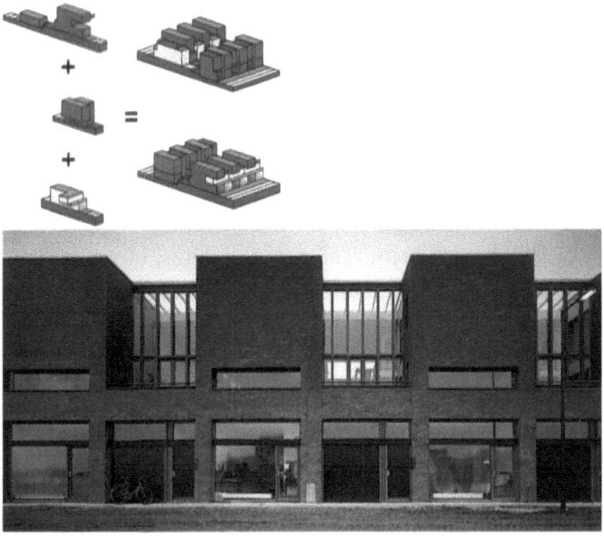

Fig. 5.52 Borneo Sporenburg, Neutelings Riedijk Architects, Rotterdam, NL. Usage modules and pier view of the house type

4 Urban Component Trade and Industry

Planning and building for trade and industry makes high demands: More than with other planning tasks, economic efficiency is a high priority. Commercial buildings must be flexible and adaptable, technical installations must enable complex operational processes. In view of these use-related requirements and the focus on the functional building envelope, other aspects have often been neglected in the past. As a result, industrial estates have been created that lack urban design and architectural quality. New technologies and the expansion of the service sector are forcing us to rethink, away from pure function and toward new quality.

In addition, planning and building for trade and industry is a factor within the construction industry and the urban fabric that should not be underestimated. In recent years, the construction industry turnover in Germany for commercial and industrial construction has reached about the same level as the turnover for residential construction, which illustrates the necessity of qualifying this Urban component (see also, Supreme Building Authority in the Bavarian State Ministry of the Interior 1996).

High-quality architecture, cost-conscious planning and execution, ecological compatibility, space-saving concepts, integration into the urban development context and social quality are criteria that areas for trade and industry must fulfil. Commercial areas have now been recognised as elementary building blocks of the city: They are a working environment for employees, living space for residents and an important factor influencing the natural balance.

Planning and design solutions must show how these different requirements can be taken into account, with the aim of creating urban places of work or designing commercial and industrial areas as independent and at the same time integrated parts of the city.

Historical Development

It was not until the second half of the nineteenth century, when industrialisation progressed, that the development of areas for industrial and commercial settlements gained in importance, making it a very young planning task. Until then, the place of work was allocated in the immediate vicinity of the dwelling and thus integrated into the urban structure. Businesses were mostly able to expand through extensions and conversions in the inner city block inner areas.

It was not until the middle of the nineteenth century that companies began to relocate to the outskirts of the city. The new buildings, however, generally did not develop any formal language of their own, the urban structures of the inner city development were transferred to the periphery in the form of space-intensive halls and story buildings. Decisive for the building configuration were the respective production processes and their rationalisation. Urban design structures played no role in the planning of commercial and industrial sites at that time.

It was only with the criticism of the inhospitable housing conditions in the cities due to the growth of industry that new models in the organisation of the urban fabric were developed. Large-scale entrepreneurs such as Owen in England (1817) and Krupp in Germany (1873) built workers' housing estates in spatial allocation to the factories.

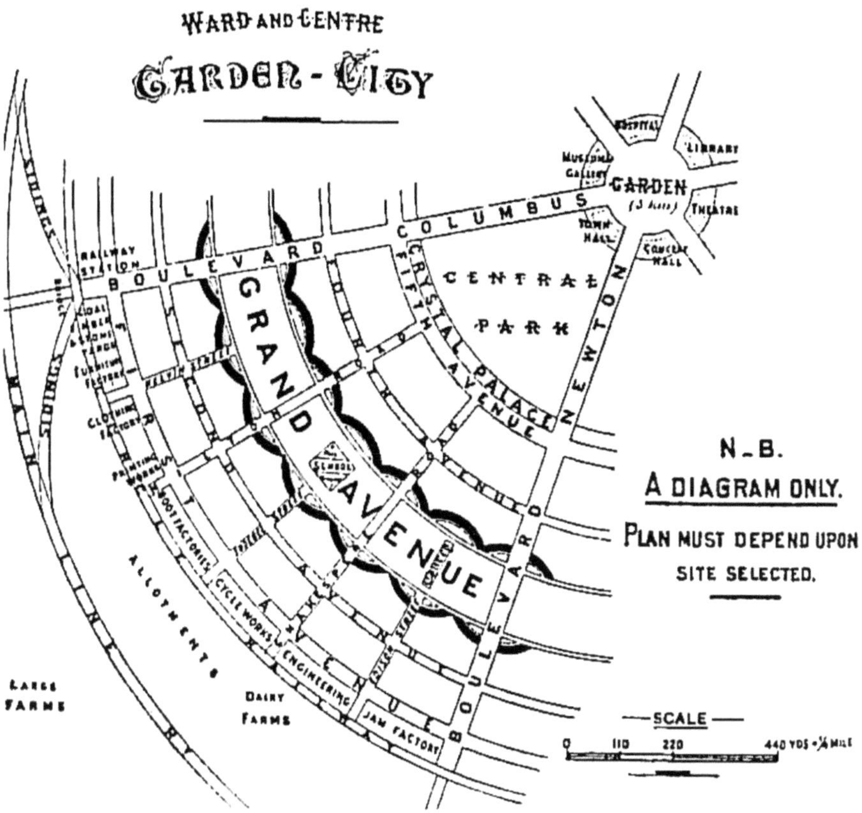

Fig. 5.53 Ebenezer Howard: Garden City 1898

The garden city movement of Howard, who first presented his "Garden-City" concept in 1902, contrasted the big city with the idea of a manageable, green garden city structured according to its functional areas of living, working, community facilities, and recreation.

Today's Requirements, Designs, Planning

Recent economic developments show that the needs of trade and industry are constantly changing. The individual business must achieve its competitiveness by adapting as far as possible to market conditions, and accordingly urban design and architecture must allow this flexibility: "Industrial estates need a design vocabulary that takes account of urban planning requirements and at the same time responds to the needs of businesses and local authorities in dealing with the sometimes rapidly changing business structures". (Schmidt et al. 2006, p. 36–37)

The ongoing structural change in the economy is most clearly reflected in the progressive shift from the production sector to the service sector, the so-called tertiarisation, accompanied by changes in the organisation of work, the division of labor and the application of new production, transport, and information technologies.

Proximity to the customer, accessibility and the urban design environment are important location criteria for many service-oriented companies today, as is the image value of the location, which also shapes the identity of a company through architecture and urban design qualities. A good urban design concept for industrial estates is today part of local and regional economic development.

In addition, the demands of new technologies are becoming increasingly differentiated. New building typologies are currently emerging, especially in the area of "innovative" technologies.

The dissolution of traditional working hours and working conditions, primarily in the tertiary sector, makes a shift toward mixed use sensible. A spatial coexistence of living, working, and other life functions puts the advantages in the foreground. When integrating commercial operations into residential and mixed-use areas, the applicable requirements under immission control and planning law must be observed.

Fig. 5.54 City of Leverkusen with Bayer plant. The city of Leverkusen is essentially shaped by the Bayer plant. Right up to the social and cultural life, the plant and the city were connected with each other and entered into an almost ideal-typical relationship, which not only significantly determines the city structure

Types of Business

Today, there is a wide range of coexistence, from small craft enterprises to highly technical research and development units. Functionally, they can be distinguished in production and manufacturing, storage, marketing with associated social area, development, exhibition, and sales. This functional allocation requires different sizes and requirements for the development and building structure, which is why the following types of business can be distinguished from an urban planning point of view:

- Craft enterprises and small to medium-sized manufacturing companies
- Large industrial enterprises
- "Clean industry" (service companies, research and development companies) and
- Special types with special location requirements (e.g., shipping companies).

In the case of larger industrial estates, social infrastructure facilities may be required, for example, shopping facilities, gastronomic, or cultural facilities as well as kindergartens and day-care centres.

Design Elements

The urban design concept for an industrial estate develops from the coherent consideration of the individual structural features. The following design elements determine the planning:

- the usage structure underlying the arrangement of the different functional areas of an industrial estate
- the building structure, which is derived from the usage structure and the overall urban development concept
- the parcelling out, which takes account of different farm sizes,
- the development of an area,
- the green order, the green and open spaces, water bodies with their surroundings and the planting in the street space in the urban context (see also, Supreme Building Authority in the Bavarian State Ministry of the Interior 1996).

Fig. 5.55 Cologne-Vogelsang industrial estate, Reinhard Angelis Planning, architecture, design, Cologne. Urban concept (below) and design layers (right)

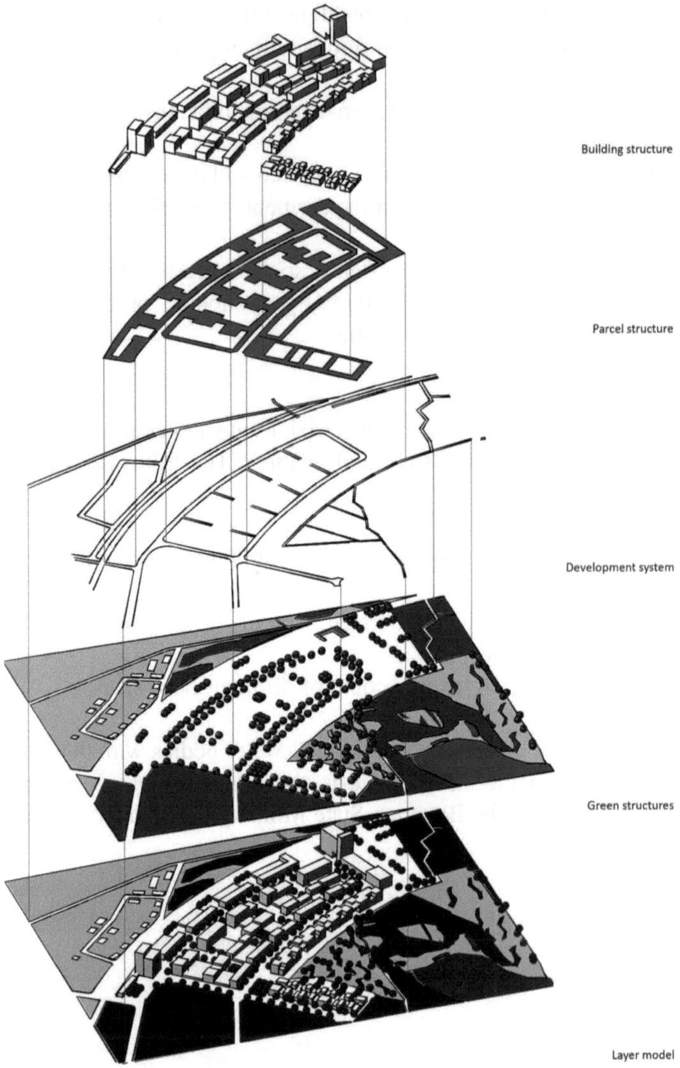

Fig. 5.55 (continued)

Use and Zoning

How a commercial area is to be used for different purposes depends on the one hand on the types of business intended for settlement and on the other hand on the specific requirements of the business and its organisation. The following units of use are to be taken into account, with different degrees of intensity depending on the type of business:

- Administration
- Production areas
- Storage areas
- Delivery and shunting areas
- Parking spaces for employees and visitors
- Infrastructure areas
- Green open spaces (incl. rainwater retention)
- Reservoir areas for possible expansion of operations

Units of use can be combined and arranged on the property in such a way that a uniform zoning in the area is possible. This zoning can be specified in the development plan. The advantage of this additive building structure is the flexibility of the system, as only individual buildings need to be replaced or reconstructed when certain areas are modernized. However, different uses can also be combined in one building structure, so that the built-up area is significantly reduced.

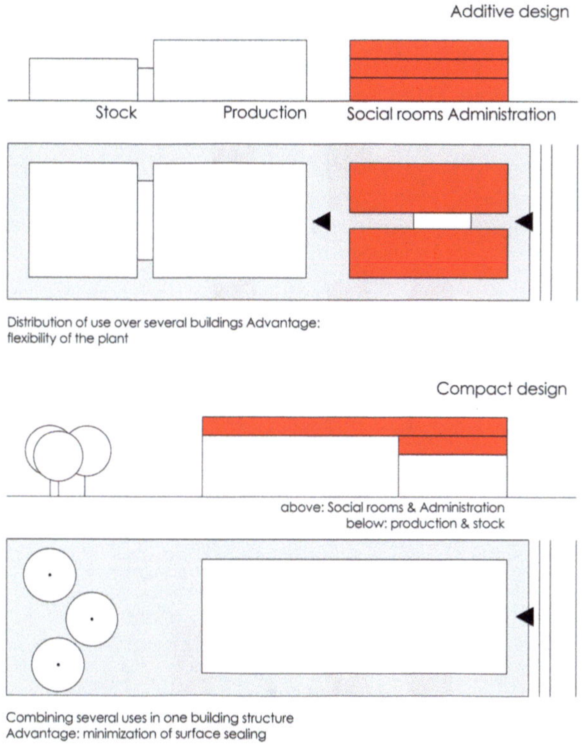

Fig. 5.56 Exemplary zoning on the property

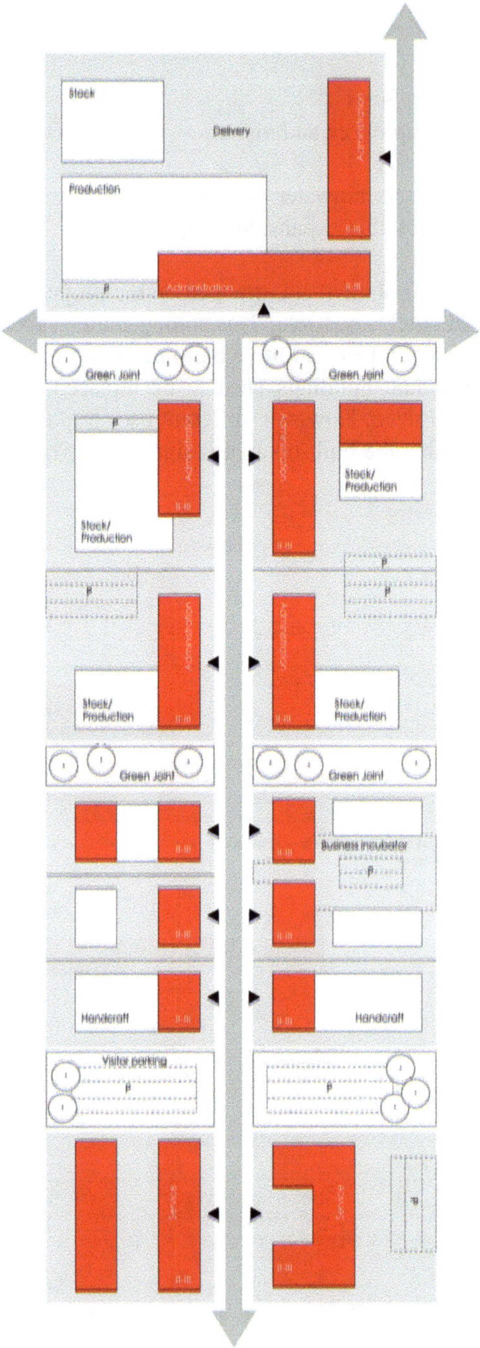

Fig. 5.57 Exemplary use zoning

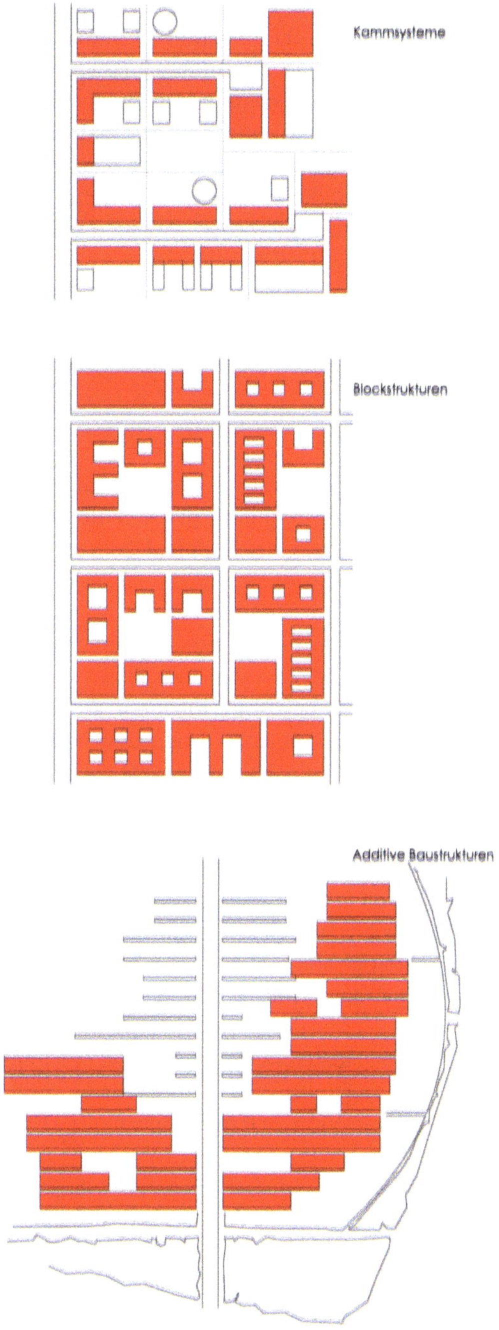

Fig. 5.58 Building structures for commercial sites

Building Structure

The building structure results from the selection and arrangement of the different functional areas of an industrial estate. The building structure is to be developed in Dependence on the functional requirements of the companies, the urban development and urban space goals of the overall concept, and the specifications of the location, for example the surrounding buildings, the topography and the development. Even if the requirements of the businesses to be settled are known at the start of planning, the building structure should be flexible enough to allow room for changing requirements.

In the sense of a resource-conserving handling of soil the buildings should be compact and in the urban solution a high density must be read. Low energy construction, passive solar energy use and waste heat recovery are also mandatory in commercial construction.

Building

Since—unlike in residential construction—in commercial construction every company needs a solution tailored to its requirements, it is difficult to classify buildings by type. For this reason, the urban development concept must allow sufficient leeway for the individual company requirements, at least when the future users have not yet been determined.

Characteristic structures in commercial construction are as follows:

- Linear structures in which the operating functions are connected in series, for example, administration—development—production—warehouse.
- Modular systems in which individual building parts are combined with each other in a modular system. The buildings can be flexibly extended by adding individual modules.
- Commercial yards, where the operational functions are divided into several buildings and grouped around an inner courtyard.
- "Spine facility" systems, in which a central development and supply axis connects individual buildings and building sections (see also, Supreme Building Authority in the Bavarian State Ministry of the Interior 1996, p. 26).

Different structures can be created by combining the structures:

- Comb systems enable a uniform external appearance. The rear areas can be adapted to different functions.
- Additive building structures provide a linear growth pattern. The individual elements are standardized and can be combined in different ways within a grid (modular system).
- Block structures allow the training of different block types which can be adapted to the needs of the companies.

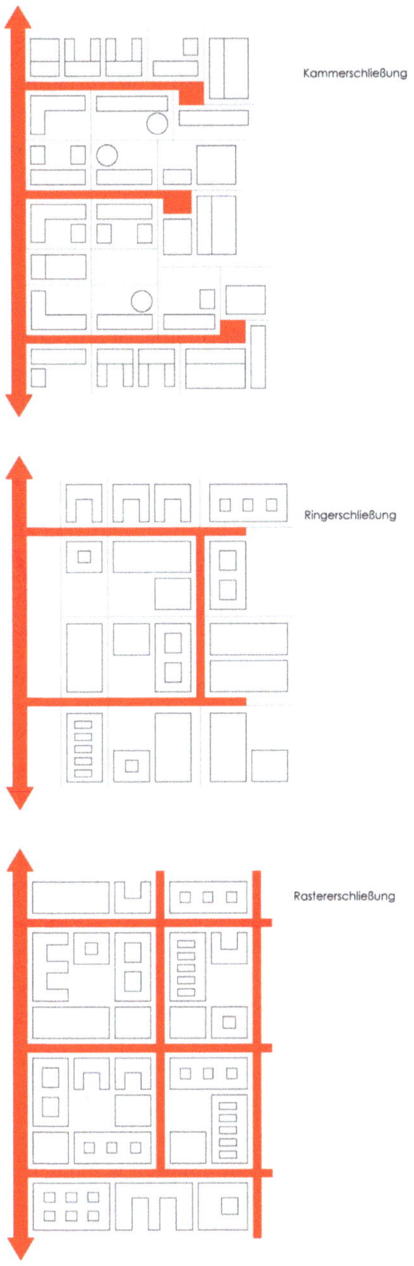

Fig. 5.59 Access systems

Parcelling

With the definition of building structures and development systems, the framework conditions for the parcelling out are fixed. The variability required in industrial estates can be achieved, for example, by defining coherent building areas that can be divided flexibly and according to requirements. In this way, the possibility of joint use of plots by two neighbouring businesses (joint depots, internal development, storage areas) is granted.

Access

Internally, an industrial estate should be given its own urban identity through a clear basic structure and a simple and legible principle of order for the buildings. The basic grid of streets and paths creates the basis for this by differentiating between main and secondary roads, by forming focal points in the area and by topographical features (see also, Lorenz 1993, p. 67).

An ecologically compatible transport development has the following objectives:

- Reduction of the space required for the traffic facilities
- Open-space-compatible development system as the basic framework for parcelling and usage structure
- Positive design of the working environment.

In road development, car and lorry traffic, foot and cycle traffic and, where appropriate, rail traffic are to be given equal consideration. Particularly in the case of the settlement of job-intensive companies, attention must be paid to an attractive network of cycle paths and footpaths, or even integration into the local public transport network. The aim must be to limit the proportion of motorised private transport to the necessary extent.

The internal development system was to allow for demand-oriented site layouts for different commercial uses and operational requirements and to enable the successive use and development of the areas. The multiple development of buildings should be avoided, and the space required for traffic facilities and parking spaces in public areas should be reduced to the necessary extent.

The following development systems are available in industrial estates:

Chamber Access

The access roads branch off from a main road like a comb. Between the stitches, manageable plots are created. A good orientation is made possible. The rear sides of the industrial estates face each other without disturbance, but the stitches only allow a limited depth of development.

Fig. 5.60 Open space structure of the Mönchengladbach Sports Park

Annular Access

The development of the commercial area is from the main road in the form of a loop. This means that only two intersections are necessary on the main access road. With the development loop, many plots are connected in an economical way. The rear sides of the farms are oriented toward the open landscape. The rear sides of the plots to the feeder road can be problematic. This can be remedied by planting or green protective walls.

Raster Access

With grid development, regular building blocks are created which can be approached from four sides and are, therefore, usually overconected. The flexibility in the parcelling is limited by the rigid development system. Only certain company sizes fit into this development scheme.

The variety of development possibilities shows that factors such as access and view sides, arrangement of green spaces or the settlement of additional uses, but also the expandability of individual businesses must be taken into account as early as the urban planning design stage. New commercial enterprises to be established give the freedom to a holistically coordinated concept in terms of urban design and traffic engineering. (see Lorenz 1993)

Parking Spaces

A sufficient number of parking spaces must be demonstrated for the destination and source traffic in the industrial estate. In the industrial estate, public parking spaces are generally only created for parts of the visitor traffic. The necessary parking spaces for employees and customers have to be proven by the tradesmen themselves on the sites. The number is usually based on the usable area of the business, the number of employees and, in the case of customer-oriented businesses, the number of visitors. Parking spaces for visitors should be located in the front areas of the business and should be easily accessible with a direct connection to the entrance. Parking spaces for employees can be located to the side or behind the buildings.

It depends on the urban development objectives of the planning concept how the parking spaces on the sites are to be arranged. Traffic development concepts should contain proposals for decentralised parking space accommodation in compact units surrounded by greenery and should show site-appropriate alternatives for parking space accommodation on private plots. Large car parks should be avoided.

Open Spaces and Green Spaces

An attractively designed industrial estate is always a good address to advertise and enhance the image of the company and the location. The implementation of environmentally compatible objectives in industrial estates also contributes to a high-quality working environment and links economic development progress with improvements in the environmental situation. For this reason, the development of open space structures and open space design should play a special role in planning and execution, but also in the further development of existing commercial structures.

Green spaces should be an integral part of the planning of industrial estates. Green planning measures are particularly necessary for the townscape and as a contribution to the balance of nature in the industrial estate and make an important contribution to urban ecology. The measures are manifold: creation of biotope areas, green and forest areas, large-scale networking with green corridors and stepping stone biotopes, maintenance and development of nature and landscape in the vicinity of industrial estates, near-natural greening of roads and parking spaces, greening of roofs and facades, and the creation of recreational and recreational areas for employees.

Commercial and industrial areas are an integral part of the city and must be integrated into the urban structures in terms of design and ecology. As a motor for settlement, high-quality open space design must be ensured. Commercial areas should fulfil the following criteria:

- Street space design for address formation, for example, through avenues (also to improve the local climate),

- Connection to the city foot and cycle path network,
- Permeability of the area,
- Providing green spaces as recreational areas for the workforce and as ecological compensation areas,
- Roof greening to minimize sealing and rainwater runoff.

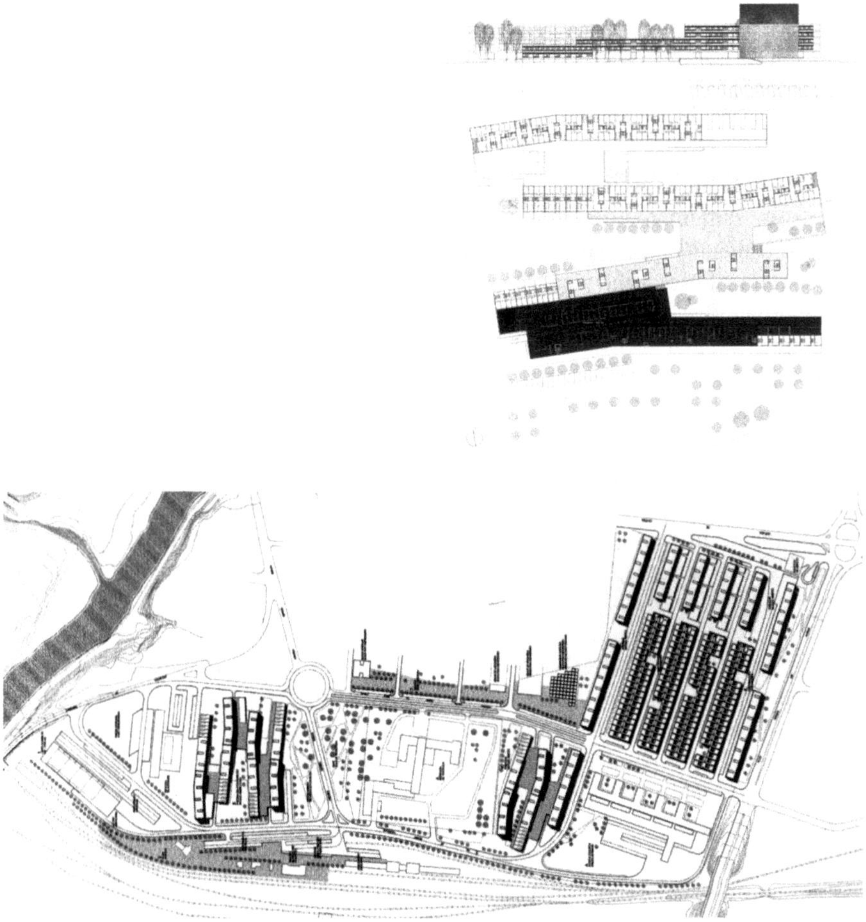

Fig. 5.61 Trade Aranda de Duero, Spain, Competition Europan 4. The design is limited by the railway line and the extensions along the main road and therefore cannot revolutionize the existing structure. The existing stock conditions the design, which formulates the contextual urban development in simple terms and not only does not exclude changes over time but accepts them as enrichment

Further Development of Industrial Estates

Further development and redensification of already existing industrial estates offer a significant development potential for municipalities, as the need for new designation of areas is reduced. On the other hand, by promoting an already established area, one achieves a safeguarding of the stock. Functional and design measures are used to enhance the value of the location and thus increase the attractiveness of existing businesses. A further advantage of development of established locations is the existing integration into the urban fabric and the coexistence of living and working. The inventory oriented and step-by-step further development of an existing structure is therefore not only a space-saving and cost-effective, but also a neighbourhood compatible alternative for the settlement of businesses.

New Designation of Industrial Estates

The redesignation of industrial sites should not take place without examining alternative settlement possibilities and a demand-oriented demand analysis, on the basis of which the type and structure of the area is determined. The aim must remain an economical use of the resource soil as well as a mixture of functions in order to create urban places of work and design industrial estates as independent and at the same time integrated parts of the city. If a section-by-section realisation is certain from the outset, each construction section should be designed with regard to the functional unit in its final state. In the absence of expansion, however, the retention areas should also be usable separately.

The urban design concept guides the development and the later structural formulation of an industrial estate. For this reason, these areas must be given just as much commitment in terms of quality and landscape and open space-relevant characteristics as is usual for office locations and residential areas. The keywords here are: “good address,” “image building” and “adding value.”

The former PHOENIX West blast furnace plant in Dortmund-Hörde is to be developed into an innovative technology park. A business location for microtechnology, information, and communication industry and various services will be created on the approximately 1000 ha development area. The landmark of the former blast furnace plant as well as other historical buildings will be preserved as identity-giving structures.

The urban design concept for the area is derived from the surrounding landscape. An urban linear structure functions as a superordinate system of order: “External order and internal complexity” create a robust and flexible basic framework in which the most diverse building configurations can be realized. Building stripes are defined as a system of order, which are clearly delimited from the access roads (building lines) and generate a multitude of building possibilities in the interior.

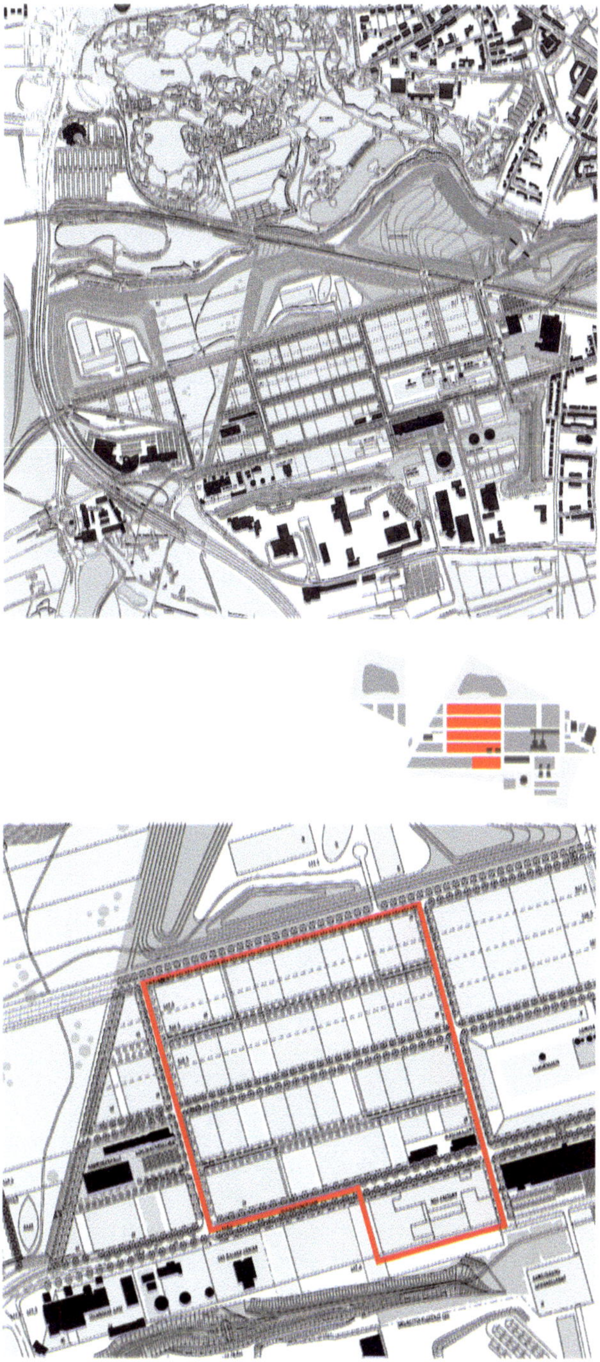

Fig. 5.62 Future location PHOENIX West, Dortmund _ stegepartner, Dortmund

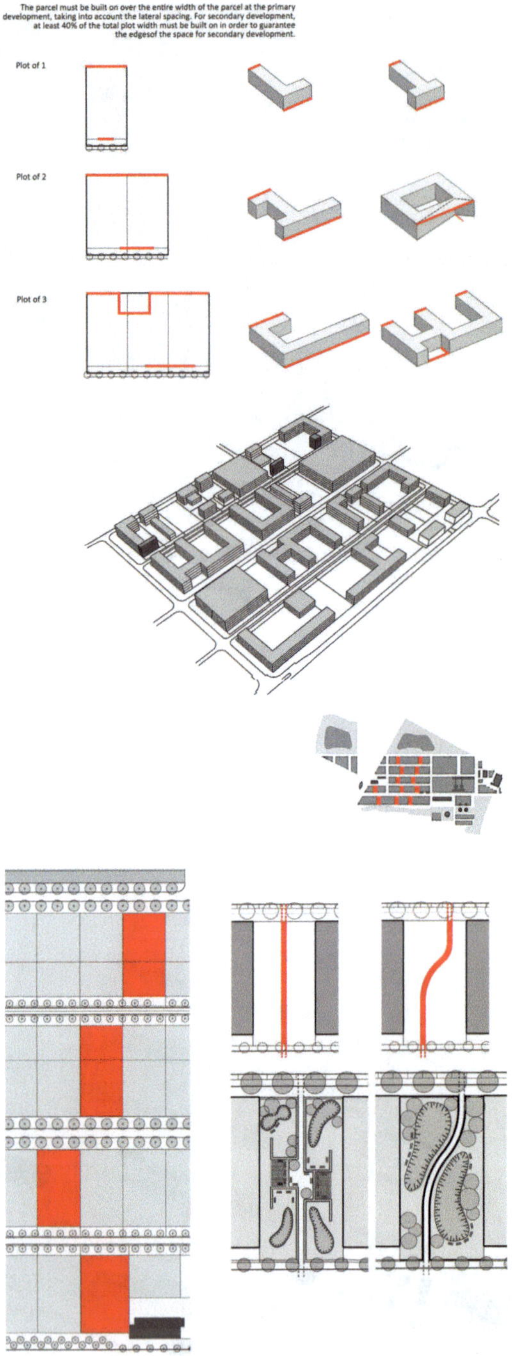

Fig. 5.62 (continued)

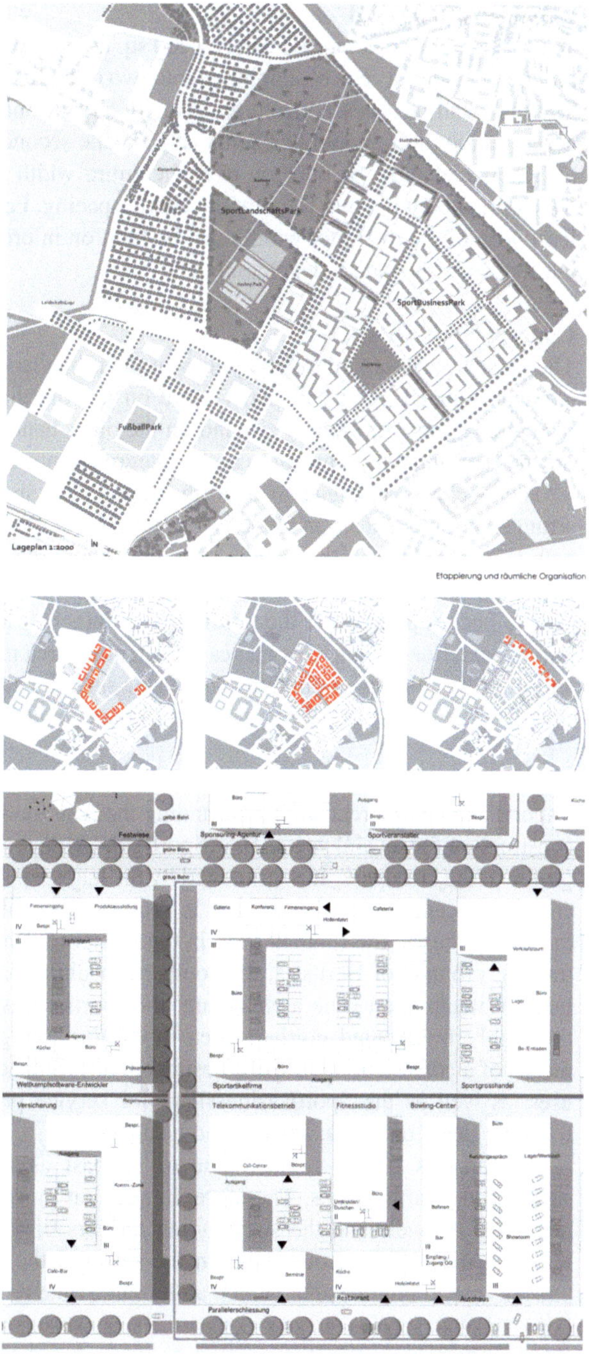

Fig. 5.63 Sportpark Mönchengladbach, RHA Architekten und Stadtplaner

Construction Area B

The basic module of approximately 1900 sq. m can be divided into two half plots in plot no. 5, and plots can be combined into two or three modules to create larger units. The plots can be accessed from the primary and secondary roads. The delivery and access to the plots should be exclusively via the secondary road.

The parcel must be built on over the entire width of the parcel at the primary development, taking into account the lateral spacing. For secondary development, at least 40% of the total plot width must be built on in order to guarantee the edges of the space for secondary development.

Concept Pocket Parks

The pocket parks are one of the pillars of the urban landscape planning design. The areas designated as pocket parks serve on the one hand as a meeting point and recreational area for employees, and on the other hand the necessary retention area for the accumulating rainwater can be integrated into the pocket park design.

The pocket parks lead to a north-south green space network. The staggered arrangement guarantees an attractive footpath through the park. As a rule, the public path is to run through the middle of the pocket parks, so that two green spaces of equal size are created.

The pocket parks offer the possibility to integrate the necessary retention areas into the exterior design. This makes the retention area more attractive for employees and improves the microclimate. The water surfaces can be produced as basins or troughs. The design as a flat lawn area is conceivable.

On the site of the former Ayrshire Barracks North, Nordpark Mönchengladbach, a sports and commercial area surrounding the new football stadium, has been under construction for several years. The "SportPark" Mönchengladbach describes the new profile of the Nordpark. It is structured by the landscape axis and the city axis, which cross at a central point. The area is thus divided into four areas with different thematic focuses: the football park in the western "quadrant" with the stadium and training grounds of Borussia, the southern media park with creative companies in listed Wilhelminian-style architecture, the sports landscape park, which combines the highly fragmented northern area (hockey, parking spaces, cemetery, etc.) by means of a consistent open space design and offers space for scenic leisure-oriented uses, as well as the SportBusinessPark for service providers and administrations from the fields of sports, leisure, and health.

The area is highly isolated due to its past as a barracks location and the surrounding traffic routes. Therefore, in addition to an internal reorganisation, it is necessary to overcome the surrounding barriers of the "SportPark" and to better integrate it into its surroundings. In the sense of the overarching theme of sport, "tracks" are to link the area, which at the same time represent a wayfinding system.

5 Urban Component Social and Educational Infrastructure

Päivi Kataikko-Grigoleit

Museums, theatres, hospitals, schools, and universities are important buildings or ensembles that shape the cityscape; as a social and cultural infrastructure they are an indispensable element in the city. They are not necessarily public institutions, but they do fulfil original public tasks: Thus, the first schools and universities were mostly founded by churches and even today the church is still the sponsor of many denominational educational institutions. In addition, in recent years more and more private schools, private kindergartens, and private universities have been founded. The increase in educational institutions underlines the growing importance of education: More and more people are spending more and more time gaining and imparting education, because with the transition from an industrial to a knowledge society, education and knowledge are becoming the most important raw materials for the future.

This does not mean, however, that the only thing to be done is to build more schools and universities. Existing educational institutions must also be adapted to their new role and the challenges they face: New forms and formats of learning are changing everyday life in schools. More and more educational institutions are becoming "all-day businesses." Educational buildings are increasingly being combined with other social or cultural facilities and should also be available for non-school purposes with a view to their efficient use. For universities, close links with private research institutions, business start-up and technology centres remain important; their campuses are increasingly to be "posturbanized" with leisure, sports and cultural facilities or residential buildings for university staff, so that they can develop into vital science districts of the city.

Schools are sometimes social stabilisers in difficult, often disadvantaged urban neighbourhoods. There they act as centres of local community by combining education and training with social and cultural district work. In addition, schools are also increasingly providing impetus for urban redevelopment and expansion. In these cases, they are not so much follow-up housing facilities but rather attractors for new residents who settle in those urban areas that are supposed to have the best schools. The same applies to universities, albeit at the regional level.

Accordingly, the fields of activity for architecture and planning are very diverse: existing educational facilities must be renovated and spatially reorganized; new educational facilities must be planned and built that can be meaningfully integrated into the surrounding city. Schools and kindergartens should increasingly be linked with each other and with other institutions to form educational landscapes—in other words: educational buildings and landscapes in particular will be formative elements of the "city in the knowledge society."

Historical and Current Developments

Educational institutions have a tradition as urban places that goes back many centuries. From the very beginning, they have usually been at the centre of towns and villages: The classic village or elementary schools, for example, together with churches and town halls, formed the characteristic buildings of a place. Larger cities had universities. The first universities were founded in the German-speaking world in the fourteenth and fifteenth centuries, and even today some of these towns are known mainly as university towns (Tübingen, Göttingen, Marburg, Freiburg, or Heidelberg). At that time, the universities comprised only very few faculties, mostly theology or medicine, and were therefore limited to a few buildings, usually located in the middle of the medieval cities. Even the elementary schools, which went back to denominational schools of reading, writing, or arithmetic, usually consisted of one building, which in times of war was often used for other purposes (e.g., as a military hospital). The number of elementary schools grew only after the introduction of general compulsory education and the associated goal of actually providing a nationwide school education. The brick buildings from the Wilhelminian period became the dominant type of school construction.

Fig. 5.64 Main building of the RWTH Aachen

In the course of the twentieth century, not only did the number of educational institutions continue to increase, but they also required more and more space due to their internal differentiation (more subjects, more types of schools, more faculties).

The campus became the common spatial model for universities. The same applies to schools: a wider range of offers, larger catchment areas for secondary schools—all this leads to a leap in scale from individual buildings (plus gymnasium and sports field) to extensive school complexes which, where different schools share certain facilities (sports hall, canteen, library), approach the (university) campus model.

In the 1950s to the 1970s a large number of schools and colleges were built, which are now in urgent need of renovation. This need for refurbishment coincides with a necessary conversion of the educational buildings: New forms of learning and teaching require new spatial principles (moving away from buildings that focus on instruction and teaching to flexible "learning landscapes" in which self-responsible learning becomes the standard). As all-day facilities, schools must not only provide learning spaces but also spaces for "non-learning," for leisure and regeneration (refectories, play and leisure rooms, etc.). Many university campuses will also not only be renovated in the coming years but will be comprehensively redesigned. The monofunctional special areas are to become urban city spaces, especially where university campuses were built on greenfield sites in the 1960s and 1970s.

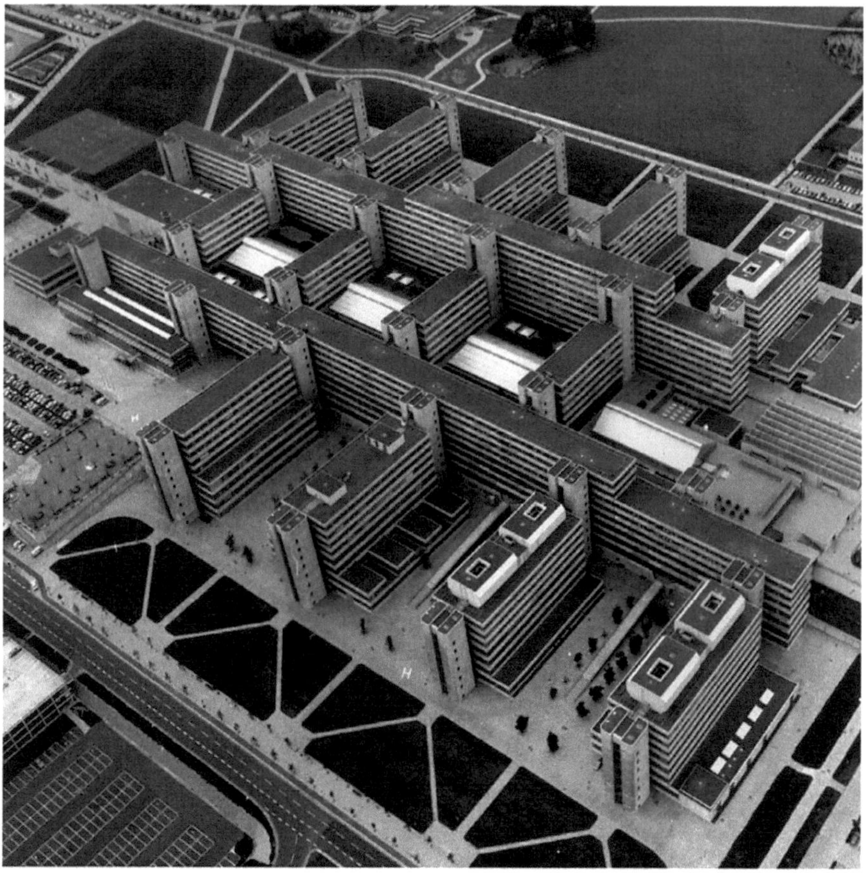

Fig. 5.65 Bielefeld University, Planning Association Bielefeld University Köpke, Kulka, Topper with Siepmann and Herzog, construction period 1972–1974

Fig. 5.66 City Campus, Cambridge and Karlsruhe

Fig. 5.67 Green Campus, ETH Zurich and Constance

Planning Handling

The locations of public educational institutions are usually determined on the basis of long-term infrastructure planning (e.g., school development or university development plans). There are various concrete guidelines or recommendations for the construction of educational institutions (e.g., on minimum sizes and equipment features of buildings and outdoor facilities), which are usually regulated at the state level. In Germany, the federal states are responsible for education. In addition, some cities have also drawn up municipal school construction guidelines.

Typologies

Schools and universities are already very different design tasks because of their different scale: Schools or school centres, which are organized as educational landscapes, are usually designed for 100–2000 students, while universities can have between 500 and more than 30,000 students, that is, they can be the size of an urban district. In such cases, universities must also be planned and developed like urban districts—with concepts for use, development, open spaces and access, based on a flexible yet clearly identifiable spatial and settlement structure. It must be possible to develop such areas in several stages over a longer period of time. The most widespread typologies are the urban campus ("City Campus") and the campus on the outskirts of the city ("Green Campus"). The campus on the outskirts of the city is mainly chosen where there is not enough space available in the city centre, but where good transport connections can be guaranteed. In recent years, however, many cities have created considerable potential on former industrial, railway, or military sites, which can be used for the construction or expansion of new universities. There, historical buildings are usually converted and integrated into the campus. Central facilities of a university (library, Audimax, refectory etc.) are either located in the centre or in the entrance area.

Unlike in the 1960s and 1970s, schools are hardly ever built on the outskirts of cities, unless they serve as a prelude to a new urban neighbourhood to be developed at that location. In most cases, schools are located in the middle of a neighbourhood or district, often combined with other public, social, and cultural facilities. Because of their required size, they are usually divided into several buildings or at least parts of buildings: Depending on the pedagogical concept, decentralized classrooms or learning houses are grouped around the communal areas of an educational institution (library, cafeteria/hall, sports hall, sports field, etc.) The building typologies on the one hand depend on the concrete spatial program, and on the other hand on the location and its urban design conditions. While one- to two-story buildings dominate in suburban environments, higher-story buildings and more compact designs are also common in inner city locations. In the case of very cramped plots of land, under certain circumstances parts of the building can be arranged underground or open spaces can be "stacked" (as for example in Hannes Meyer's design for the Petersschule in Basel from 1926).

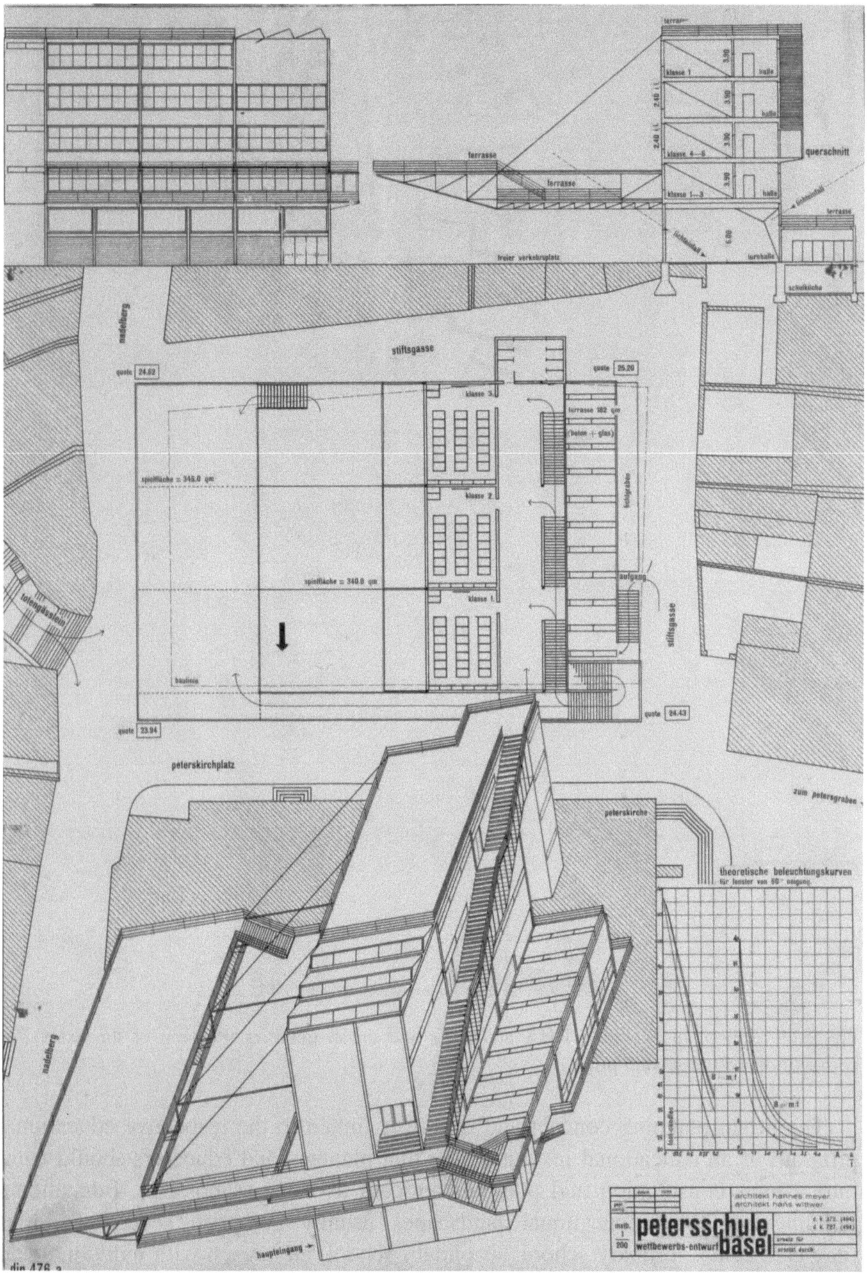

Fig. 5.68 Competition design for the Petersschule in Basel, Hannes Meyer, Hans Wittwer, Bauhaus Dessau 1936

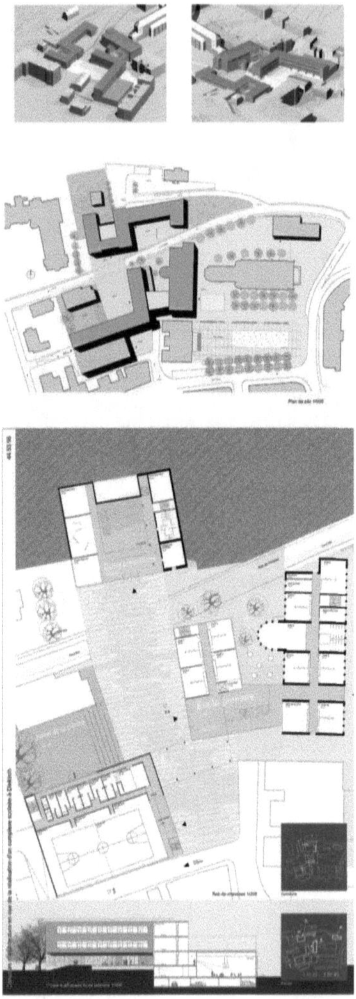

Fig. 5.69 School in Diekirch, RHA architects and urban planners. Phasing of the extension, competition design second prize

The urban planning concept and design is linked to the respective educational program of an educational institution, so that planners and educators should enter into a joint consultation and planning process as early as possible. Educational buildings or local educational landscapes usually comprise several facilities (day-care centre, primary school, secondary school, etc.), each with individual and communal areas for use. As so-called "district schools" they take on additional functions and are also available to nonschool user groups from the district. However, the extent to which educational institutions open up into the district is answered in different ways, especially in socially disadvantaged neighbourhoods: In some cases, schools their function more as sanctuaries and retreats where children and young people can develop new values and attitudes.

Access/Design of the Open Spaces

For educational institutions a very good connection by public transport is especially important. This applies to universities, which should generally be served by efficient underground or suburban trains, as well as to schools, for which trams or buses are usually sufficient. Equally important is a pedestrian- and cyclist-friendly connection of the school and university locations, which should be continued in the design of the open spaces within the areas. Concrete parking space requirements for passenger cars depend not only on the size and profile of the educational institutions, but also on the quality of development through local transport or integration into local cycle networks.

Buildings and open spaces have different degrees of publicity. This must be reflected in the zoning of the open spaces, their allocation to the respective buildings, and also in the orientation of the buildings to the surrounding urban neighbourhood. If, for example, individual buildings are used for extracurricular activities, they should have their own entrances or access points.

Fig. 5.70 Kamp-Lintfort, development of the new university location. Urban design competition. first prize (top): .pbr Planungsbüro Rohling second prize (bottom left): Müller Reimann Architekten, Berlin third prize (bottom right): Marcus Patrias Architekten, Dortmund

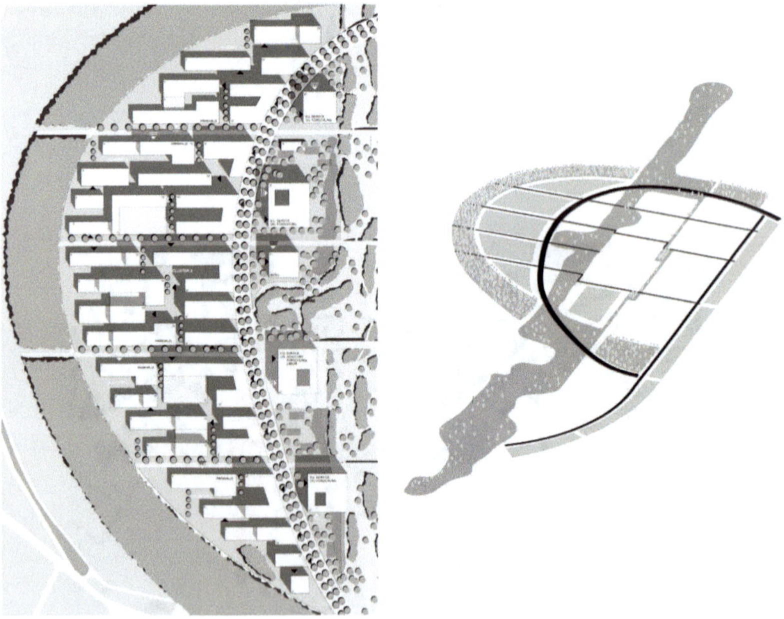

Fig. 5.71 RWTH Aachen University. Future location of the university areas within the city of Aachen

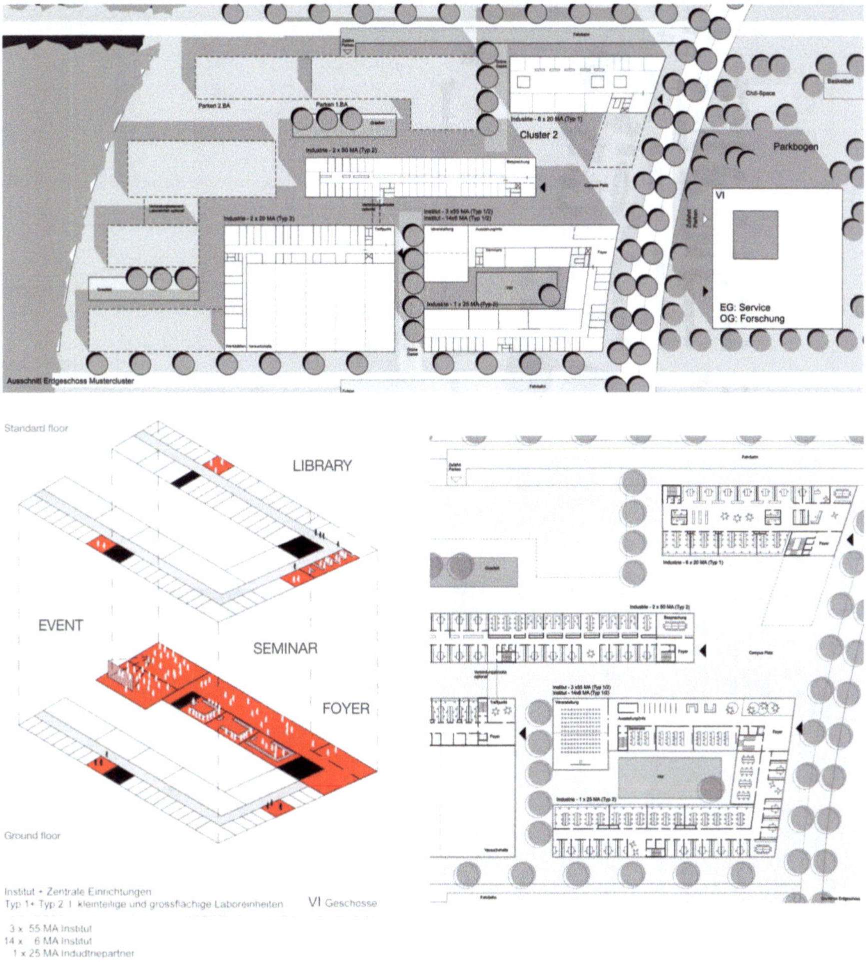

Fig. 5.71 (continued)

RWTH Aachen University will massively expand its university locations in the coming years; the largest university campus in Europe is being built here. The urban design provides the framework for the further development of the university location in the west of the city of Aachen. The concept is based on a concise landscape concept with a central park and a green boulevard as its backbone. The individual clusters have their address to public space. They are flexible in their use and in their building structure.

Fig. 5.72 University of Kassel, urban design concept for the North Campus, spacetime—Läufer, Tausch, Tuczek Architects Berlin, kl Landscape Kuhn Klapka, Berlin. The urban design is the basis for the future development of the Campus North of the University of Kassel. The basic idea is an "educational landscape" with individual blocks between the city and the park. The building blocks are polygonally shaped and as a striking element are intersected by a central development axis

6 Urban Component Trade and Supply

Angela Million

As an Urban component, trade shapes the appearance of cities and landscapes to an extent that should not be underestimated. In many places it is a basic element of the urban environment and influences the overall urban and even regional context in its development and structural appearance. The urban landscape is, always, also a shopping landscape, in which the retail trade represents uses within the urban structure (above all as shops), but also claims space as independent Urban components, whether in the form of a department store, a shopping centre, a market hall, or a supermarket. Whereas once retail was mainly concentrated in the city centre, today commercial buildings can be found in urban structures of all ages, including suburban locations and industrial estates.

Whereas until a decade ago the focus in Germany was on settling the periphery with shopping centres, consumer and specialist stores, today the interest of investors and consumers is once again concentrated on the centres. One consequence is the realization of retail space in inner city (integrated) locations. Above all, shopping centre projects and food markets are on the advance into the city and town centres. This development has positive aspects for the city, such as the strengthening of the inner cities and secondary centres and the securing of the network of local supply, which has become more extensive over the years. A major problem, however, is the insufficient functional and urban design integration of retail typologies into the urban structure and space.

With its requirements for generous sales areas and a large number of parking spaces, the retail trade often goes beyond the scope of the urban environment. The standardized, in-slung, flat market buildings do not take sufficient account of the urban space and context. Despite the high quality of the hurdles set for the design of inner city retail trade projects, streets are degraded to delivery roads, public pathways are built over, existing building lines and heights are negated. The task of planning is to treat the peripheral location requirements of the retail typologies in such a way that a high-quality integration into the urban context is achieved.

Historical and Current Developments

In its long history, trade has gone through various phases: from trade on a free, undeveloped square in front of and within the city (market square), as an integrated, subordinate component of the city in residential and commercial buildings (shops), to independent commercial buildings (market halls, shopping markets, department stores, shopping centres, etc.). In the process, older urban structures were often destroyed in order to realize the large-format commercial buildings—that is, already during the construction of market halls and department stores at the end of the nineteenth century—and the urban ground plan was quasi adapted to the trade.

Due to its integrated location in the urban fabric, the retail trade has always had a formative influence on the cityscape: as a component and user of public space, as a user of the ground floor zones in buildings, but also via the façade design through large shop windows and advertising signs. Already at the beginning of the twentieth century, forms of organisation in the retail trade, such as chain stores and cooperatives, favored a uniform shop window and façade design in the grown shopping locations and city centres. At the same time, sales concepts were always reflected in the construction. The introduction of the self-service concept, combined with technical innovations and a changed shopping behavior of the population, demanded larger sales areas and finally resulted in the operating and construction form of the supermarket. The discount principle can in turn be seen as the result of strong competition in food retailing. Its introduction was initially reflected, among other things, in a reduced shop fittings and design and later in the choice of location and standardized building typologies.

Last but not least, urban development models of past planning epochs can also be read in the building structures. Already in the garden city and cooperative housing development, shop areas and department stores were an integral part of overall urban development concepts. Centralised (marketplace, shop and shopping centres) and decentralized location concepts (solitary locations in head buildings or individual solitary food markets or similar) were pursued in the settlements. Especially in postwar urban design, trade was then promoted as an independent, monofunctional building typology in settlement construction. Freestanding shopping markets, shop, and shopping centres were built in the settlements—later, with the increasing motorisation of the population in the 1950s, also on the outskirts of the city. This development was only counteracted by the amendment of the Building Utilisation Ordinance (BauNVO) in 1977, a planning instrument with the aim of strengthening business locations close to residential areas and city centres in comparison to large-scale settlements of shopping markets and shopping centres on the outskirts of the city.

The main development trends in the retail trade today are the expansion of sales floor space, declining number of outlets, a leap in the scale of building typologies, increasing corporate concentration, the emergence of new service concepts and operating forms and the resulting changes in location structures. Experience, entertainment, festivalisation of shopping (Urban Entertainment Centre, Factory Outlet Centre, Shopping Centre) are contrasted with a discounting of retail (building typologies reduced to warehouse buildings). At the same time, planning law in conjunction with retail trade and centre concepts provides planning control for the settlement of large-scale retail trade projects (from 800 sq. m sales area), again to strengthen central business locations in the cities. This has been visible, among other things, in the construction of new food markets and the realization of shopping centre projects at central locations since the mid-1990s. It is also driven by demographic change and other economic, ecological, and sociocultural developments. In view of rising mobility costs, an ageing population and the increasing professional activity of women (who, incidentally, still do the shopping in many partnerships despite their professional activity), supply concepts are supported that are easily accessible even without a car. The presence of shops in the neighbourhood today offers flexibility and relief in everyday life.

However, in order to achieve a high-quality integration of retail trade buildings in central locations—while at the same time ensuring trade-geographic compatibility, which is not the subject of the following comments—urban design has the task of identifying development and space options in the centres (see Table: Exemplary area sizes of retail trade buildings) and to insert the module trade in form and design into the architectural context of the city.

Operational form	Offer priority	Sales area[a]
Self-service shop, specialist shop	Food (supplementary intake), non-food	By 200 sq. m
Food discounter	Food, everyday consumer goods	700–1000 sq. m
Supermarket	Food, everyday consumer goods	800–1600 sq. m
Supermarket	Food, everyday consumer goods	from 1600 sqm
SB-Warehouse	Extensive range of food and non-food products for short, medium and long-term requirements	from 5000 sq. m
Department store	For example, textiles	at least 1000 sq. m
Department store	Supply-side swings in the clothing, footwear, household goods, sometimes food and catering sectors, among others, long-term demand	4000 sq. m
Hardware store	Among others, building materials, tools, machines, gotten hardware and furniture, plants	4500–5000 sq. m, mostly over 10,000 sq. m
Neighbourhood centre (magnet tenant supermarket, consumer market, discount store)	Fast moving consumer goods	3000–8000 sq. m
Municipal and district shopping centre (additional magnet and leading companies)	Wider and deeper range of goods and services as a neighbourhood centre	8000–15,000 sq. m
Regional shopping centrecentres (magnet tenants department stores, SB warehouse, specialty stores)	Comprehensive range of offers (retail, services, gastronomy, including leisure activities)	Larger than 15,000 sq. m, mostly 30,000–40,000 sq. m

[a]Area used for sales, excluding storage areas and customer parking spaces

Fig. 5.73 Schocken department store, Stuttgart 1926–1928, Erich Mendelsohn

COMPANY FORM	MAIN FOCUS OF THE OFFER	SALES AREA
Self-service store, specialized store	Food (supplementary purchase), non-food	at 200 sq m
Food discounter	Food, goods of daily use	700 to 1.000 sq m
Supermarket		800 to 1.600 sq m
Consumer market	Food, goods of daily use	from 1.600 sq m
SB Merchandise Market	Comprehensive range of food and non-food products for short, medium and long-term requirements	from 5.000 sq m
Department store	e.g. textiles	at least 1.000 sq m
Warehouse	Main focus of the product range, including clothing, shoes, household goods, sometimes food and gastronomy long-term requirements	4.000 sq m
DIY store	Among others building materials, tools, machines, garden hardware and furniture, plants	4.500 to 5.000 sq m mostly over 10.000 sq m
Neighbourhood center (magnet tenant supermarket, supermarket, discounter)	goods of daily need	3.000 to 8.000 sq m
Municipal and district shopping center (additional magnetic and guiding drives)	Broader and deeper designed goods and services as a neighborhood center	8.000 to 15.000 sq m
Regional shopping centers (magnetic tenants department stores, hypermarkets, specialty stores)	Wide range of offers (retail, services, gastronomy, etc.)	larger than 15.000 sq m, mostly 30.000 to 40.000 sq m

*Area used for sales, excluding storage areas and customer parking

Fig. 5.74 Exemplary area sizes of retail trade buildings

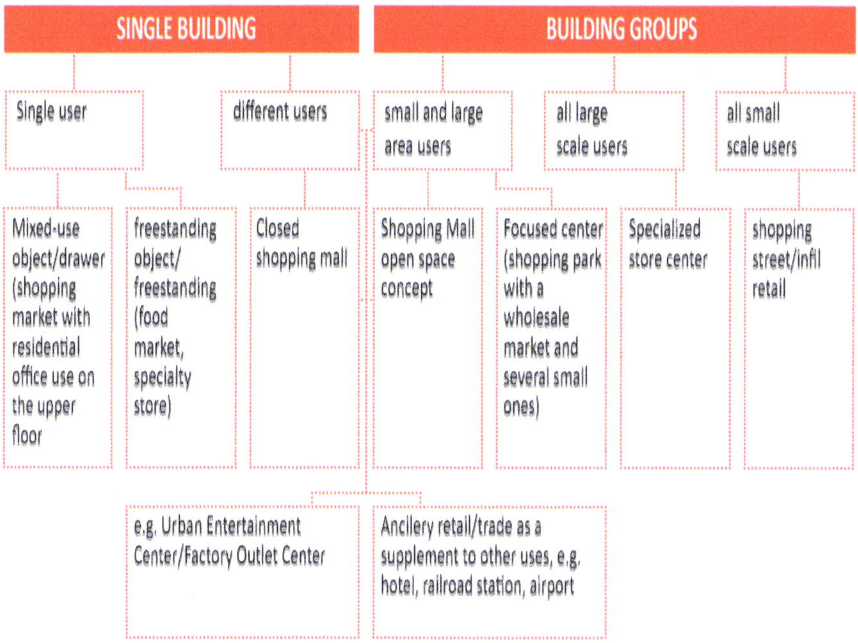

Fig. 5.75 Building typologies in retail trade

Design Principles

Small plots of land, existing cultural heritage worthy buildings, mix of uses and a dense and differentiated network of public spaces require a stronger structural treatment of retail buildings in central, grown locations than peripheral locations. Due to the scale-breaking character of large-scale retail trade projects, various options for integration should be examined. Freestanding market buildings (**"Freestander"**) with ground-level parking spaces can seldom be qualitatively combined with the multistory, enclosed construction of city and town centres. It should, therefore, be examined to what extent these can be pushed under multistorey buildings according to the **"drawer"** principle and then develop into the depth of the plot alongside the road. A further principle is conceivable in the style of cinema buildings of the 1920s (**cinema principle/market hall principle**). Here, there is only a narrow entrance on the street next to other shops, the market also develops into the depth of the block.

A distinction is made between closed and open or partially open shopping centres. A **closed centre** can be designed as an elongated mall or in a star shape. An **open centre**, on the other hand, picks up more strongly on the typical structure of the European city. It is distributed over several buildings that line the existing street structure ("open space concept").

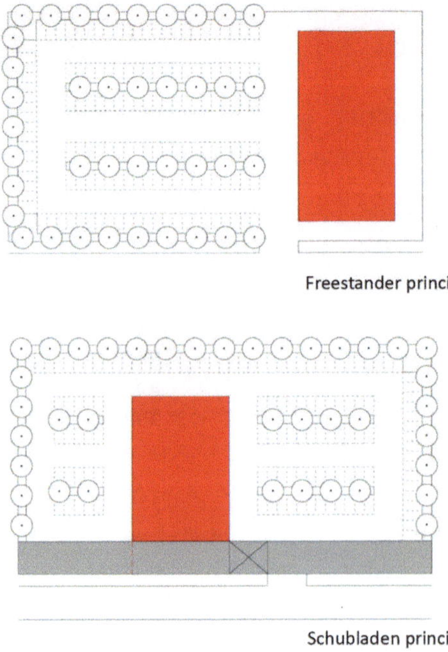

Fig. 5.76 Principle sketches for market buildings

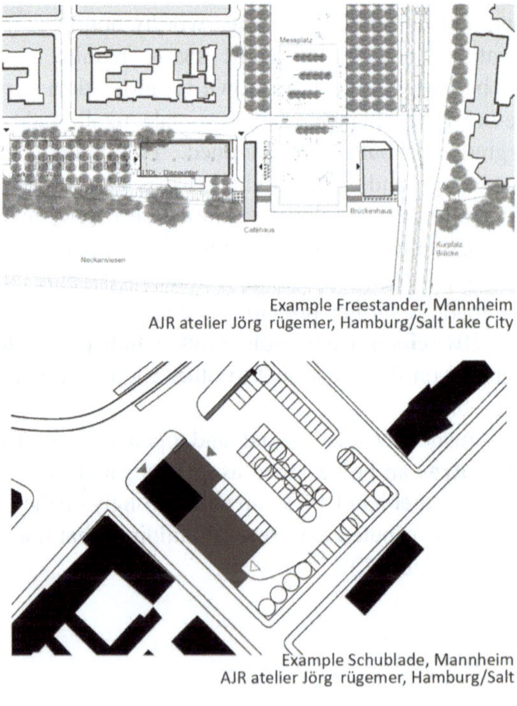

Fig. 5.77 Built examples of retail trade buildings

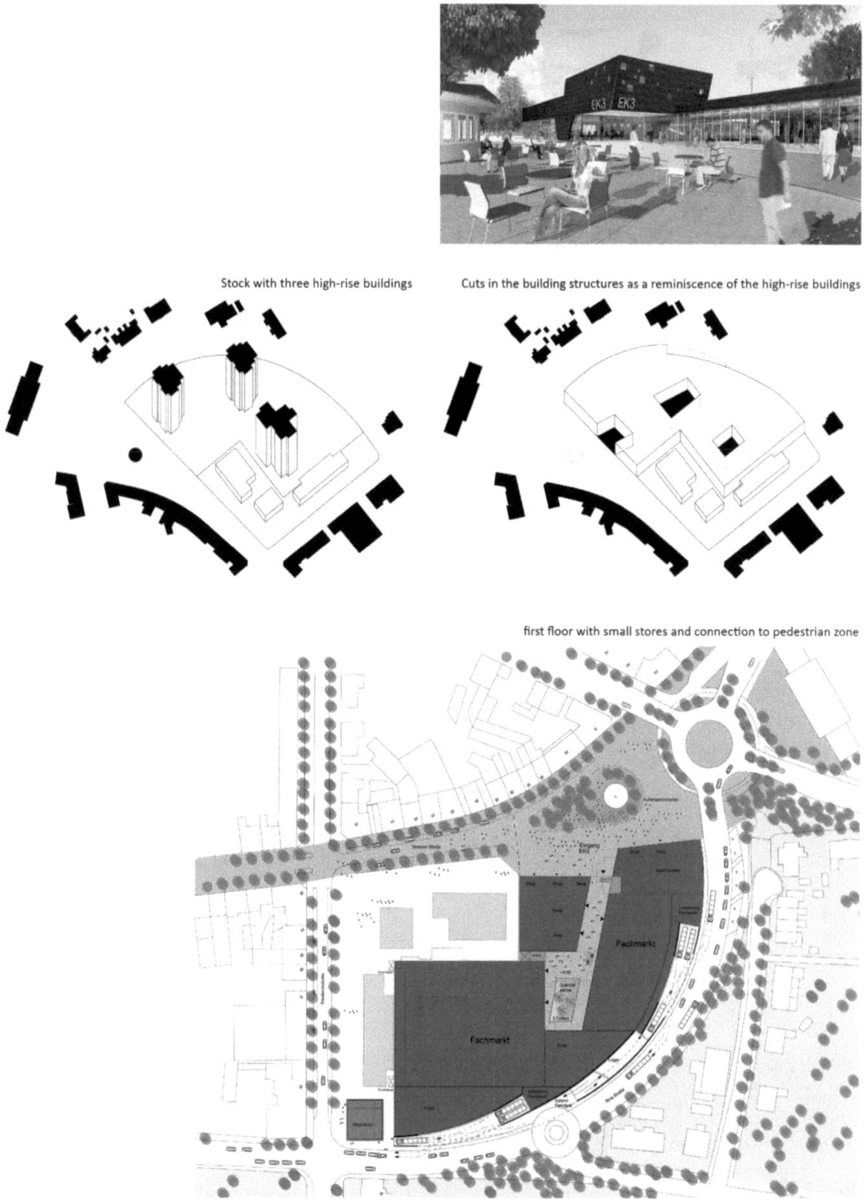

Fig. 5.78 Oaks shopping centre in Kamp-Lintfort, bob-architecture, Cologne

Fig. 5.78 (continued)

On an inner city site, three high-rise buildings are giving way to a new shopping centre. The project is an example of how quality can be created by connecting to public space, designing the roof landscape and creating advertising despite the leap in scale.

Urban Design Requirements for Retail Trade Buildings
In addition to the aspects described above, the following essential requirements for retail trade projects in urban design and architecture are to be discussed:

- Recording of urban structures and path relationships
- References to the available free space
- Visual relations to traditional business situations
- Formation of spatial edges

A retail trade project is not an isolated building block. It must enter into a dialogue with its surroundings and establish references to the existing structural and spatial structure and the available open space, thus making a positive contribution to the urban development situation. This can be achieved by, among other things, taking up or interpreting the construction method, the building lines and the materials of the surroundings, as well as by continuing open space structures and existing path relationships.

Since the quality of public and private spaces is largely determined by their unambiguous formulation and delimitation, which are mainly perceived visually, spatial edges must be formed along streets and also to parking areas. In many urban design situations, multistorey buildings are necessary to form spatial edges. A spatial effect can be created by the arrangement, cubature, and height of the building, but also by other structural elements (perforated walls) and partly by space-creating green plantings. As a rule, advertising boards cannot create the desired effect.

The conversion of (historic) building structures also contributes to a retail trade project becoming an integral part of the urban structure and at the same time acquiring a distinctive character. This is important because shopping centres and grocery stores in central locations, for example, sometimes also take on an important function as places of encounter and communication. This sociocultural function can also be promoted through the usage structure of the retail trade project, which avoids monofunctional usage exclusively through retail spaces. Targeted mixed-use objects (trade combined with housing, public facilities, office use, etc.), a compatible level of additive uses (gastronomy, services) as well as the multiple use of the parking space enhance an otherwise monofunctional location.

In the case of retail trade buildings, the all-round treatment of the building structure, the entrance and delivery situation and the advertising facilities play a special role. Through the facades, the building enters into an active exchange with its surroundings. The opening of the façade, especially to public space, also prevents the creation of anxiety dreams. In contrast to the public urban space, transparency, openness, and communication in the façade design must therefore be given priority. This can be achieved by using appropriate glass surfaces that allow views into the interior and by structuring the façade. The presence and visibility of the retail facility is also supported. The respective façade theme must consistently address all sides of the building structure. This also applies to the delivery area, which, in addition to its functional location, should also respond to directly adjacent neighbouring uses by making noise protection, visibility, areas for old materials, and recyclable material containers a design theme. The materiality of the façade should be justified by the structural context.

Advertising installations serve to identify and attract attention to a retail facility. However, they must also positively support the character and the respective atmosphere of the urban space and subordinate themselves to the context. Therefore, advertising should be reduced to a necessary level and be classified in terms of design. The type of advertising installation (signs, banners, gable logo . . .), its size and location must be coordinated with the appearance of the surroundings and the respective space.

The **entrance situation** has to meet the needs of pedestrians and cyclists. Even a car driver becomes a pedestrian when he moves to the entrance of the market. Therefore, the entrance should be barrier-free and safely accessible for all user groups in addition to the aspect of orientation. This is usually the case if the entrance is located at the public footpath or if areas for pedestrians are marked on the car park. The larger the car park, the more clearly the areas should be marked or separated by high kerbs. In addition, parking facilities for bicycles are just as necessary as seating that meets the needs of older people.

The location and design of the **traffic** must not only ensure a good flow, but above all keep the impact on adjacent residential uses to a minimum. This can be achieved by the actual location of the entrances and exits, if necessary supplemented by structural measures. In the design of the **delivery,** turning-free entrances and exits of the delivery traffic are functionally advantageous. Since delivery areas usually also have to have containers for recyclable materials and waste, the aim should be to achieve limited visibility from the main façade.

Parking spaces are to be realised depending on the location and character of the site and the adjacent uses. If the planned retail trade project does not impair the function of other higher-level centres or retail locations, no restrictive specifications are recommended for integrated locations. Analyses of the number and utilisation of parking spaces in food markets have shown (Uttke 2009) that the following orientation values can form the basis of a discussion on the number of parking spaces appropriate to the location;

- for grocery stores up to 700 sq. m sales area: approximately 1 parking space per 30–40 sq. m sales area,
- for food markets of 700 sq. m UK and more: approximately 1 parking space per 15–25 sq. m UK,
- for combined locations of two food markets and other retail businesses: approximately 1 parking space per 20 sq. m UK.

An additional demand could be justified, for example, by the accessibility of the site by the consumer and multiple use of the parking space.

In addition to the number of pitches, their arrangement and accommodation are of even greater importance. The variety of possibilities must be addressed, for example, ground-level parking spaces in front of, next to and behind the building, open parking decks, multistorey car parks. In shopping centre projects, the distribution of parking spaces within the urban fabric can also ensure that visitors to the shopping centre also visit neighbouring shopping locations. Parking spaces for peak occupancy (overflow parking) are particularly suitable for such a decentralized parking concept. Another positive aspect is the multiple use of parking spaces by different users during opening hours and after closing time.

Green spaces can not only increase the integration of the entire complex into the urban design, but in the case of ground-level parking spaces they also provide shade and improve the ecological situation on site (e.g., dust absorption, creation of infiltration areas). Parking facilities at ground level in particular can be structured and "humanized" by means of green structures—especially tree plantings. The number of trees results from the parking concept. It is recommended to plant trees in a continuous green strip at the head or side of the parking space. If the green strip is planted with trees at intervals of 5–10 m, this arrangement will result in one tree for every three to four parking spaces.

Plant beds and tree discs must be dimensioned and irrigated with sufficient soil contact to allow species-appropriate plant growth. A 2 m wide planting strip should be a minimum (from the inner edge of the kerb to the opposite inner edge of the kerb). This results in the requirement that tree slices should have a minimum size of 4 sq. m in order to provide the planted trees with suitable growth conditions. Necessary retention areas for precipitation water must be integrated into the open space design. At least on the car parking spaces, seepageable coverings, for example in the form of paving, are possible. Areas for "overflow parking" can be designed more economically and ecologically as multilayered gravel surfaces or as gravel lawns compared to a complete expansion. Rainwater should—if infiltration on the property is not possible—be channelled into adjacent green areas or into the green strips in the edge or middle area of the car park. A trough or infiltration channel and other combined methods are conceivable. It is also possible to design the roof areas as green roofs or to install photovoltaic systems.

Literature

1. Urban Component Open Space

Aminde, Hans-Joachim (Hrsg.) 1994: Plätze in der Stadt. Ostfildern-Ruit: Hatje

Giseke, Undine; Spiegel, Erika (Hrsg.) 2007: Stadtlichtungen: Irritationen, Perspektiven, Strategien. (Bauwelt Fundamente). Basel, Boston, Berlin: Birkhäuser.

2. Urban Component Public Space

Havermann, Antje; Selle, Klaus (Hrsg.) 2010: Plätze, Parks & Co. Stadträume im Wandel – Analysen, Positionen und Konzepte. Dortmund.

Knirsch, Jürgen 2004: Stadtplätze. Architektur und Freiraumplanung. Verlagsanstalt Koch.

Reicher, Christa 2009: Der öffentliche Raum. Berlin: jovis.

3. Urban Component Living

Harlander, Tilman 2007: Stadtwohnen: Geschichte Städtebau Perspektiven. Ludwigsburg.

Reicher, Christa; Schauz, Thorsten 2010: IBA Emscher Park – Die Wohnprojekte 10 Jahre danach. Essen.

4. Urban Component Trade and Industry

Lorenz, Peter 1993: Gewerbebau - Industriebau. Architektur. Planen. Gestalten. Leinfeld-Echterdingen: Koch.

Oberste Baubehörde im Bayrischen Staatsministerium des Inneren 1996: Gewerbegebiete. München.

Schmidt, J. Alexander; Bremer, Stefanie: Orte der Arbeit - Gestaltungsmöglichkeiten in Gewerbegebieten. In: Europäisches Haus der Stadtkultur e.V. (Hrsg.) 2006: 5 Jahre Landesinitiative StadtBauKultur NRW, Gelsenkirchen.

5. Urban Component Social and Educational Infrastructure

Hochbaudepartment der Stadt Zürich, ETH Zürich (Hrsg.) 2004: Schulhausbau - Der Stand der Dinge. Basel, Boston + Berlin. Birkhäuser.

Kees, Hoeger, Kerstin (Hrsg.) 2007: Campus and the City. Urban Design for the Knowledge Society. Zürich: gta.

Wüstenrotstiftung (Hrsg.) 2004: Schulen in Deutschland -Neubau und Revitalisierung. Stuttgart, Zürich: Karl Krämer Verlag.

6. Urban Component Trade and Supply

Uttke, Angela 2009: Supermärkte und Lebensmitteldiscounter: Wege der städtebaulichen Qualifizierung. Dortmund: Rohn Verlag.

Reading the City

6

1 Seeing and Perceiving

The perception of laws and characteristics of a place are a basic condition for the individual steps during a design process. Wolfgang Goethe's statement "You only see what you know" is also relevant for the perception of the built city.

Perception... Is more than a Visual Process

Everyday experience proves that perception goes beyond the mere process of seeing. The functionally significant sections of the environment are conveyed to a living being via its sensory organs. Stimulation, understood as chemical-material and energetic processes, triggers individual sensations and perceptions and thus the experience.

Perception intervenes via the biological process in layers of meaning. However, this insight only became established in the nineteenth century. Until then, the "postulate of a conception of space given to man, independent of empiricism" was valid. The researcher Johannes Müller assumed that "the development of spatial perception in the womb begins with the self-induced body movements and the sensation of our inner organs". The ability to perceive was seen as something that was already immediately and completely present at birth.

In the first half of the nineteenth century, various developments, especially in the new natural sciences, led to a restructuring of vision. Theoretical questions of vision and perception were among the burning issues in science. How does spatial perception work? What is involved in the visual experience? From the physiological basis of the process of perception, knowledge of the psychological components of it is increasingly gaining ground.

In his contribution "Vorschule der Ästhetik" (1876) Theodor Fechner dealt with the experimental investigation of perceptual relationships, stimulus, and sensation. An external purpose lying in the built thing becomes an aesthetic one through subjective associations. At the centre of all associations is human nature. In a "philosophy from below" observations and experiences of impressions and effects

C. Reicher, *Urban Design*, https://doi.org/10.1007/978-3-658-34370-5_6

are systematically collected. Helmholtz recognized spatial perception as being learned on the basis of experience and as an "ongoing interactive learning process between the individual and the outside world". According to this, the process of seeing is "not in the least determined by inherited or innate anatomical and physiological mechanisms". Theodor Fechner, Ernst Brücke, Hermann von Helmholtz and others developed different approaches and theories on these topics, which also had a decisive influence on urban design.

Perception... Is more than Passive Observation
Perception includes active participation. It is precisely this participation in the architecture, in the building, that no one can escape.

We do not only perceive what we perceive sensually, it is rather embedded in emotional situations that correspond to memories and wishes from which it stands out. This is the essential difference to literature (a book can be closed if it is too boring) or music (a radio can be switched off without further ado if you do not like the music anymore), but the shape of a room, a building, especially its exterior, the facade, always influences us, in negative and in positive ways.

Perception... Is a Subjective Matter
Both objective and parameters of personal constitution are connected with the visit of a place. Reference system research assumes that these subjective perceptions can also change in everyday life. Wohlwill and Kohn point out that people who move from a small town to a large city often perceive the big city as louder, dirtier and narrower than the long-time city dwellers. If they return to the small town after a certain period of time, they see it as quieter and cleaner than before they moved to the big city.

> All perception and thinking goes back to action. Conceptual cognition is also a kind of inner action. The development of ideas and concepts prepares the perception of (external) reality (Jean Piaget).

Fig. 6.1 Ophthalmotro, Knapp. Apparatus for measuring eye movements

Perception... Is Supported by Conspicuousness
In architecture and urban design, planners often look for tools to create eye-catching features, to draw attention to themselves. However, eye-catching does not necessarily go hand in hand with quality. Colour, protrusions and recesses, material breaks, mass, height development and much more are used to direct the eye, to intensify perception, without us immediately recognising the interests behind it. Perception is consciously controlled.

Perception... Is the Prerequisite for Cognition
The ideas we develop about spaces are primarily visual ideas. Our recognition is bound to the space and to the view, as Karl Jaspers has already expressed in a similar form. That conceptions are primarily pictorial conceptions is confirmed by psychology. However, this statement is only valid with restrictions. The degree of pictoriality also depends on the type of person in question. There are certainly very abstract and non-representational thinking people in urban design—that is, the discipline to which priority is given in the context of this observation. Some of the ideal urban designs of the Renaissance, but also many of our traffic systems, bear witness to this.

The perceived image contributes to an insight; at the same time, we process the gained image individually. Imaginations of images are "not the things themselves, but their sensually received images lie there (in memory) ready for the call of the active mind that remembers them" (Augustine). The cognition is followed by the individually shaped process of implementation.

Perception... Is an Offer for the Viewer
Gibson assumes that the perceived world for an observer consists of objects with certain surfaces. The surfaces "make offers" to the observer: the soft moss floor invites him to lie down and rest. A steep surface stops the movement, overlapping and hidden objects make curious and invite to look behind. According to Gibson, perceptual-psychological interactions between object and observer should be criteria for architects, for example, when designing house fronts. They determine the functional statement of a building.

The various theories of seeing and perceiving our surroundings make it clear that urban design thinking also requires a spatially three-dimensional view. Urban design goes beyond the mere technical solution of problems; the aesthetic quality of space seizes our innermost consciousness through perception.

"You only see what you know": The greater the knowledge about the development history of a city, a neighbourhood, a planning area, of social-spatial contexts, the more you perceive it. It is, therefore, necessary to determine the first guiding ideas and framework conditions of the design as early as the step of stocktaking and analysis, that is, to carry out a problem-oriented stocktaking.

> There is more at every moment than the eye can see and the ear can hear—there is always a background or a view waiting to be explored. Nothing is experienced through itself,

everything is connected to its surroundings, to the sequence of events leading to it, to the memory of past experiences (Kevin Lynch).

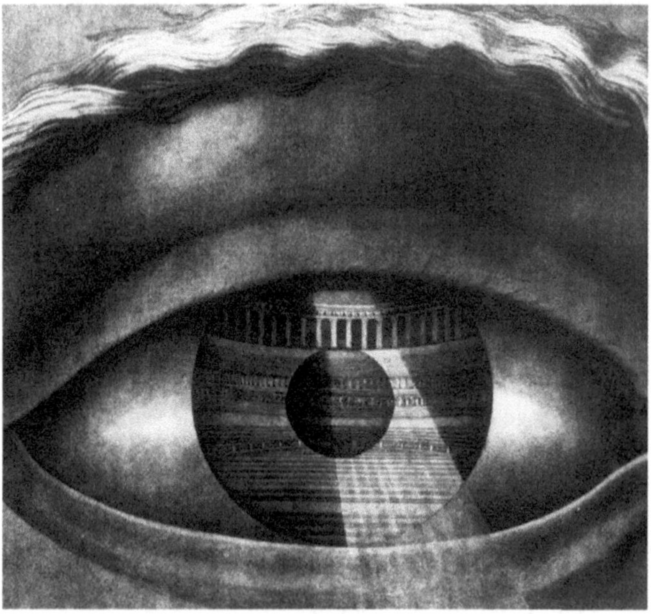

Fig. 6.2 Theatre de Besancon, Claude Nicolas Ledoux

Fig. 6.3 Urban planning thinking requires a three-dimensional view

Fig. 6.4 Campaign LEARN TO SEE of the State Initiative CityBuildingCulture NRW. Between August 2008 and October 2010, the 7 m × 7 m × 12 m SEHSTATION by architect Andy Brauneis—in collaboration with Nicolette Baumeister, communication modules, and Christian Schüller, structural design—toured the cities of North Rhine-Westphalia, supplemented by 10 oversized color frames. The wooden construction, reminiscent of the bellows of a pit camera, is intended to direct the public's attention to the successful and unsuccessful, the existing and the future. The SEHSTATION is both a visual aid and a place for discussion. SEHSTATION and oversized frames were an offer to interested citizens to acquire a new perspective of the built environment in a playful and sensual way and to actively enter into dialogue with experts

2 Site Survey and Analysis

Every process of urban design is based on the intention to intervene in existing conditions, to initiate a development process that can best serve future needs and requirements. Profound knowledge and an intensive examination of the planning area, its problems and possibilities are important prerequisites for urban design. The site survey and evaluation is the basis of urban design and thus an essential step in the overall context of planning considerations and concepts. The aim of site surveying and analysis is to identify the specific local conditions, problems and focal points of action, and to ascertain and evaluate them with appropriate depth.

A problem-oriented approach is essential here. A complete recording of all points and investigation criteria is neither practicable nor desirable: Limiting the analysis to the essentials and thus to the individual objectives of the respective project is helpful and important so that an analysis following the inventory is also effective. Because: every site survey must also be evaluated. A problem-oriented site surveying is guided by responsibly formulated framework conditions, guiding principles and the specific problems of the respective project.

> Going through an iterative cycle of information/analysis, synthesis and evaluation leads from one stage of concretization to the next. In order to finally arrive at the realization of a planning concept, it usually requires numerous feedbacks to previous stages of the design process (Dietmar Reinborn).

In order to familiarise oneself with the planning area and to record the actual state for comparison with the later planned target state, an on-site visit with map material for notes, surveys, first sketches as well as the recording of urban development and design features with the help of a camera is helpful. The evaluation of existing data bases in the form of plans, maps, statistics, etc., is also elementary. Social processes, for example, demographic change, which are not necessarily reflected in one's own spatial inventory, can also be relevant for planning decisions and must, therefore, be taken into account.

The problem-oriented inventory is followed by an analysis of the information collected. The inventory analysis describes the process of critical examination of building and spatial structures, their use and interdependence as well as their design and quality, and concludes from this that there are deficiencies, opportunities, restrictions, and special features of the planned area. There is a connection to the type and scope of the analysis derived from the purpose and significance of the task. An evaluation and assessment of the existing situation, that is, an analysis of the initial situation, is carried out to shed light on the specific problem relationships.

Based on the analysis results, the respective need for action is derived and goals are formulated. Analysis and the resulting objectives are closely intertwined as an exchange of arguments and conclusions.

However, since planning is not a linear process and new information must also be incorporated in later work phases, feedback on analysis and formulated objectives should take place in every stage of the design process. Data gaps may have to be filled in later. External references (from experts, residents, etc.) also lead to the analysis being extended and completed.

Planning tasks are seldom the same and can be clearly delimited, and planning areas are never the same, depending on the project and task. And as different as the planning tasks and planning areas are, so are the individual analyses. Despite these differences, certain analysis methods and analysis contents have developed that are often used for specific tasks.

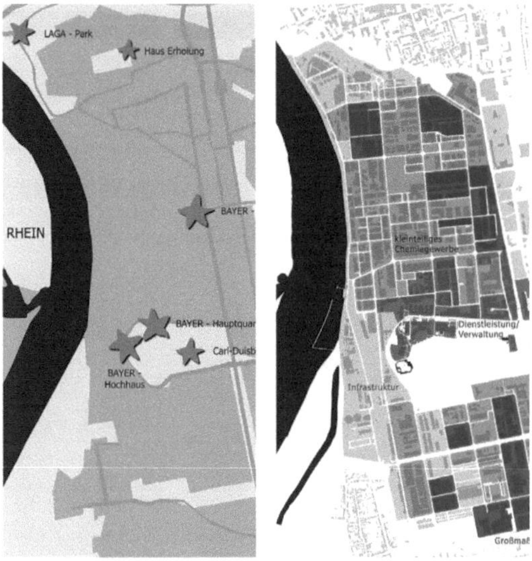

Fig. 6.5 Analysis plans Leverkusen, ideas competition Cultural Circle of German Business, BDI. Identity carrier and usage structure

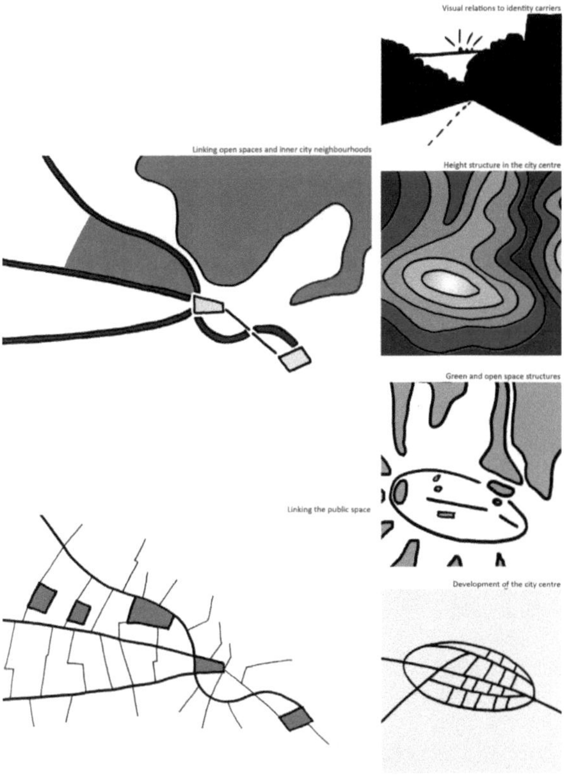

Fig. 6.6 Analysis and concept pictograms, RHA architects and urban planners with Oberhausen planner group. Stocktaking of Remscheid city centre. Linking open spaces and inner city neighbourhoods. Visual relations to identity carriers. Height structure in the city centre. Green and open space structures. Linking the public space. Development of the city centre

	CRITERIA AND ASPECTS	SOURCES AND METHODS (SELECTION)
Historical development	History development of the planning area, Historical uses and functions, Buildings, Structures, Historic preservation, Economic developments, Demographic developments, Historic classification of urban development	Maps on settlement and urban development, historical maps, statistical data, urban development concepts, Land use plans, GEP, expert opinions, historical photos, photos of present-day traces of history, texts, publications, scientific papers, lists of monuments, building files, archives
Natural specifications	Natural specifications Topographical elements, soil, fauna and landscape worthy of protection, water bodies, climate	Thematic maps, statistics, expert reports, environmental reports, landscape plans, green space plans
Landscape structure, green and open spaces	**Large-scale:** Regional green and landscape systems, green corridors, worthiness of protection, legal requirements, characteristic features, type, uses. **Small-scale:** Green space structure of the planning area, public green spaces, private green spaces, parks, accessibility, design, proportion of sealed surfaces, worthiness of protection, networking with urban/regional green systems.	Maps at regional, city and district level, development concepts, master plans, photos, landscape protection maps, aerial photos, green space plans, landscape plans
Land law requirements, planning law requirements		Land use plan, recreational map, thematic maps, site visits, internet
Uses and usage structure	**Uses:** Culture, leisure/sport, social, education, supply, housing, economy/ industry/ commerce, transport. **Structure of use:** forms of use, distribution and focus of use, structure of use,history of use, centre structure, catchment areas, networks, qualitative equipment, quantitative equipment, needs, planning	Land use plan, recreational map, thematic maps, walk-throughs, internet
Building structures, typologies	Forms of arrangement, urban layout, plot structure, street layout, orientation of buildings, building lines, building density, number of storeys, number of floor areas, settlement density, building epoch and age, condition and design of buildings, height development, storeys, roof forms, visual relationships,landmarks, landmarks.	Figure-ground plan, photos, sections, perspective sketches, historical plans, aerial photos, site inspections
Public Space	Types, overall structure, links, structural setting, design qualities, equipment, furnishings, vegetation, user groups, intensity of use, conflicts	Figure-ground plan, photos, sections, perspective sketches, observations, interviews, site visit
Access, transport	Hierarchies, transport systems, types of transport, pedestrian and cycle traffic, supra-regional and urban connection of the planning area, hierarchisation and function of roads, design/appearance, public transport stops, barriers, danger points, usage history	Route plans, master plans, cycle path register, cycle maps, hiking maps, NRW cycling network, various internet sources
Environment	Air emissions, noise emissions, contaminated sites, suspected contaminated sites, soil contamination, use history	Plan work, atlests cadastre, land use plan, thematic maps, photos, historical information
Social structure	Age structure, educational level, socio-spatial distribution, elderly people, young people, single parents, families, migrants Social use of space: meeting places, playgrounds, event spaces, common spaces, recreational areas, conflicts of use Overarching developments: Demographic change, social change, lifestyles	Statistical data, surveys, observations

Fig. 6.7 Aspects and criteria of site survey and analysis. Total city. Plan area and surroundings. Detailed considerations

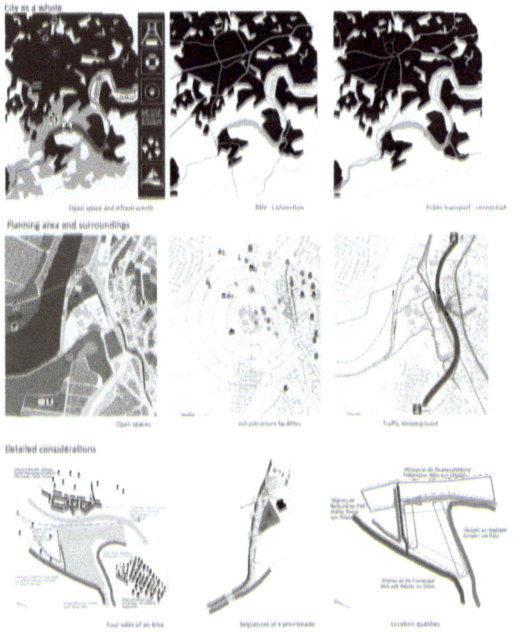

Fig. 6.8 Viewing levels for the analysis of the Essen Kupferdreh district

3 Morphological Analysis

In order to be able to plan and design in an urban space, it is necessary to grasp and understand the interrelationships and inner logic of the building and open space structures. The morphological analysis or morphology of a site provides clues to this on a structural level.

The Greek word "morphe" means shape, form, appearance. "Lodge", too Greek "lógos", means "scientific investigation, teaching". "Morphology" is, therefore, the study of form and shape, the science of changes in form. Transferred to the urban context, morphology means the connection between external form and shape and the formation of shape.

The morphological analysis is based on the understanding that urban design is concerned with the continuation, linking and re-creation of structures. The spatial order and the spatially determining elements of the urban ground plan form the basic framework for the further development of a site. The analysis looks for the underlying regularities of networks, areas and building structures that have formed in the course of the development history of a site. Why does a structure look the way it looks? What developments, breaks, stability, flexibilities, bonds, freedoms, permanence, changes have existed and continue to exist? These are questions that the analysis is intended to answer. Building structure and open space structure are of equal importance. The forms that can be read in the urban structure analysed in this way are called the morphology of the city.

Every urban structure is characterised by the given topographic framework and the specific historical development and change processes. In spite of the permanent change of individual components over time, certain structural characteristics remain. They have a persistent character, an inertia, and ensure that basic structures remain recognisable in their development.

Curdes defines morphology as "two- and three-dimensional shape in the sense of the architectural-physical structure of the city". If one looks at the basic urban structural elements, one finds that these are interrelated and cause mutual dependencies at different levels of scale. Humpert describes four spatial levels:

- Region and city
- City and city layout
- City fields
- Parcel structure

Within each of these levels there are elements and structural principles which can also be found in subordinate levels and which correspond to each other.

Urban morphology results from the addition, division and iteration of these principles and elements. The structural fabric of a city is formed by the readable forms of the networks, the urban fields, the plots and the buildings as well as the logic of their creation. There are determining elements (e.g., infrastructures, water bodies, special building blocks) and properties (qualities, defects, stability, homogeneity, etc.). All parts together provide clues for the urban design in one place.

In the understanding of the founders of the modern morphological considerations from the "Italian school" (Saverio Muratori, 1910–1973), the method of analysis assumes that in the urban structure, chronologically subsequent developments refer to the previous ones and have given rise to the system city. The interest here is not primarily historical, although the city ground plan becomes the historical source and the lines of development can be read by studying historical planning materials and city ground plans. It is the basic topographical files—fortifications, trade routes, gate and network structures on various scales, historical events, inventions, new technologies, etc.—which, in interaction, provide information about the development and have influenced the structural development of the city.

If, for example, an individual building block is modified, the interventions are carried out within a certain modular system, namely within a defined development and plot structure. This structure also has a preserving effect on the overall structure, so that there is a retroactive effect of the urban structure on the selective changes of the building.

The following example of the development of the city centre of Dortmund shows this system by means of a sequence of individual historical steps.

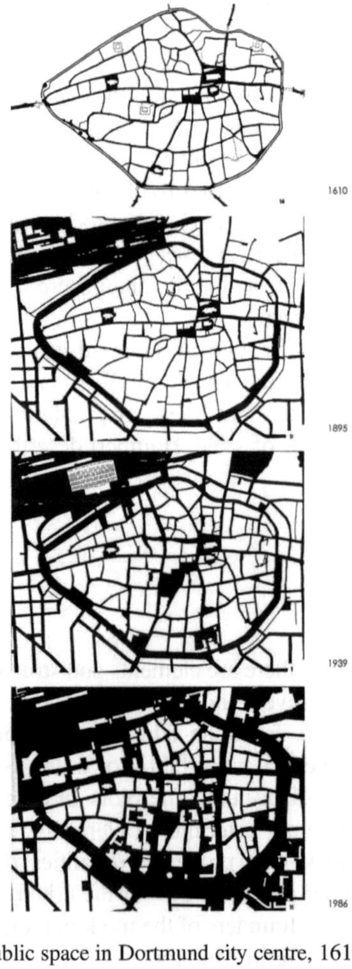

Fig. 6.9 Development of public space in Dortmund city centre, 1610–1986

Fig. 6.10 Development of buildings and building structures in Dortmund city centre, 1610–1986

4 Subjective Analysis

Atmospheric Mapping

A new field of analysis in spatial planning is the description and mapping of atmospheres of urban spaces. The focus is on the aesthetic perception of spaces, their moods and their affective, that is, emotional qualities and effects on people. The atmosphere of an urban space, a building or a landscape is often only perceived unconsciously; atmospheric mapping aims to systematise such perceptions, to make them readable and discussable, so that they can be incorporated into planning and design in a further step. Future planners should become more sensitive to these aspects of spatial planning and design, because the creation of specific atmospheres is a design goal that will become increasingly important because the affective qualities of spaces, the way they appeal to senses and feelings, will become even more important in the future, especially in the competition between cities.

Atmospheric mapping can include different forms and methods of representation:

Texts

These can be subjective recordings, also literary descriptions of moods and atmospheric qualities of a room; also possible are, for example, interviews or surveys to be able to record the affective effects of a room on other people.

> Beyond six rivers and three mountain ranges stands Zora, a city that no one who has seen it once can forget. But not because, like other memorable cities, it leaves an extraordinary image in the memory. Zora has the quality of being remembered point by point with its succession of streets and houses along the streets and doors and windows on the houses, although it does not have any particular beauty or rarity. Its secret is the way in which the eye glides over figures that follow one another like in a score where you are not allowed to change or swap notes. The person who knows by heart what Zora is like walks through its streets at night, when he cannot sleep, in his thoughts and remembers the succession of the brass clock, the striped awning of the hairdresser, the fountain with the four rays, the glass tower of the astronomer, the stand of the watermelon seller, the statue of hermits and lions, the Turkish bath, coffee at the corner, the crossroads leading to the port. This city, which one does not erase from one's memory, is like a scaffolding or a network in whose fields everyone can place the things they remember (Italo Calvino).

Sketches

Freehand sketches (or watercolour drawings, comics, elaborate paintings, etc.) are very suitable methods for visualising the moods and effects of a room. Here too, the graphic recording should be preceded by the most precise observation possible.

Mental Maps

The development of cognitive maps goes back to Kevin Lynch, who in his studies on the image of the city examined the subjective understanding of space and corresponding spatial perception patterns. Such maps can contain real structural elements (paths, mountains, buildings, etc.), but they can also spatially represent their sensory qualities or affective effects.

Photographs, Films
They are the most frequently used methods for documenting the atmospheric effects of spaces; however, audio portraits or so-called auditory collages are becoming increasingly important because they provide long neglected access to the mood of an urban space.

Fig. 6.11 Mental map. Recording of spaces and atmospheres on a daily route from the Dortmund city centre to the TU Campus South

Fig. 6.12 Sketches for recording atmospheres

Fig. 6.13 Pedestrian zone in Ahaus. By night and by day

Fig. 6.14 Atmosphere at Malecon, Havana, Cuba

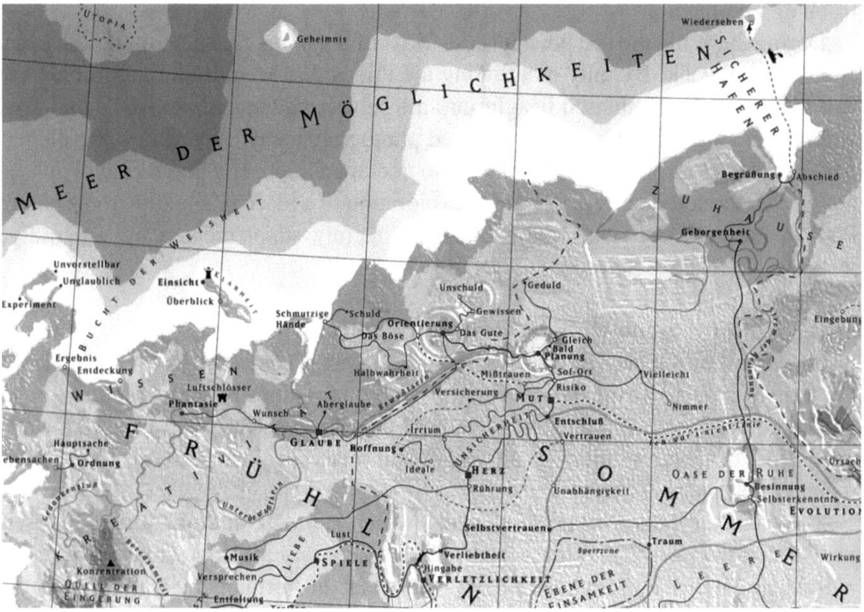

Fig. 6.15 “Map” of atmospheres and moods

Path-analysis (“Méthode Itinéraires”)

The path analysis was developed in the 1970s by Jean-Petiteau under the name “Méthode des Itinéraires” and was mainly tested in French cities. What initially began as an analysis and effect of architecture on the individual, developed over the years into extensive studies of spatial experience and social relationships in the urban district (see Petiteau, Pasquier 2001, p. 74). The core of this method of analysis is that the researcher and the test person exchange roles; the researcher leaves the testing of a space to an outside person.

The method of analysis is divided into a three-step procedure consisting of a combination of qualitative instruments of empirical social research.

Step 1: The Pre-interview ("Pré-Entretien")

An initial contact (the so-called pré-entretien) between the researcher (or the designer) and the test person is intended to familiarise the latter with the aim of the project and the method. In the course of the conversation, the test person learns the detailed intentions and requirements of the project, the exploration of a condition or situation.

Step 2: The Day of the Walk ("Journée de l'Itinéraire")

The core element of the method consists of the walk. On the day of the walk, the researcher and the test person change their respective roles. The researcher briefly communicates the question again and otherwise limits himself to the role of a listening companion.

During the walk, the test person is the focus of attention. She determines the exact course of the walk. By only mentioning the things that seem important to him, the researcher gains an unbiased insight into the subjective view of the individual. The walk is recorded on tape and documented photographically. After the walk, the test person's stories are overlaid with the photo documentation, creating a kind of "photo novel". Methodologically, this type of exploration of a place or situation represents a kind of participatory observation, which is merely supplemented by a narrative interview in relation to the question posed.

Step 3: Confrontation with the Results ("Confrontation")

At the end of this method of analysis, the test subject is confronted with the results in the form of a photo novel. The researcher can ask questions about ambiguities or go into more detail about issues that were only addressed selectively during the walk. This type of confrontation is an important methodical step to avoid possible misinterpretations.

The chance of this method of analysis, which is based on de-ligating the observation to a test person, is that it provides insights into the subjective view of the individual. Studies in recent years have shown that this form of city walk, if prepared and carried out carefully, is a promising instrument for a sensitive reading of urban spaces.

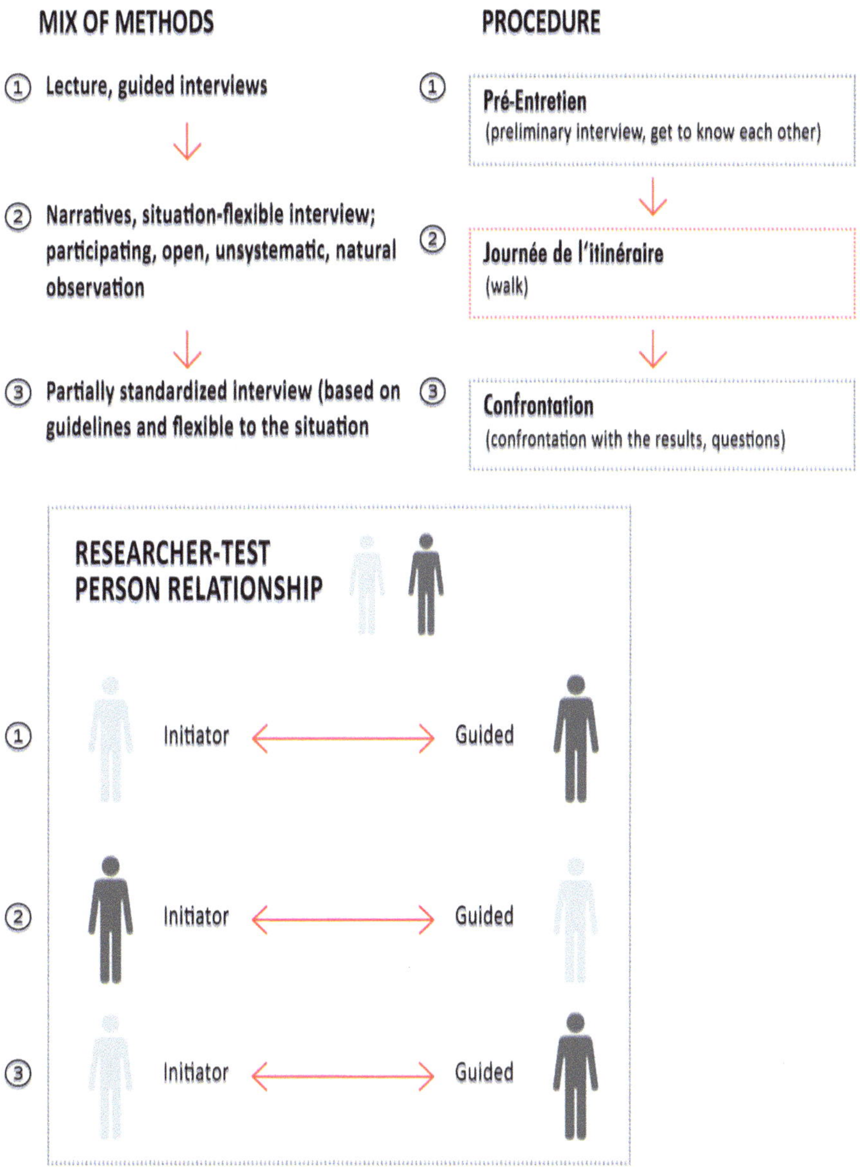

Fig. 6.16 The path analysis

Literature

1. Seeing and Perceiving

Fechner, Gustav Theodor 1876: Vorschule der Aesthetik. Leipzig: Breitkopf und Härtel.

2. Site Survey and Analysis

Curdes, Gerhard 1995: Stadtstrukturelles Entwerfen. Stuttgart, Berlin, Köln: Kohlhammer.
Humpert, Klaus 1997: Einführung in den Städtebau. Stuttgart, Berlin, Köln: Kohlhammer.

3. Morphological Analysis

Curdes, Gerhard 1993: Stadtstruktur und Stadtgestaltung. Stuttgart, Berlin, Köln: Kohlhammer.

4. Subjective Analysis

Petiteau, Jean-Yves; Pasquier, Elisabeth 2001: La Méthode des Itinéraires – récits et parcours. Grosjean, Michèle; Thibaud, Jean-Paul (Hrsg.): L'espace urbain en méthodes. Marseille: Ed. Parenthèses.

Urban Design

7

1 The "Layer Method" and its Design Steps

Peter Empting, Päivi Kataikko, Ilka Mecklenbrauck, Lars Niemann, Jan Polivka, Angela Uttke, and Mehdi Vazifedoost

The layer method as a systematic approach to urban design is based on the complex understanding of urban life described by Jane Jacobs. The urban structure is understood as a multi-layered fabric made up of identifiable components, elements and layers. This structure can be dissected into individual layers. The dissection serves as an aid to recognise site-specific laws and to create a new image, an initially single-layered, concise statement.

The relationships between the different levels are revealed. By looking at individual layers, regularities, site-typical arrangement and design principles become apparent. By tracing the rules of the game and the laws of the existing, clues and specific principles for the new can be developed.

The layer method describes steps related to each other, which are sorted according to the following laws:

- according to chronological sequence of individual analysis and design steps,
- according to thematic and content levels and layers,
- according to the scale of the context under consideration.

C. Reicher, *Urban Design*, https://doi.org/10.1007/978-3-658-34370-5_7

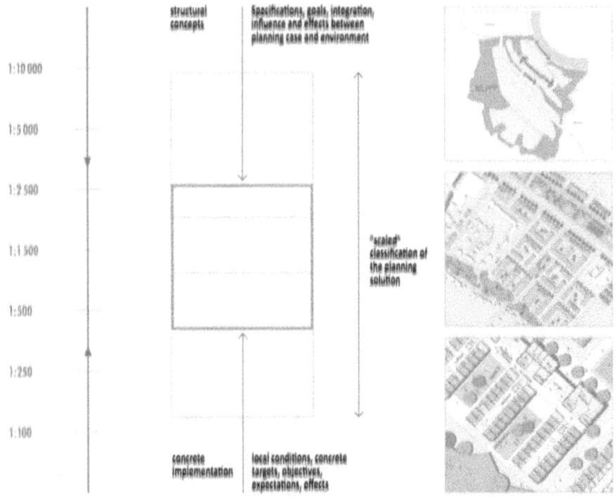

Fig. 7.1 Scale and time levels of urban development planning

1.1 Site Survey and Analysis

The approach and scale of the site survey and analysis depends on the planning area and the environment associated with it (see Chap. V. "Reading the City"). Site survey and analysis can result in different planning representations or photographs.

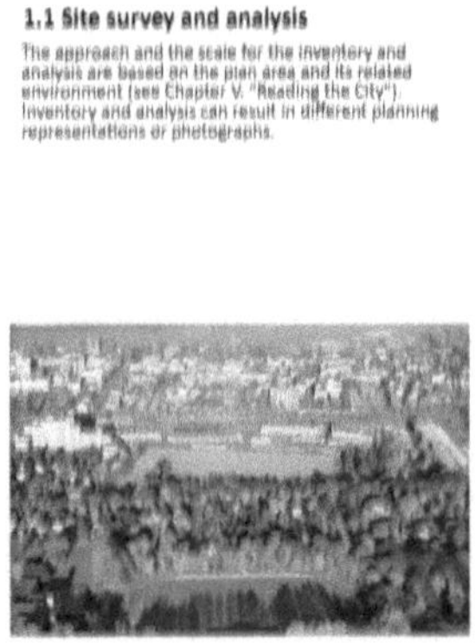

Fig. 7.2 Plan area in the city of Wesseling

1.2 Vision

In the context of urban design, the vision is understood as a pictorial idea of a spatial goal (in contrast to the programmatic vision). It is independent of scale and can refer to both local and regional context.

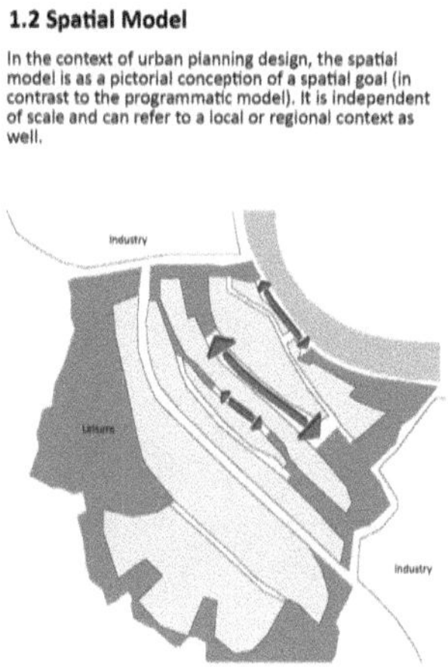

City of Wesseling

1.3 Structure Plan

The structure plan provides an idea of zoning and organisation of the planning area by presenting rough structures. It can refer to different spatial scale levels: to the region, to the city or to the neighbourhood. The scales range between M 1:25000 and 1:5000.

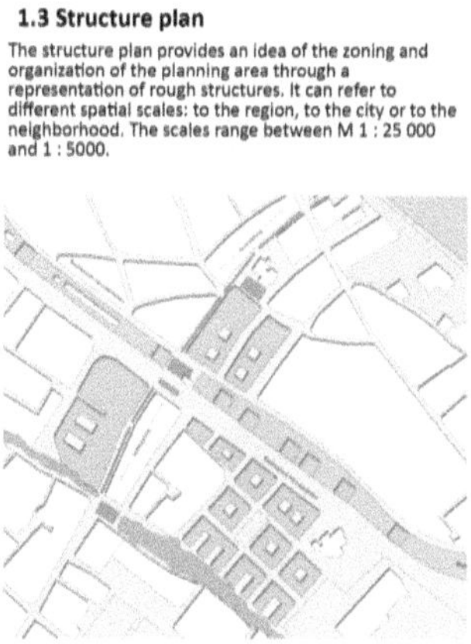

Structure plan

1.4 Framework Plan

The framework plan provides the "framework" for the future development of a district or neighbourhood. It is created for the extended planning area and embeds it in the spatial context. The reference to the surroundings of the planning area is taken up and represented. The scales range between M 1: 5000 and 1: 2000.

City of Wesseling

1.5 Design Plan

The design plan translates the conceptual statements into a proposal for a concrete building and open space structure. It refers to the actual planning area and shows functional, spatial and design principles of the plan. The scales range between M 1: 2000 and 1: 1000, depending on the size of the planning area and the desired depth of expression.

Plan area in the city of Wesseling

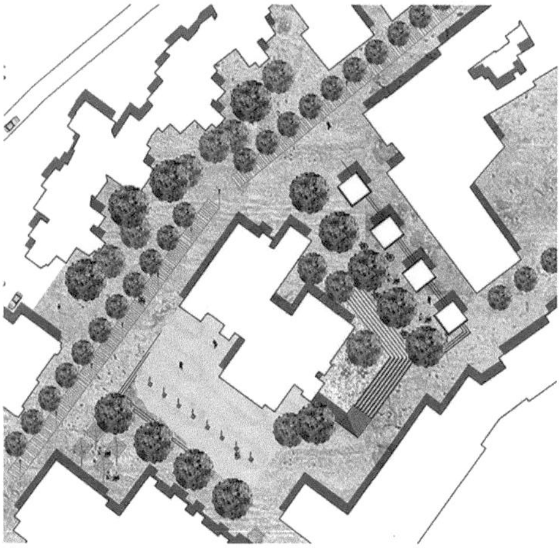

Town hall surrounding in the city of Wesseling

1.6 Urban Design Detail

The urban design detail serves to explain and deepen individual areas. It shows the design principles for particularly meaningful subareas. As a rule, these can be, in particular, the representation of the qualities of use and design of public space as well as the transitions between private and public spaces. The scales range between M 1: 500 and M 1: 200.

An excursus on further design methods is intended to illustrate that there are methods that propose similar steps in urban design, but also set their own priorities and sometimes work with completely different concepts.

Rhine bank in the city of Wesseling

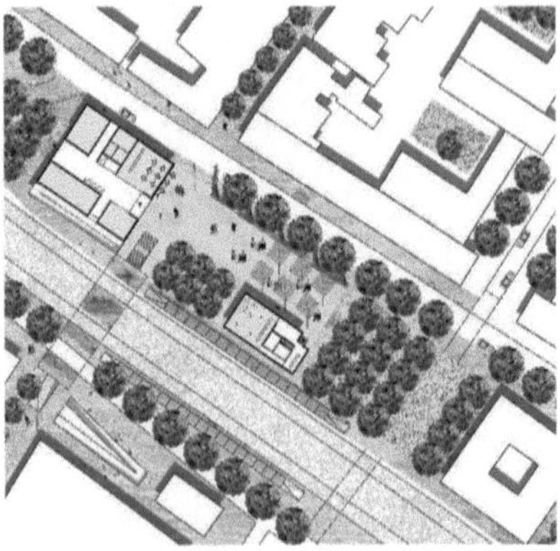

Station environment in the city of Wesseling

2 Vision

Today, urban design still uses a definition of urban design model initiated by Gerd Albers. According to this, the concept of the urban development model means a "pictorial concretion of complex objectives, which gives individual drafts, planning concepts and personal design preferences a common background and integrates them into an overarching consensus on values, which forms the basis for a comprehensive show of desirable spatial order" (Durth, Gutschow 1988, p. 214).

The understanding of planning at the respective time of origin has always been reflected in the urban development models. However, the term "Vision (Leitbild)" was not introduced into spatial planning terminology until the second half of the twentieth century. In the phase of reconstruction after the Second World War, guiding principles developed into a much-noticed phenomenon in the social science disciplines. They were initially used intensively in regional planning as feasible and at the same time ideal objectives, and soon afterwards also in urban planning. Particularly with the successful constructional-spatial implementation of the probably best-known model of urban planning of that time, the model of the "structured and loosened up city" by Göderitz, Rainer and Hoffmann (1957), models gained considerable professional significance.

The concept of the vision is understood and discussed quite differently—especially in connection with projects and designs. The understanding ranges from a pictorial idea of a future state of a spatial situation to a clear theoretical vision of how to approach a problem. There is no uniform understanding of the concept of a vision. Nevertheless, there is a consensus: Visions are—as the word already expresses—pictorial, striking ideas of future states. The word syllable "picture" makes it clear that a theoretical goal is not yet sufficient. A vision is always cross-cutting, realistic and contains objectives.

Task of the Vision

Urban design concepts are created in a design process. Such a process is a constant search for a convincing solution for the task at hand. Once a solution has been found, it is challenged by new findings.

> Urban design is not a linear, finite trace from program to implementation, but a basically endless process of developing, conceiving, rejecting and replacing spatial proposals, a permanent survey of the situation (Erik Pasveer)

In this process-based approach, it is precisely the vision that can provide assistance in the sense of an individual "image" function. It accompanies us on the way to a consolidating objective in the sense of a corrective, which does not itself represent a rigid dogmatic pattern. In the design process, it is the task of a vision to provide guidelines for further work and solution finding. These offer orientation, coordination and motivation. A vision thus takes on a signpost function throughout the entire process, without prescribing final states. It can be modified, and its content can be checked, supplemented and open to changing content. The vision compresses the contents in a more striking way and thus makes an important contribution to the illustration of a spatial and content-related central idea.

Following Albers' description, the spatial model represents a **visual concretization of the goal.**

We expect the following statements from this vivid presentation of the vision:

- The vision should convey the basic features of planning in an abstract, rough form.
- The vision is not a faithful reflection of reality, but evaluates it by means of an individual evaluation and point of view.
- The vision refers to a long-term considered state in the sense of a vision.
- The vision represents a consensus or agreement between individual planning levels or partial drafts.

By grasping the fundamental connections of an imagined future, but not defining them in detail, the vision creates scope for individual interpretation and imagination.

Visions Have an Integrating Function!

Especially in the context of interdisciplinary cooperation, visions can make an integrating contribution. They bring together abstract concepts and discipline-bound ideas in a common reference picture.

Visions Serve to Communicate Planning!

Not only are pictures much easier to remember than terms, but pictures also make it easier and more comprehensible to communicate. Visions are at the interface between idea and reality, principle and reality. They are precise enough to be discussed and too general to provide detailed planning. The more complex and theoretical planning is presented, the more important the function of the spatial vision as a medium of communication becomes. This is why they are playing an increasingly important role in coordination processes, public procedures with a large number of involved actors, in conveying thoughts and objectives. Thus, at an early stage, a consensus can be developed in the definition of objectives, which generates a successful process. Urban design models, with which the claim to implementation is connected, are no longer created behind closed doors of the administration; they are publicly available for discussion and debate:

> Via the pictures, the path leads to the emotions and thus to passionate participation in the city's projects (Detlev Ipsen)

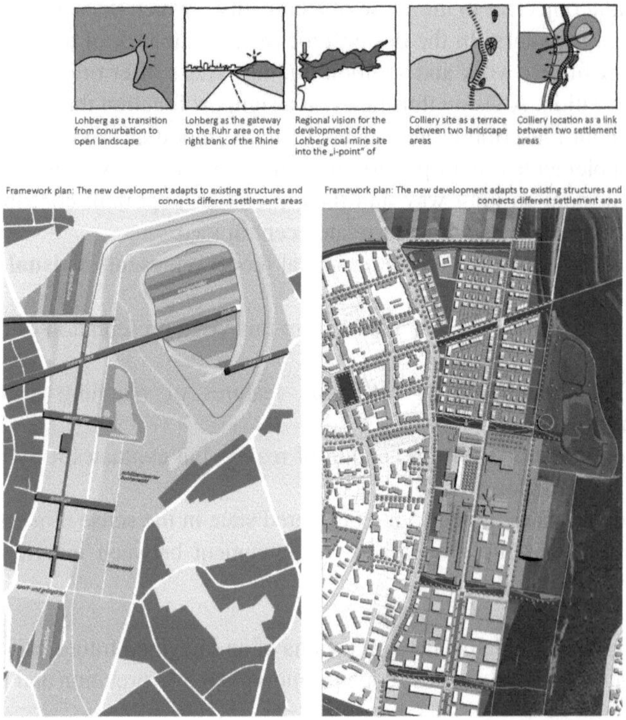

Fig. 7.3 Development of the Lohberg site, Dinslaken, architects and urban planners, Oberhausen planning group. Thematic visions. Lohberg as a transition from conurbation to open landscape. Lohberg as the gateway to the Ruhr area on the right bank of the Rhine. Regional vision for the development of the Lohberg coal mine site into the "i-point" of. Colliery site as a terrace between two landscape areas. Colliery location as a link between two settlement areas. Structural plan: Creation of connections between landscape areas through joints and linear parks. Framework plan: The new development adapts to existing structures and connects different settlement areas

Content of the Visions

A vision can be a motto, a title or a slogan that reflects contents in a striking way. It can also be an image in the real sense—in the form of a logo, a sketch or a collage.

A **programmatic vision represents** the design guiding idea less as an image, but rather in terms of content. The individual goals are derived thematically from a higher-level vision. With the help of a programmatic vision, objectives can be concretised down to the level of measures or implementation. Programmatic visions are designed for topics which, due to their complexity of content and objectives (e.g. in the context of urban development concepts), must be presented in a comprehensible manner.

The **spatial vision** mediates between the abstract motto and the statements of the structural plan in terms of content and presentation. It shows spatial outlines of the planning in a striking and abstract form, but is not a true reflection of reality. The spatial vision formulates value-setting principles. It reveals the point of view of planners, from which the individual tasks and problems in the planning area are viewed and solved. Later planning proposals and measures are to be measured and checked against this value standard.

Fig. 7.4 Development of the Nordstad, Luxembourg, RHA Architects and Urban Planners. Thematic visions. Landscape as connection. River as backbone. Urban band as braces. Urban development framework plan: The river is the backbone of the new development, which creates a connection between existing and new structures

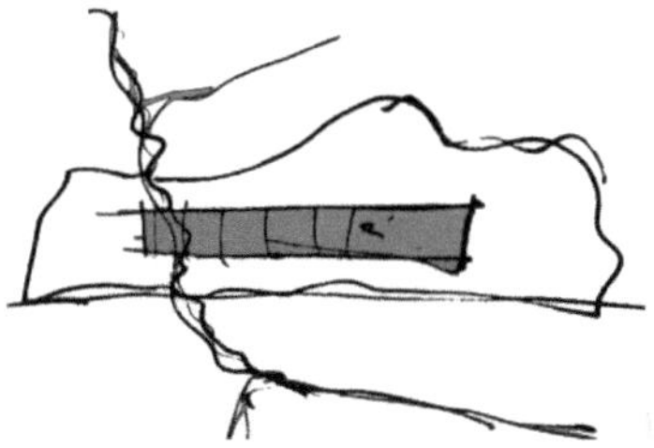

Fig. 7.5 Sketchy model for the development of a brownfield site, Düsseldorf Gerresheim (for an urban development concept, see section on brownfield development)

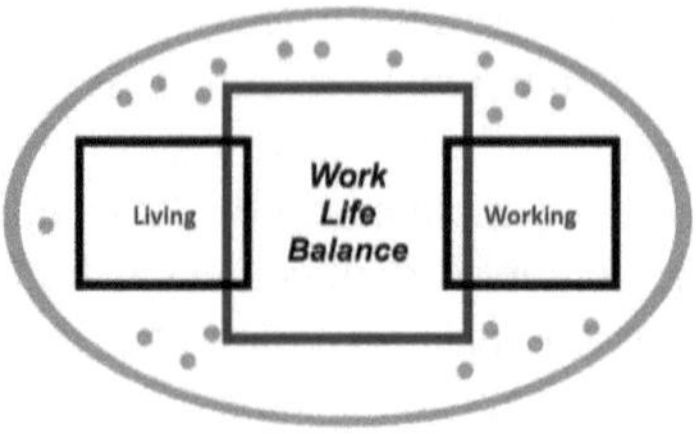

Fig. 7.6 Programmatic vision Nordstad, Luxembourg

Fig. 7.7 Large-scale guiding concept "Roman Road experience area"

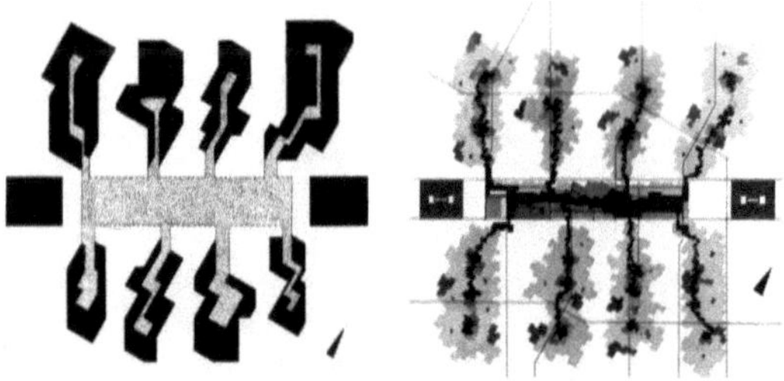

Fig. 7.8 Vision in the context of the exhibition "City of the Present" 1962 in Dortmund. The entire city complex with main access roads and a fine interlocking of the district edges with the green zones

3 Structural Plan

The structural plan deals with spatial development of superordinate structures of a neighbourhood or district—open and landscape spaces, settlement areas, transport infrastructure. It thus moves to the scale level of the city as a whole or even of a region and addresses cross-city or regional issues in urban design.

Task of the Structural Plan

Structural planning usually represents longer-term concepts for action, which are not concerned with a detailed spatial solution approach, but with objectives and rules for establishing spatial relationships. These rules, spatial target perspectives and corridors provide an orientation framework both for further detailing and for decision-makers in city administration and politics. Future (and current) sub-projects are thus placed in a broad spatial-thematic context with the help of the structural plan. However, the structural plan can also include approaches to solving prototypical problems.

At the same time, the strategic approach and the scale of a structural plan (M 1: 50000, M 1: 20000 or 1: 10000) requires a fuzziness and openness of the plan, which is certainly desired. It is not only a direct starting point for spatial detailing and concretisation, but must also be able to take up additions with further fields of action and measures resulting from urban development. In this way, the structural plan comes close in its intention to the formulation of a spatial model and differs significantly from land use plans according to the German Building Code.

As with the framework plan, the preparation of structural plans does not follow a standardised planning procedure. Due to the large planning area, numerous actors have to be involved in the preparation through information events and workshops. A legal and content-related obligation of the plan for a municipality arises everywhere through the adoption of the plans by a council decision. This is a self-obligation of the municipality.

Content of the Structure Plan

The base of spatial structure planning are analyses that deal with the formative elements of an entire city or a region:

- Which elements of cities/regions are combined into a structure and how?
- Which of these elements are particularly durable, which are in a state of change?
- Which elements are particularly useful for the orientation and legibility of regions or the body of cities?
- Which formative natural and topographical conditions (river systems, canals, mountains, slagheaps) determine the shape of regions and cities?

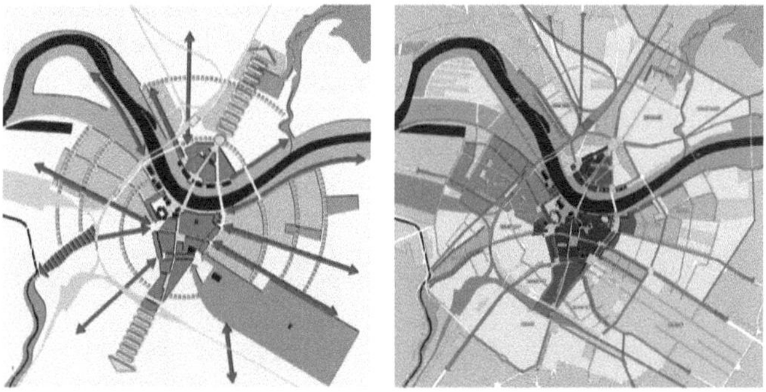

Fig. 7.9 City centre development Dresden, pp I as, Herdecke, Stuttgart. The primary goal of the urban development is the regaining of inner city urbanity from the existing buildings, taking into account high-quality buildings and the historical town layout. The aim is to create an urbanity with a graduated density and a network of upgraded public spaces and private green spaces. Presentation of the central elements of the development strategy (left) and of spaces and neighbourhoods (right)

Conceptually, the structural plan shows spatial-functional links between landscape and city or city and region that are to be secured or created in the long term.

Depending on the reason for planning, the subject and content of structural planning may include the spatial development of

- Centres of cities and regions (stock of centres, potential for new centres, classification of new centres/shopping centres)
- Regional traffic routes, main roads and arterial roads in cities (accentuation of the spatial structure/marking of transitions, nodes and panoramas),
- Watercourses and water locations (upgrading, development and opening up of riparian areas),
- Landscape areas and open space systems (creation and completion of open space composite systems),
- Altitude development and landmarks (setting new altitudes and formulating determinants for altitude development),
- Industrial sites (integration of the areas as well as potential for new developments/ definitions) and
- Large-scale, supra-local and regional significant individual projects.

Integrated structural planning combines various topics in a map or action concept (scale of representation: M 1: 50000, M 1: 20000 or 1: 10000), supplemented by in-depth maps and concept sketches for impulse-giving or prototypical intervention sites as well as a detailed textual explanation.

The map display is mainly area-based. Hatching and arrows are used to show connections and links. The thickness and size of the arrows indicate the focus and intensity of measures. Building contours are not shown—except in the case of ensembles or individual projects of highest conceptual importance.

At present, **overall urban structure planning** is mainly applied in processes of urban planning models of large cities, which formulate the principles of structural development in maps, sketches and texts. Examples of this are the "Planning model for the city centre 2007" in Dresden, the "Emscher Landscape Park "and the "Urban design strategy for Hamburg".

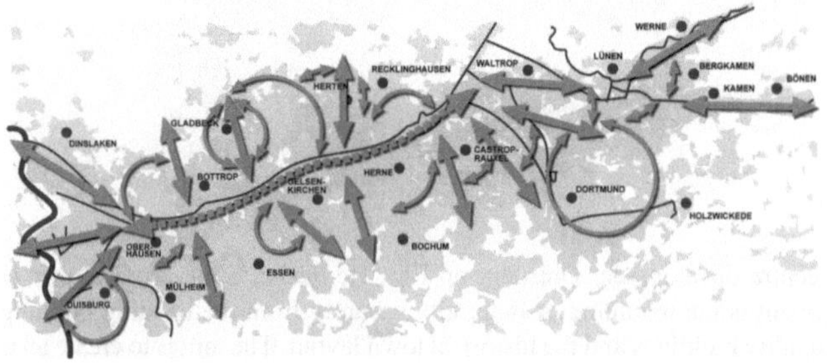

Fig. 7.10 Emscher Landscape Park structure plan. The Emscher Landscape Park was the central lead project of the International Building Exhibition (IBA) Emscher Park 1989–1999, with the aim of creating a new urban cultural landscape. This basic idea of a coherent regional landscape park was continued after the end of the building exhibition with the ELP 2010 master plan. Here: Green Rings and new connections

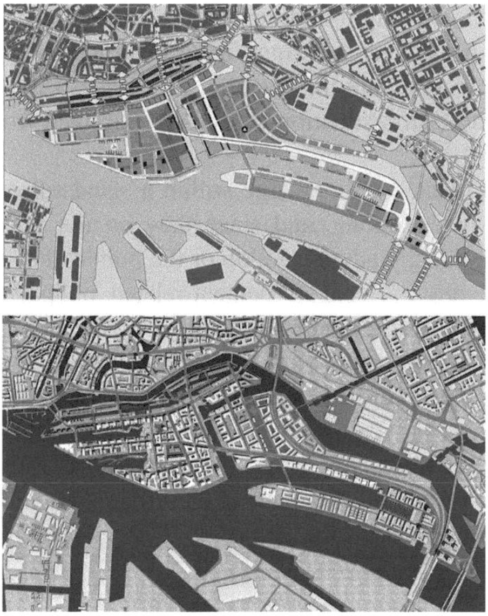

Fig. 7.11 Structural concept and urban master plan Hafen-City, Hamburg, ASTOC Architects and Planners, Cologne/KCAP Architects and Planners, Rotterdam/Hamburgplan AG. An urban piece of Hamburg is being created on the former port area. The different dimensions and restrictions of the port basins result in a variety of location qualities and characters. The links between the new Hafen-City and the evolved city centre are achieved through urban development structures and public space

4 Framework Plan/Master Plan

Urban development framework plans, including master plans, are informal planning instruments. This means that they have no legally binding character and are not subject to a fixed procedure in their preparation. Nevertheless, they provide guidelines for administrative action. They usually represent urban subspaces and different contents.

Task of the Framework Plan

Urban framework plans are used as mediators between large-scale, overall urban planning (vision, structural plan, land use plan) and detailed planning (design plan, development plan). The standard scales are 1: 5000 or 1: 2000, which provide a framework for the actions of actors such as the city administration and for the development of concrete measures. Framework plans show development potential and present perspectives for the future. They are urban development plans and communicate clearly readable, spatial order patterns for an urban area. They serve as a basis for planning based on the framework plan.

They describe the formation of spaces or spatial units and urban development contexts, establish relationships within the urban structure and clarify the relationship between existing buildings and planning, e.g. urban development upgrades, closing gaps between buildings, new construction, demolition, etc. Framework plans often also integrate the sectoral planning, so that statements on open space planning or the traffic situation can also be derived from a primarily urban framework plan.

Since framework plans do not emerge within a standardised planning procedure such as urban land-use planning, and since they do not go through any binding coordination steps in the city administration or the public, they have a certain degree of flexibility. There are no legal or content-related obligations, but nevertheless frameworks that guide the development of a city. Citizen participation and discussion rounds in the administration are process elements that are often used in the development of a framework plan in order to make the result as consensus oriented as possible. For municipalities, these are reasons for a widespread use of this planning instrument.

Framework plans are drawn up for specific subspaces of a city and basically make statements about the future development of the urban space. They can also provide information on concrete options for action.

Fig. 7.12 Urban development framework planning for the Hohlmühle neighbourhood, Bayreuth, RHA Architects and urban planners. Development: streets, parking and squares. Open space: eighbourhood squares and common areas. Rainwater: altitude profile, water bodies. Green order: trees and shrubs

Zoning: parcelling out, ownership. Uses: Housing with social infrastructure in the south/southwest.

or make suggestions for concrete implementation measures, development priorities or time schedules.

There is no binding and uniform definition of the contents or the form of presentation of framework plans. As a rule, importance is attached to a high degree of readability, that is, to clearly understandable statements on the corresponding contents. Thus, a framework plan is also suitable as a mediation instrument between the administration and citizens or municipal interest groups.

One of the strengths of the framework plans is their information character. Benefiting from the creative freedoms, "readable" plans can be drawn up for citizens, tradesmen, commerce or other interest groups, which are not bound to the requirements of urban land-use plans (e.g., planning sign ordinance).

The term master plan is often used as a synonym for framework plan. The statements are similar, there is no sharp definition of the term. In planning practice, however, it is apparent that the master plan has a more binding character and makes more statements on more concrete planning. The framework plan remains at a more fundamental depth of statement and offers greater flexibility for interpretation.

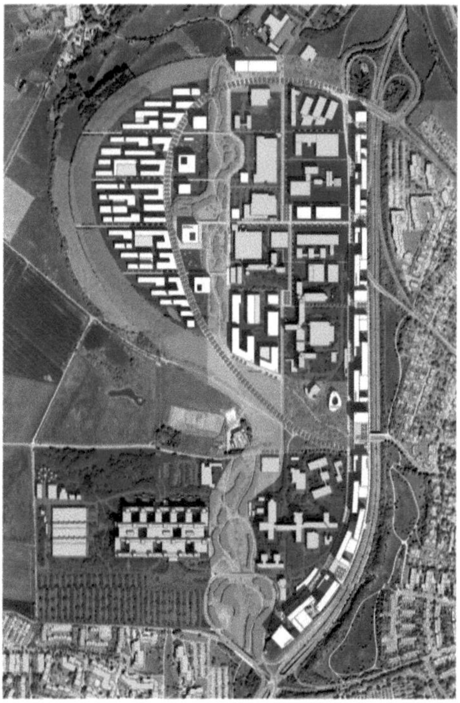

Fig. 7.13 Urban development framework plan RWTH Campus Melaten, Aachen, RHA Architects and urban planners. Urban design competition, first prize

Content of the Framework Plan

Framework plans can be drawn up for different occasions and contain a variety of core statements. The depth of the respective statement depends on the size of the planning area and the content-related examination of the planning area. Furthermore, intersections with the structure plan are formed.

Cross-municipal framework plans include the territory of several municipalities, with the individual cooperating municipalities jointly determining the land use of the adjacent areas.

General urban framework plans stand alongside the traditional, formal land use plans. Due to their flexibility and relatively short coordination phases, they are often used as action-guiding plans.

Partial spatial framework plans are drawn up—especially in larger municipalities—in order to develop perspective ideas for specific districts or neighbourhoods.

Project-oriented framework plans focus on a sub-area and deal with individual projects for major urban development projects such as the conversion and subsequent use of brownfield land or urban expansion areas.

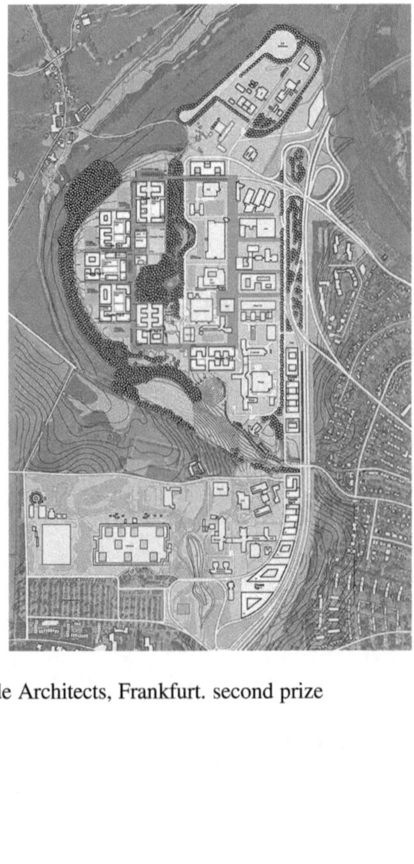

Fig. 7.14 Ferdinand Heide Architects, Frankfurt. second prize

The majority of urban development framework plans focus on the building structure and its character as well as its use. Contents can be:

- Key objectives
- Location and character of existing and new development structures
- Closure of gaps between buildings
- Future uses
- Development concepts for open space and landscape
- Renewal areas in the portfolio
- Infrastructure with spatial impact
- Traffic concept
- Urgencies, priorities, implementation phases

Partial aspects or thematically in-depth considerations within a framework planning contribute to the overall concept.

This includes, for example, special aspects of use:

- Conversion of buildings
- Demolition/partial demolition of buildings
- Reorganisation of commercial uses
- offshoring

Aspects of open space planning:

- Links to open space, green and landscape areas
- Measures to upgrade/reorganise green areas
- Interventions at neighbourhood and block level
- Greening and unsealing of the inner block areas
- Creation of public green and play areas
- Planting of street trees
- Consideration of a rainwater concept.

Aspects of traffic planning:

- Representation of the street hierarchy
- Design of the road space
- Consideration of stationary traffic
- Deconstruction and expansion of the road network as well as the everyday and leisure cycle path network and public transport.

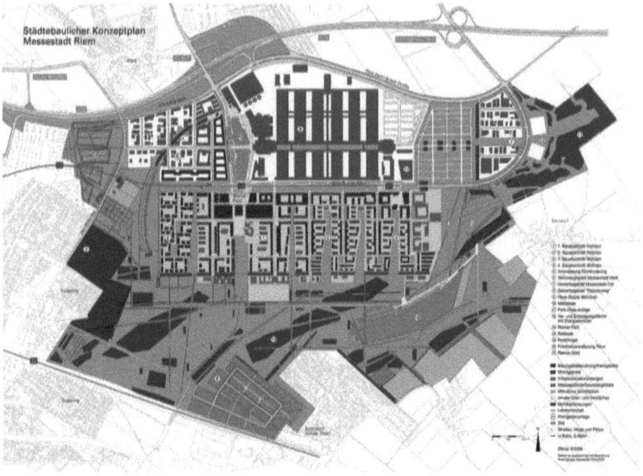

Fig. 7.15 Urban development framework plan Trade fair city Riem, Munich, Valencia + Valencia. compact—urban—green

5 Design Plan

The term "design plan" illustrates that on the one hand—as the name suggests—it is about design and on the other hand it is something projected into the future, something new or changed. This can involve both structural changes to districts and neighbourhoods and also new housing developments and industrial sites. Even small changes in urban space, such as the design of streets, paths and squares or the redesign of open and green spaces with their axes of movement, design elements and vegetation can be focal points.

Task of the Design Plan

The task and purpose of the design plan is primarily to make the planning intention as plausible and uncomplicated as possible so that not only experts can read these design plans, but they can also be understood by citizens.

The design plan is the transporter of planning intentions. It must elaborate in a comprehensible way how a spatial situation can be changed in a positive way. And all this can only be achieved by means of a two-dimensional representation, which can achieve plasticity, that is, three-dimensional readability, through special types of representation and methods, in order to make it easier for the viewer to grasp and understand the planning contents. The general viewing habits must be taken into account.

Fig. 7.16 Design plan Campus II, Heidelberg, RHA Architects and urban planners. Campus II will create space for research institutions and science-related services on the Heidelberg Bahnstadt development site

Content of the Design Plan

The contents of design plans can be as varied as there are possibilities for planning cases. It is essential to depict the cubature of main and auxiliary buildings as well as their orientation with their development sides and uses. The different uses can be represented by different colours, structures, inscriptions or similar markings. The spatial height development of the buildings, trees, bushes, walls, masts, hills and excavations are also important contents of design plans. The use of different presentation methods creates a plastic effect. To further clarify the building typologies used, the representation of the roof shape is important. For this purpose, various line representations, colour transitions and structures are available, depending on the respective roof shape.

In addition to the spatial development, the design plan must also define undeveloped open spaces and their uses. The uses of areas can be made clear by selecting different, specific colours and/or structures. The differentiation of areas into private and public open spaces should be visualised as well as the special forms of vegetation, planned floor coverings or, if the sharpness of detail of the scale allows it, the use of furniture.

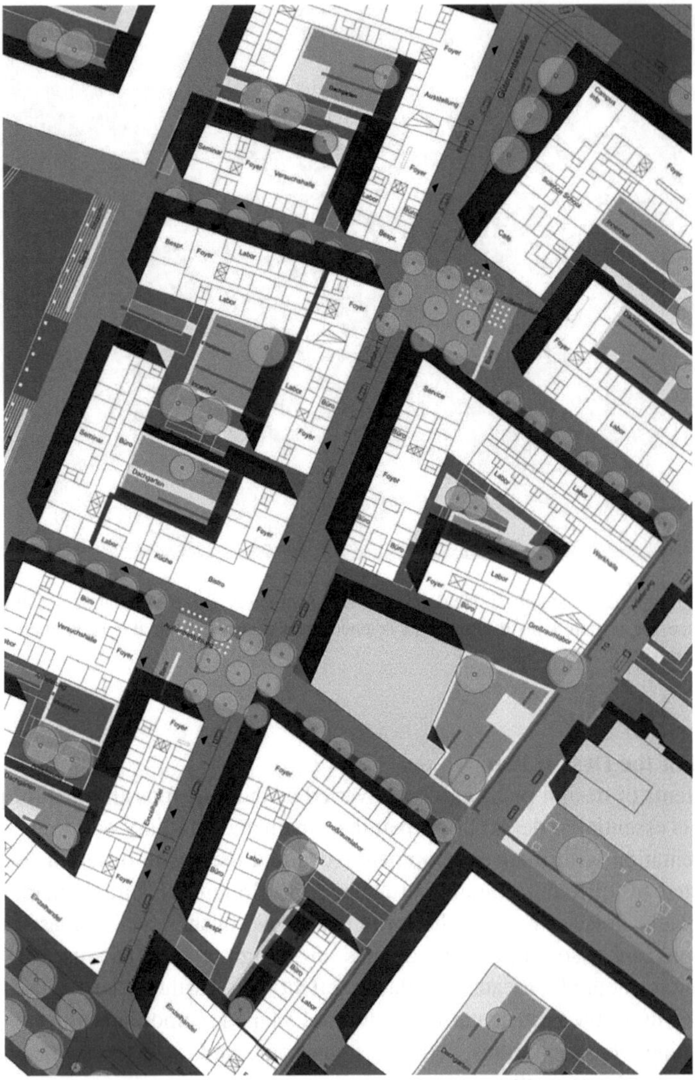

Fig. 7.17 Urban development specialisation Campus II, Heidelberg RHA Architects and urban planners

Traffic areas, their structure, use and surface design are also presented in design plans. The different areas of use—carriageway, cycle lane, footpath, areas for stationary traffic, mixed traffic areas—are illustrated with help of different colours (if possible, from one colour family), structures and textures. Due to the different dimensioning of the cross-sections, the viewer can quickly grasp a hierarchy of the traffic areas.

Roadside greenery, distance green spaces and trees in public spaces structure the traffic areas. A picture of the planning should be created that is as realistic as possible, since at this point an important functional control up to traffic safety will also take place in and on the design plan.

Even if the presentation of public transport routes in the design plan is less useful, stopping points should be drawn in. This visualisation is important both for recognising the functionality of the traffic area and for showing that neighbourhoods are sufficiently connected to public transport. It is important to show bridge and tunnel facilities, including underground car parks, with their entrances and exits.

Presentation Forms

Depending on the planning case, content or target group, different forms of presentation of design plans are possible and necessary. Design plans do not necessarily have to be in colour, they can also be displayed in black and white or grayscale or even in a half-coloured version. A coloured display makes it easier to quickly understand and record the planning contents. It also enables an extreme abundance of content and information. The choice of colour is of great importance for comprehensibility and legibility.

The colour scheme of some uses is obvious—open spaces, green spaces and vegetation are shown in different shades of green, water areas in blue. When using colours that correspond to the habits of vision, many things are self-explanatory. In order for the design plan to give the observer a realistic picture of the intended situation as accurately as possible, the plan should create a more naturalistic image. This also includes the representation of the height development. Higher elements should plastically protrude, and lower ones should recede. In order to achieve this spatial effect, it is useful to imagine the height development of the design plan in individual height layers and to break it down into individual levels.

From bottom to top for example:

- Traffic areas: roads, paths, squares,
- Green spaces, gardens,
- Bushes, hedges, fences, outhouses,
- Trees and buildings according to their height.

Another method of achieving plasticity in the design plan is to use different line widths. The thickest line widths are used for the highest elements, e.g. buildings and trees, and thinner line widths are used for the lower elements. Additional plasticity can be achieved if the thinner lines do not follow the thicker ones, but stop about one to two millimetres before them, i.e. do not touch them.

Shadows can be displayed to achieve a particularly strong depth effect and three-dimensionality. The length of the shadow cast by buildings, trees, bushes, masts, hills or similar things protruding from the earth is represented in a dark colour, grey or black, depending on their height. The uniform angle of all shadows must be taken into account.

In black and white design plans, there is no way around the use of hatching and textures. Even with coloured design plans, it may not be possible to use additional colours for additional content. In this case, using hatching and textures is also possible. Hatching and textures are usually two-dimensional representations and are mainly used for additional information.

Special Features and Additions

All maps and plans are to be north orientated if possible. However, there are also planning fields, plan sections and plan formats which make it sensible and economical to rotate the map section in order to make better use of the plot and paper format. In such a case, special attention should be paid to the legibility of the north arrow.

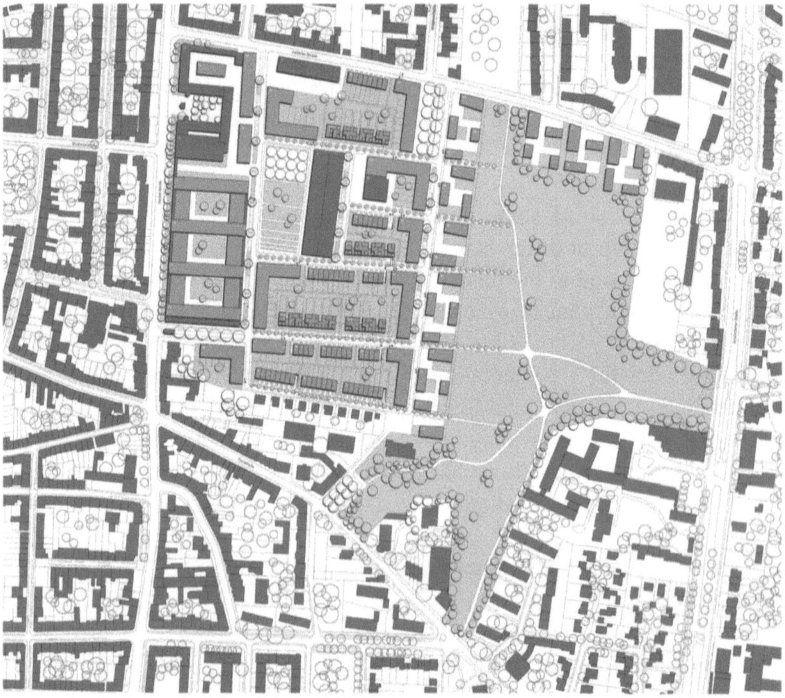

Fig. 7.18 Design plan for the Clouth urban neighbourhood, Cologne-Nippes, scheuvens + wachten, Dortmund

The signs, symbols, strokes, colours, hatching and textures that appear in a design plan are usually explained in an explanation/legend despite the plan being clearly legible. It is a relief if special features are not only found by searching in the legend, but are made visible by writing directly into the plan at the direct location.

Fig. 7.19 Structure plan and design plan Attendorn, RHA Architects and urban planners

6 Urban Detail

Contents of the Urban Detail

Starting from the design plan, selected areas or aspects of the design are presented in detail in a separate plan to specify the urban and open space design idea. The in-depth presentation of a section of the plan is referred to as an "urban detail". In this detailed plan, for instance, important public squares and street spaces or exemplary transitions between buildings and open spaces are presented on a larger scale and linked with statements on spatial formation, spatial division, building typology, materiality and individual design elements. The urban detail provides further insight into the planning and form dimension of the design and serves as a basis for planning steps that follow in the course of realisation (approval planning, implementation planning). In detail, this includes the following design elements:

Detailed overall display of a selected plan area—scale 1:500 - 1:200

- Building typologies with assigned open spaces
- Development
- Public and private open spaces
- Detailed representation of the surface of the square with design elements, fixtures, surface materials and formats, furniture, lighting, seating
- Representation of the building structures with details of roof shapes

- Function and usage information
- Street structure
- Presentation of the demarcation of public and private areas
- Detailed representation of the height conditions
- Vegetation

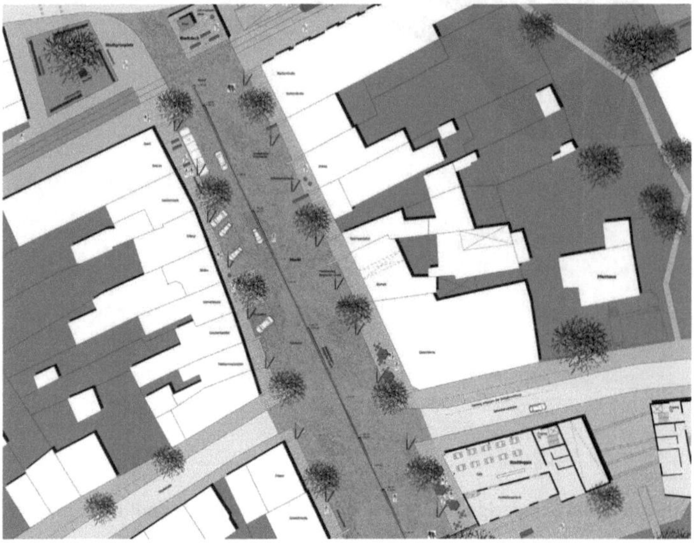

Fig. 7.20 Urban design detail Würselen, RHA Architects and urban planners. Design of the market, Würselen

Cross Sections
- Scale 1:100 - 1:20
- Proportions of the planned spaces (square, street), including building heights and street or square widths
- Street structure
- Representation of the planned vegetation
- Dimensions

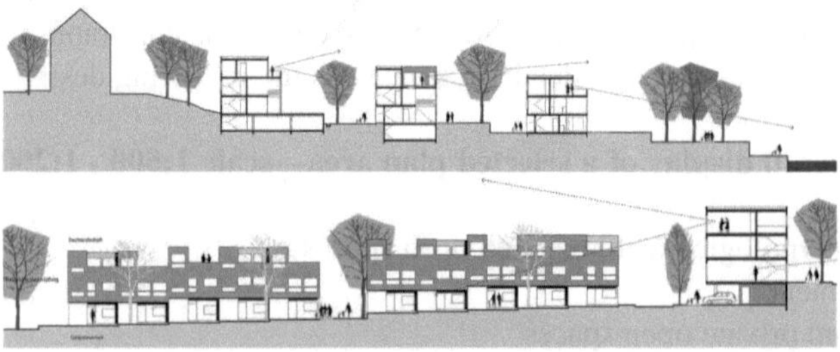

Fig. 7.21 Cross-section through a terraced residential development in Gevelsberg, RHA Architects and urban planners/ LAD+ Landscape Architecture Diekmann

Spatial Sketches/Perspective Representations

- Representation of the spatial proportions and atmosphere

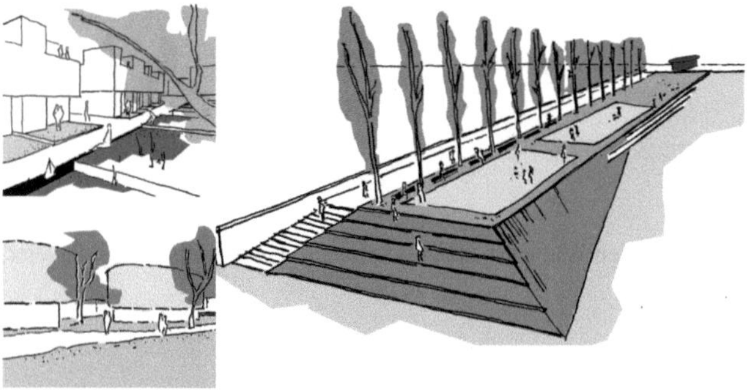

Fig. 7.22 Spatial sketches, Enneppe Park Gevelsberg, RHA Architects and urban planners / LAD+ Landscape Architecture

Fig. 7.23 Photomontage, RHA architects and urban planners. Competition slagheap Carl Alexander in Baesweiler

Details

- Scale 1:20 - 1:50
- Detailed presentation, for example, on materiality, colourfulness, vegetation or the structure of a designed street furniture
- Differentiated material specifications/material collage

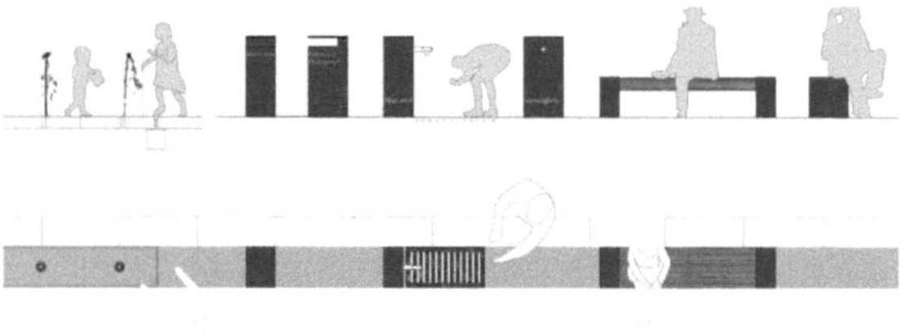

Fig. 7.24 Göttingen city centre design, pavement detail and street furniture, LAD+ landscape architecture Diekmann, Hanover

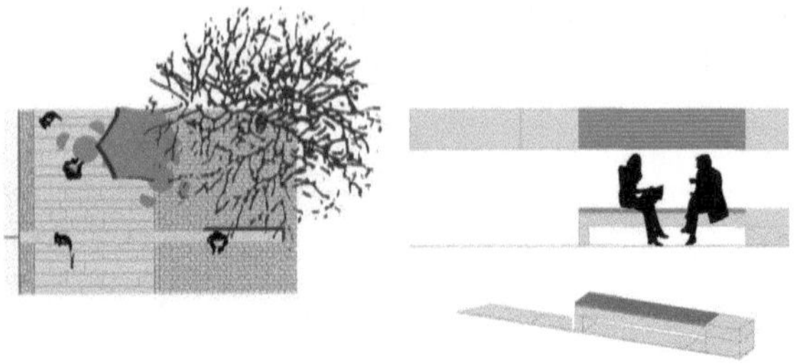

Fig. 7.25 City centre design Würselen, RHA Architects and urban planners. Pavement detail, street furniture

Reference Pictures

- Development
- Open space design
- Structure and design of the street space
- Lighting
- Street furniture

Fig. 7.26 Würselen, redesigned street space

Fig. 7.27 Ahaus, newly designed square

7 Verification

In urban design, there are various ways of examining a spatial solution in relation to its context. This involves

- interfaces between existing and the new,
- chosen scale and granula rity for the new,
- height development and its relevance to the context.

7.1 Figure-Ground Plan

The figure-ground plan—often also called "Schwarzplan"—provides a good tool for checking a spatial concept. The urban design draft is reduced to two statements: the buildings, which are shown in black, and the remaining open spaces as the so-called "ground". The existing building stock and the planning are presented in the same way and joined to form an overall picture. The reduction to this representation makes it possible to check whether and in what way the existing building stock and planning communicate with each other. The intention of the design authors can be to insert the new in the context (see Chap. VII. Urban Design and Attitude) or to consciously set it apart from the existing.

If the open space plays a major role in a design concept, it makes sense to invert the figure-ground plan to an inverse representation in which the open space is shown in black and the buildings are shown as a light background.

7.2 Model

With the help of a model (usually even a simple working model), scale and granularity, but especially the height development can be checked. Similar to the figure-ground plan, the model can be used to check the further development of existing regularities from the existing building or the development of new spatial qualities. This representation in the third dimension also allows conclusions to be drawn regarding the quality of lighting and ventilation as well as the effect of open spaces.

Fig. 7.28 Working model Gummersbach, RHA architects and urban planners. Concept development and review

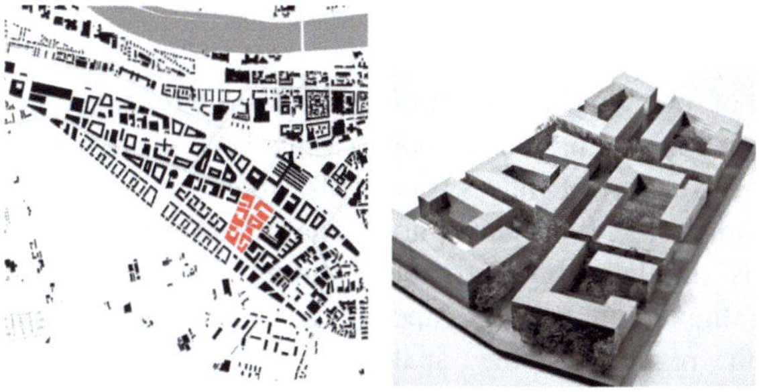

Fig. 7.29 Heidelberg development study, Campus II, RHA architects and urban planners. Figure-ground plan (left) and checking the cubic capacity

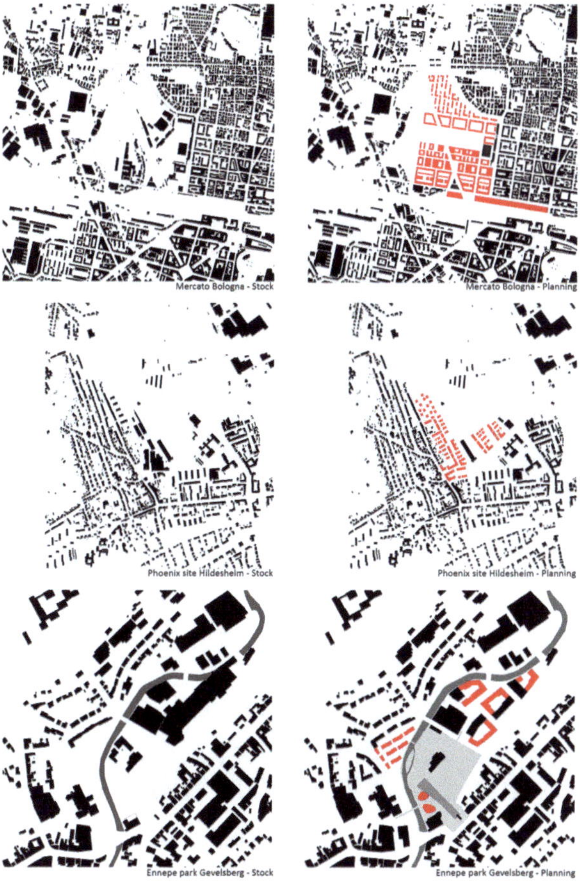

Fig. 7.30 Figure-ground diagramm of inventory and planning. The new structures fit in with the existing structures. Mercato Bologna—Stock. Mercato Bologna—Planning. Phoenix site Hildesheim—Stock. Phoenix site Hildesheim—Planning. Ennepe park Gevelsberg—Stock. Ennepe park Gevelsberg—Planning

7.3 "Zeppelin" Method

The possibility of reviewing a concept using the so-called "Zeppelin" method is only applied when the decision has been made to implement an urban design concept, but the elevation development raises questions about an overall urban or even supraregional impairment of perception. The actual height development of an urban design is simulated on site using balloons. In selected lines of sight, the impairment of the urban or landscape image is recorded photographically and evaluated with regard to the spatial consequences. This can result in recommendations for a shift of high points, but also for the cancellation of a planned elevation development.

Fig. 7.31 Zeppelin method. Verification of the planned heights with the help of zeppelin measurements/helium balloons, Campus Melaten, Aachen. View from Tour Bauduin in direction Melaten, marking of the Zeppelin locations and important landmarks

8 Excursus: Further Urban Design Methods

8.1 Network City Method

The network city method developed by Franz Oswald and Peter Baccini is a scientific instrument with the aim of analytically recording the specific system properties of cities and regions and using them as a basis for model formations (on form, supply and disposal, social and political organisation . . .). The network city method requires transdisciplinary work. In the following, the procedure and the instruments for analysing urban systems and supporting the design are described.

The network city method is based on five successive design steps:

Design step 1: First, the so-called "observation perimeter", the viewing area, and the so-called "project perimeter", the urban system to be designed, are defined. The observation of a larger space is necessary to characterize the nature of the interactions to the outside.

Design step 2: An initial analysis is performed according to morphological criteria, then according to physiological criteria. The morphological analysis serves to perceive the physical characteristics of the cultural landscape in order to understand it in its historical context. Using morphological instruments, the territorial forms, their structure and their function in relations to each other are recorded. This type of identification also serves to explain morphological characteristics in the "architecture of the territory" (such as water bodies, forests, settlements, agriculture, infrastructure, fallow land).

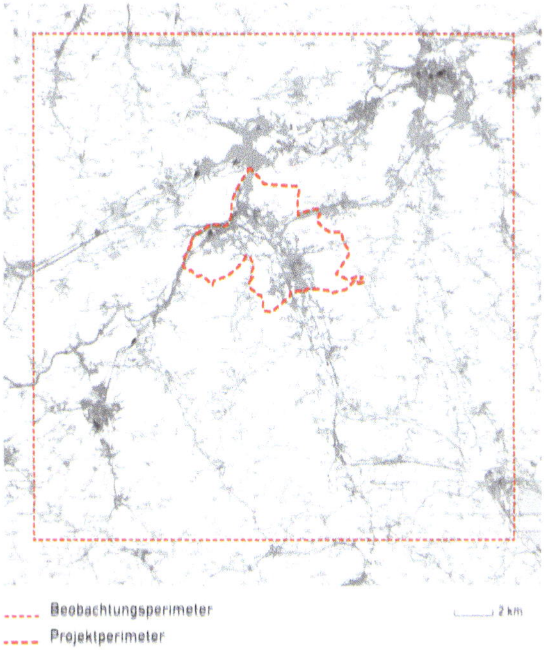

Fig. 7.32 Observation and project perimeter

Fig. 7.33 Architecture of the territory: morphological view

Six basic morphological terms are used to record and interpret the topographies found:

- Outline
- Field
- Size
- Structure
- Figure
- Hierarchy

The analysis by means of physiological instruments aims to record the relevant physical resource budget of urban systems (built and open space) and to use the results for their qualitative and quantitative description.

The following indicators seem particularly relevant:

- Population density
- Job density
- Service density
- Institutional density
- Workers (rivers)
- Students (Rivers)

Design step 3: This is followed by an initial **assessment** of the urban characteristics based on five quality criteria

- Identification
- Diversity
- Flexibility
- Supply level
- Resource Efficiency

Identification means the distinguishing features in the sense of orientation and order in space and time, which are necessary for urban coexistence. An essential criterion is the extent to which urban systems are capable of giving people distinctive images that can give them a feeling of home and security.

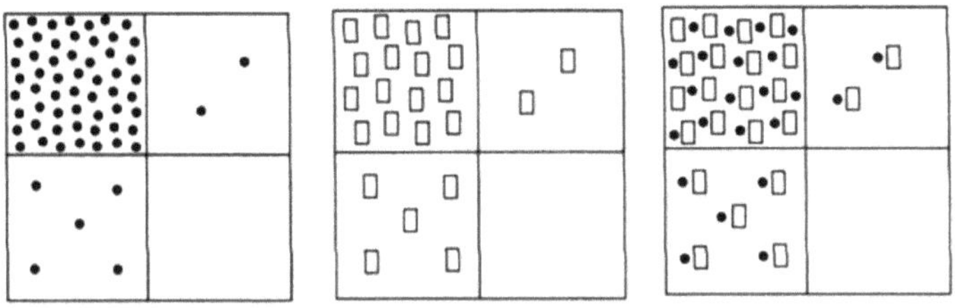

Fig. 7.34 Illustration of morphological terms

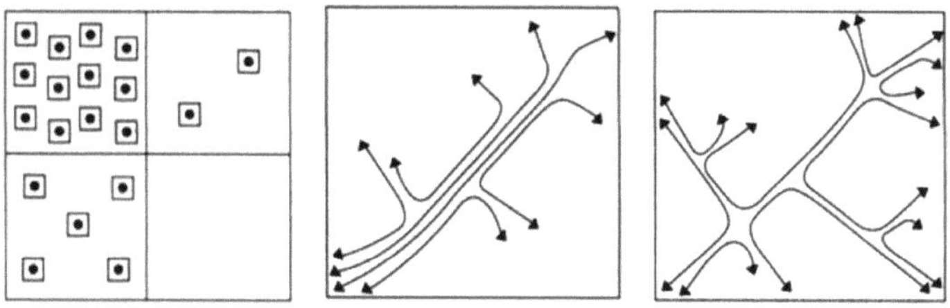

Fig. 7.35 Overview of physiological indicators

Diversity refers to the variety of ways in which a particular function can be fulfilled in an urban system, from the possibilities of transport to the availability and production of consumer goods.

Flexibility describes the property of a system to react to changes in the external and internal environment . . ., that is, to be able to adapt to different conditions and needs without endangering the system as a whole.

Supply level refers to the ratio between regional resources and the total resources needed by the region to meet its needs. If the supply (of water, food, energy sources) is high, this means that an area or region is highly self-sufficient.

Resource efficiency refers to the relationship between the useful quantity of a resource and its primary quantity, that is, between the consumption of resources and the space, energy and material required to use them. If the efficiency is high, the system is robust.

Design step 4: In a participatory process, the corresponding development goals are defined for the project perimeter.

Draft step 5: At the end of the chain there are drafts that show the target state of the rebuilding process.

8.2 Scenario Method

In spatial planning, scenarios are used to design, point out and discuss possible future developments—they are a method for representing the future or alternative futures. Starting from the circumstances of the present, they gradually show how future situations could develop as a logical sequence of events. Scenarios from the 1970s onward were used in planning as a reaction to the failure of previous quantitative forecasting methods. These methods (explorative forecasts, simulation models) were questioned and the necessity of dealing with qualitative approaches in research was recognized.

Instead of mathematically capturing the diverse interdependencies of reality and deriving forecasts from them, scenarios can clearly capture and process complicated causal relationships, show necessary prerequisites and present solutions. For the recognition and description of system interdependencies and complex, realistic structures, the methodology of scenario techniques is therefore more useful than the purely quantitative approach of forecasting. However, this does not exclude the supplementation of scenarios by forecasts or the quantification of statements in a further step. It is primarily an argumentative procedure, however, because the objective associated with the creation of scenarios is to derive and present the relationship between the present and the future in a plausible manner.

In contrast to quantitative forecasting methods, value is placed on writing scenarios:

- Vivid illustration
- Qualitative, not only quantifiable changes and elements of reality
- Diversity and interaction of developments
- Possibilities of political intervention at the stages of the development of the future
- Broadening the horizon of conceivable futures

Task of the Scenario Method
In addition to the substantive goal of designing (alternative) futures, the scenario method pursues the dialogue on topics of the future, on modes of action, interrelationships and effects, making it a "methodical vehicle of communication" (Hartmut E. Arras).

Scenarios cannot describe the truth about the future—rather, by capturing preconditions, they serve to illustrate possible development paths that need to be questioned and, if necessary and possible, changed. They thus offer a great didactic potential, which they have been using in professional and public discourse since their establishment in spatial planning, politics, economics and administration.

Content of the Scenario Method
The method presupposes that the future has its origin in the present and that it is therefore developed and influenced from the present.

Requirements for scenarios:

- Logically comprehensible in every step
- Reality-based
- Measures and their application must be clearly identified
- Comprehensible and plausible for both experts and outsiders
- Possible is the narrative style or the factual style
- Graphics serve as a striking addition

Different scenario categories are used to describe the content objectives. A distinction is made between explorative and normative scenarios. Explorative (also reactive) scenarios have a future-oriented function; lines of development and effects of events are followed in the future. The following types are assigned to explorative scenarios, although their differences in content and methodology are minor and fluid.

Status-Quo Scenario. . .
. . .describes the future state that would arise if present structures and conditions were to continue unchanged. It is therefore a very static view.

Trend Scenario. . .
. . .starts from a present state of development and continues into the future; in doing so, a possible, hypothetical course of events is designed step by step from the present into the future. However, the present is not merely continued as in the status quo scenario, but a dynamic development is assumed.

Alternative Scenario. . .
. . .serves, similarly to the trend scenario, to project future development possibilities, but the most likely development is replaced by several alternatives and variants. The assumption of changed conditions—change in social values, occurrence of economic crises, etc.,—results in different courses of the future. An alternative scenario

is used to show which developments have an effect and how, and which positive, negative or contradictory perspectives can be expected.

Trend Scenario...

...assumes, similar to the alternative scenario, changing conditions. Certain objectives from politics, economy and other areas are specified and taken into account in the process. In this way, the effects of certain objectives can be clarified, and contradictions and developments can be presented.

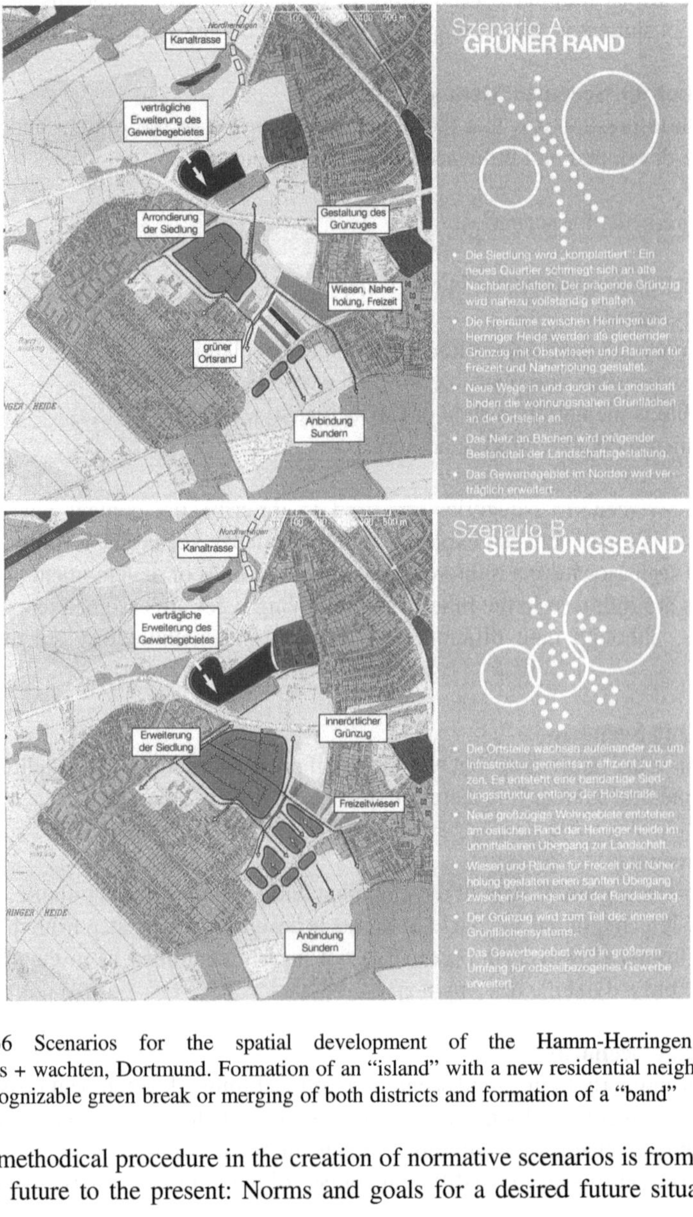

Fig. 7.36 Scenarios for the spatial development of the Hamm-Herringen district, scheuvens + wachten, Dortmund. Formation of an "island" with a new residential neighbourhood and a recognizable green break or merging of both districts and formation of a "band"

The methodical procedure in the creation of normative scenarios is from looking into the future to the present: Norms and goals for a desired future situation are formulated. Based on this vision, the necessary steps for realisation are then derived—the development is traced back from the desired vision of the future to the present. These scenarios thus have a future-shaping function.

Contrast Scenario. . .
. . .designs a picture of the future, accepts it as a future state and traces its development back to the present. A contrast scenario is often used as a strategy scenario to identify the political and instrumental approaches that can lead to the desired development, or to show how disruptive developments can be counteracted at an early stage.

The choice and application of the respective scenario category depends on the objective of the method. Explorative scenarios are intended to clarify already existing development trends, to criticize them if necessary and to take countermeasures in time; they are a common method in planning practice. If the intention is to present societal objectives, basic proposals for solutions and ideas for the future, normative scenarios are drawn. They tend to emerge in scientific contexts.

Development and Application
Scenarios are used in all fields and at all levels of spatial planning, precisely where communication takes place: Federal, state and municipal levels, in politics and administration. Therefore, they must be comprehensible—both for a scientific or planning expert audience and for laypersons (e.g., the public in citizens' hearings, etc.). It is important to define the addressees before the scenarios are drafted.

The development of scenarios follows a structured, step-by-step procedure:

1. **Basic survey, site survey, situation analysis**
   The first step is to gain and evaluate as comprehensive and detailed expertise as possible. This includes first of all a precise description of the task and definition of the problem, followed by the derivation of hypotheses and the setting of objectives. Based on these, necessary information is collected.
2. **Assumptions regarding future developments**
   After setting targets, arguments on future developments are formed in line with the hypotheses. They serve as a basis for drafting initial rough scenarios. Already in this phase possible branching points and alternative directions of development are designed. They are used to identify critical points as well as to determine leeway and further alternatives. Based on the findings of the rough assumptions on future development, these are now increasingly specified and substantiated.
3. **Creation of the scenarios**
   After the "preliminary phase", the compilation of the scenario or several alternatives in text and pictures begins. It is important that the scenario reflects itself, that is, the initial state is compared with the new state and the objective is checked against the results of the scenario. Measures and recommendations developed as a result are used for information and discussion. Different methods should be used during the entire processing phase. In the first phase, literature, statistical and document evaluations serve to gather information. Expert interviews may be useful. In the development phase, goal-finding methods such

as brainstorming, business games or workshops can be helpful. In the scenario design phase, the focus is ultimately on one's own creativity in the development of ideas and their textual and graphic implementation.

In the run-up to the new preparation of the land use plan for each city district, the city of Hamm has drawn up concepts for urban district development. They take up the results that were worked out in future conferences on the spatial development of the city districts. The adjacent variants illustrate two possible development scenarios for the urban district of Herringen.

As an "island", a new residential neighbourhood could be built adjacent to the school grounds. The open spaces could be preserved as a recognizable break between the locations and be significantly upgraded. In the form of a "ribbon", a merging of the districts would also be conceivable—with considerably more residential units and the integration of open space elements into the settlement area, thus supplementing the inner green space system. In both variants, areas for the protection and development of local trades and crafts are integrated with a connection to the planned canal route.

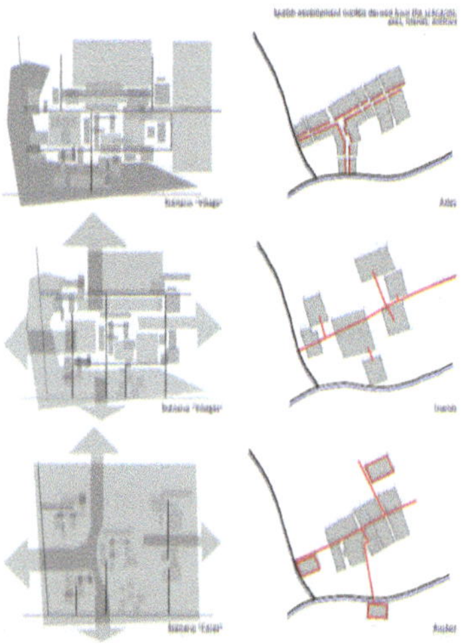

Fig. 7.37 Development of Tilbeck Abbey, Havixbeck community. Spatial scenarios: Village, villages, nuclei. Spatial development models derived from the scenarios: axes, islands, anchors. Scenario "Village". Axles. Scenario "Villages". Islands. Scenario "Cores". Anchor

9 Reading Plans and Concepts

Plans and concepts must be legible. A plan is only readable and thus comprehensible to outsiders if it clearly communicates the ideas and objectives that guide the design, from structural statements to urban planning and design details. Especially when it comes to the participation of citizens, experts and other actors in the planning process, a clear communication of the overall concept and objectives of the planning is extremely important.

The following examples are used to explain how the concept ideas can be made recognisable at different scale levels and what depth of content is required for "reading" each planning step.

Urban Design Concept Study Nordpark Mönchengladbach

On the site of the former Ayrshire Barracks North, Nordpark Mönchengladbach, a sports and commercial area surrounding the new football stadium, has been under construction for several years. The planning area is an area initially designated as a residential area, which is now also to become part of the commercial area.

The SportPark Mönchengladbach is the new profile of the Nordpark. It is structured by two axes that cross at a central point: landscape axis and city axis. The cross of the axes divides the SportPark into four areas with different thematic focuses:

The area is highly isolated due to its past as a barracks site and the surrounding transport routes. Therefore, in addition to an internal reorganisation, it is necessary to overcome the surrounding barriers of the SportPark and to better integrate it into its surroundings. In keeping with the overarching theme of sport, paths and tracks that can be used for sports activities are to link the area.

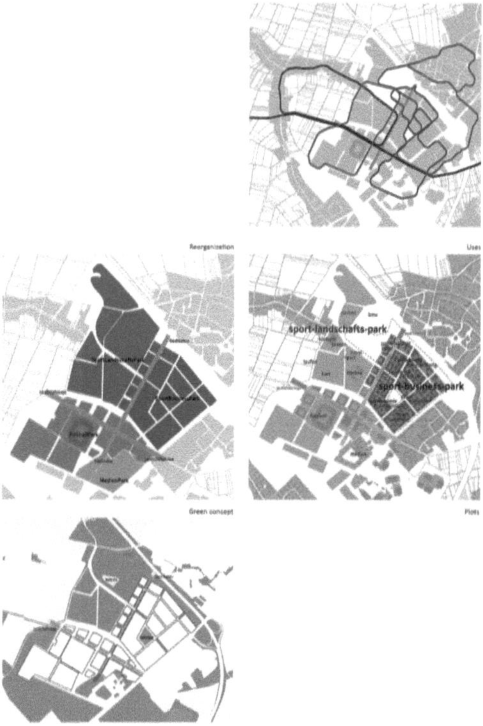

Fig. 7.38 Urban design concept study Nordpark Mönchengladbach, RHA—Reicher Haase architects + urban planners. Networking. Reorganization. Uses. Green concept. Plots

The SportBusinessPark as a “built” part of the plan area can be divided into four areas with different building and user structures. The dense development on the north side of the CityAxis continues the structure begun with the financial centre and thus forms a clear contrast to the adjacent SportLandscapePark. It does not seal off the area, but creates a dialogue between urbanity and landscape through openings to open space. The *core of* the SportBusinessPark is characterised by its small-scale, flexible building and plot structure. The *solitary strip* in the east of the area is the “jewel” of the SportBusinessPark. Solitaires embedded in the green space with alternating heights and alignments serve as representative company headquarters or office buildings for large companies. The high-rise building at the southern entrance to the neighbourhood becomes a landmark of the SportPark and is an eye catcher from the adjacent motorway.

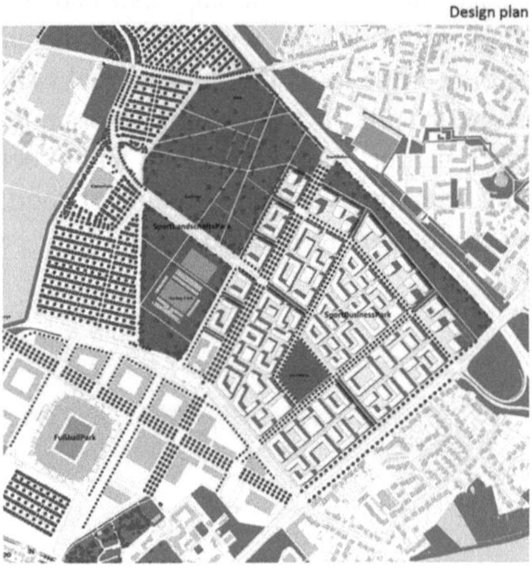

Design plan

The LandscapeLodge is located at the northern end of the LandscapeAxis at the transition from the busy pedestrian boulevard to the Land Forces local recreation area. The opposite pole is the CityBalcony at the north-eastern end of the CityAxis. The central green space element of the urban design concept is the festival meadow, which can be used as an event area for various purposes. The SportLandscapePark with its bathing lake forms a clear contrast to the densely built SportBusinessPark as a spacious open space.

The urban detail shows the intended flexible building structure to accommodate different uses. The theme of integrating the landscape into the new neighbourhood is reflected in the high design quality of the public space.

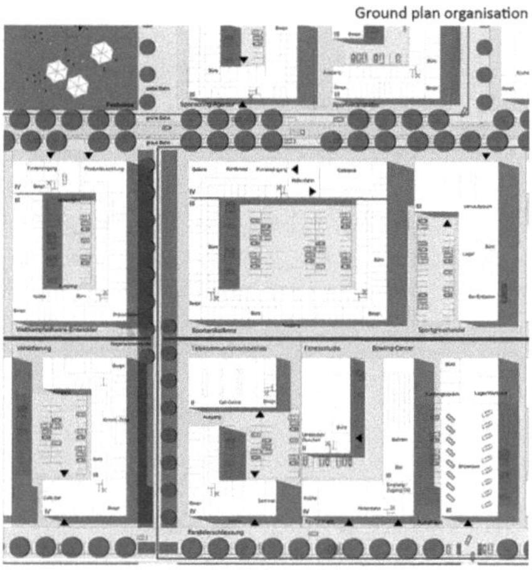

Ground plan organisation

Station Quarter in Mersch: "Le Quartier Vert"
The urban design concept for the new "Quarter Vert" in Mersch, Luxembourg, is based on the following guiding ideas and objectives:

- Develop a sustainable, future-oriented neighbourhood, integrated into the community of Mersch
- Create a close interlocking of the built environment and landscape on a local scale
- Generate impulses for large-scale urban development

The urban integration of the formerly sealed-off Alzette arch with the high-quality urban spaces of Mersch is a particular concern, which is important from both a social and tourism point of view.

In the "search for traces", elements and structures are defined which may be of importance for the site and beyond for Mersch. In the examination of the existing building, both the structural-spatial solution and the open-space concept are located. At the Quarter square, the existing building, the "Reception hall ", will later be integrated into the new structure with its use as an identity carrier.

The existing woody structures and the topography decisively determine the concept for the open space.

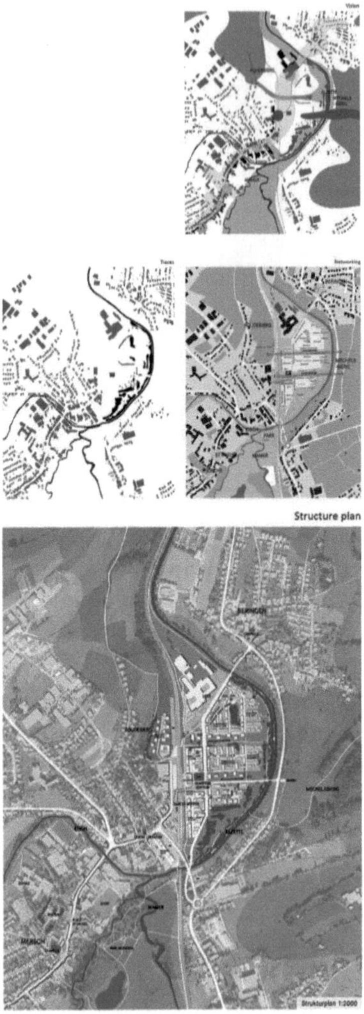

Fig. 7.39 Mersch Station Quarter, RHA architects + urban planners Traces. Vision. Networking. Structure plan

Through an appropriate density and a balanced mix of uses with different building typologies, a place with a high quality of life is to be created in interaction with public, community and private open space.

The urban design concept is based on the principle of solar construction. Especially in the transition areas to the landscape, the buildings are oriented towards the south. In the entire area, residential areas with different residential typologies and energy optimisation will be created. Two differently shaped landscape joints zone the neighbourhoods and connect with the riverbank. At the same time, they link the river and the hilltops, thus providing orientation and identity. Meadow spaces are created that offer recreation and play facilities. Attractive flower meadows alternate with playable lawns.

The development takes the interests of all road users into account. Mersch station, located between the "Quarter Vert" and Mersch, is the interface between cycle and rail public transport and motorised private transport. The bus station will be docked directly north of the new central underpass at the station square.

The objective of mixed use and the associated short distances between different uses within the neighbourhood are reflected in the urban detail.

Urban design concept, excerpt

Aerial photo

Bird's eye view

Urban design concept, excerpt. Aerial photo. Bird's eye view. Urban detail

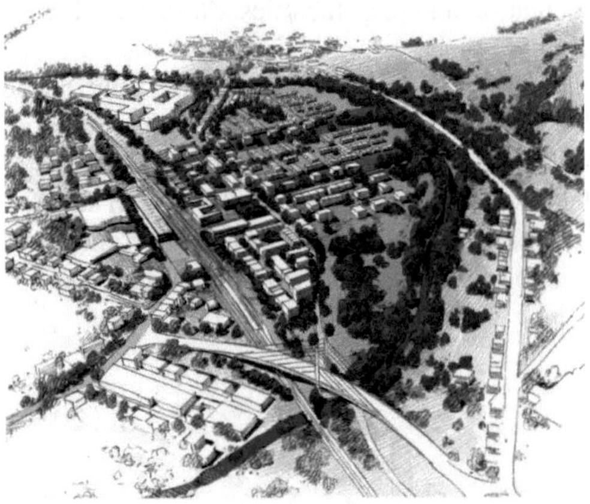

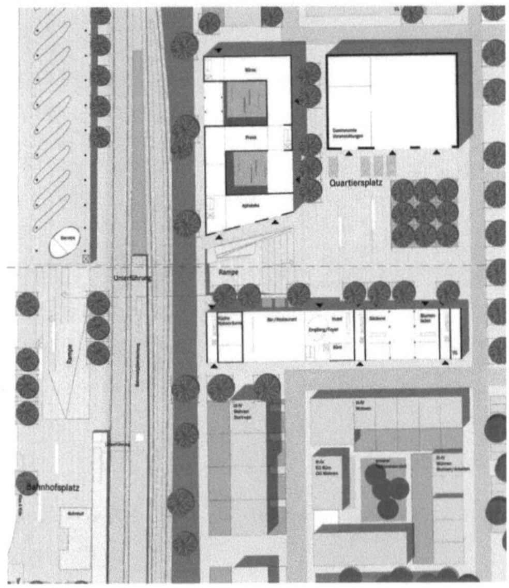

(continued)

Literature

1. The layer method and its design steps

Albers, Gerd; Wékel, Julian 2008: Stadtplanung. Eine illustrierte Einführung. Primus-Verlag: Darmstadt.

Düwel, Jörn; Gutschow, Niels 2005: Städtebau in Deutschland im 20. Jahrhundert. Ideen, Projekte, Akteure. Bornträger: Berlin.

2. **Vision statement**

3. **Structural plan**
4. **Framework plan/master plan**
5. **Design plan**
6. **Urban planning detail**
7. **Verification**

8. Excursus

Forschungsverbund Lebensraum Stadt (Hrsg.): Szenarien und Handlungswege. Mögliche Zukünfte der Stadt: Stadt, Mobilität und Kommunikation im Jahre 2020–Konsequenzen für Politik und Verwaltung. Band II. Ernst & Sohn, Berlin 1994.

Oswald, Franz; Baccini, Peter (2003): Netzstadt: Einführung in das Stadtentwerfen. Basel, Bosten, Berlin

9. Reading plans and concepts

Durth, Werner; Gutschow, Niels 1988: Träume in Trümmern. Planungen zum Wiederaufbau zerstörter Städte im Westen Deutschlands 1940-1950. 2 Bd. Braunscheig, Wiesbaden

Göderitz, Johannes; Rainer, Roland; Hoffmann, Hubert 1957: Die gegliederte und aufgelockerte Stadt. Archiv für Städtebau und Landesplanung, 4. Tübingen: Wasmuth

Urban Design and Attitude

8

1 Urban Form and Urban Image

1.1 Urban Design

Urban form is an essential theme of urban design. It encompasses both the material, physical form as well as immaterial factors such as atmosphere or diversity.

> Consequently, the elements of the urban environment include both material and immaterial factors, i.e. physical phenomena as well as economic or social aspects (...) The level of urban design is an enclosed continuum defined by boundaries. (...) with which man consciously or unconsciously stands in constant interaction. (Michael Trieb)

As inhabitants of the city, we use certain paths and can therefore only perceive parts of the urban form. Every observer stores the urban space as his or her "inner" image, as a so-called "city appearance". The environment affects us and is thus more than the sum of the mathematical and physical environment. This image is structured by the urban form and filtered through the subjective level of urban appearance. The resulting urban image is the result of an interrelation between individual and environment.

As the product of a constant interrelation, the real world is concrete reality and thus for the observer the existential performance of space. Urban form and urban appearance are interdependent and are today, at the beginning of the twenty-first century, supplemented by the level of media. This aspect leads to an expansion and new definition of the levels of perception of the *"cityscape"*.

1.2 Urban Image

Today, cities are increasingly communicated via images. In the perception of cities, the physical shape of the city plays an important role both externally and internally.

C. Reicher, *Urban Design*, https://doi.org/10.1007/978-3-658-34370-5_8

Images contribute to the understanding of reality. A picture represents something that it is not itself, it combines individual features into a whole.

Understanding Urban Image According to Kevin Lynch

According to Kevin Lynch's investigations, the image of the city or a place is composed of five typifying elements that refer to representational forms: paths, boundaries, areas, focal points and markers. The elements describe urban contexts, but also landscapes or sections of the overall urban context.

Paths describe the lines of movement in an urban context. By this Lynch means streets, footpaths, connecting paths, waterways, railway or underground connections. The observer moves through these channels habitually, occasionally or possibly rarely. They are the observer's main viewing space and connection between other elements. From here space is perceived. Paths are characterized by their characteristics (narrowness, width, façade design, composition, planting, location, clarity, etc.), continuity (length and course) and direction. The hierarchization of paths is not only based on spatial dimensioning. Guidance systems can be used to control the course of the path, that is, the primary perceptual space, especially for tourists and strangers.

Areas can be regarded as two-dimensional zoning. The observer thus divides the city into different sections that can be identified from the inside or outside by an individual character. According to Lynch, typical characteristics of areas are continuity and a uniform character, for example, in topography, social structure, the type of predominant use or urban planning and design features.

Boundaries are edges of areas that are not used as paths by the observer. They can be interpreted as elements that structure space. Here, according to Lynch, they are rather lateral directional marks and coordination axes such as topographic jumps, coasts, railway lines, walls, motorways or building site edges. They are either insurmountable barriers with a separating character or similar connecting seams, which enable the lively exchange between two areas.

Focal points are the links in the network of paths or the centre of the areas. They thus become strategic points of a city: intensively used central points, destination and starting point, intersections in the sense of a junction or concentration points in terms of use-specific densification. Many focal points are doubly occupied—nodes and concentration points at the same time. Focal points are conceptually only points in the cityscape. In reality, however, they can be large squares, or even entire cities on an international scale. At intersections, a decision often has to be made (e.g., intersection) or one stays there a little longer (e.g., square). Due to the direct link between focal point and experience, these places shape the image particularly strongly.

Markers represent optical reference points with conspicuous expression. This includes simple objects: buildings, signs, department stores or even hills. They are filtered out of context by a process of separating the individual element. Markers structure the urban space on different levels. They are macro- or micro-space oriented and thus create an orientation network. Basically, they overhang smaller elements and serve as "radial marks". These include towers, golden domes and topographic markers. Local markers are only visible from certain points, such as signs and posters, department store fronts or trees (see El Khafif 2009).

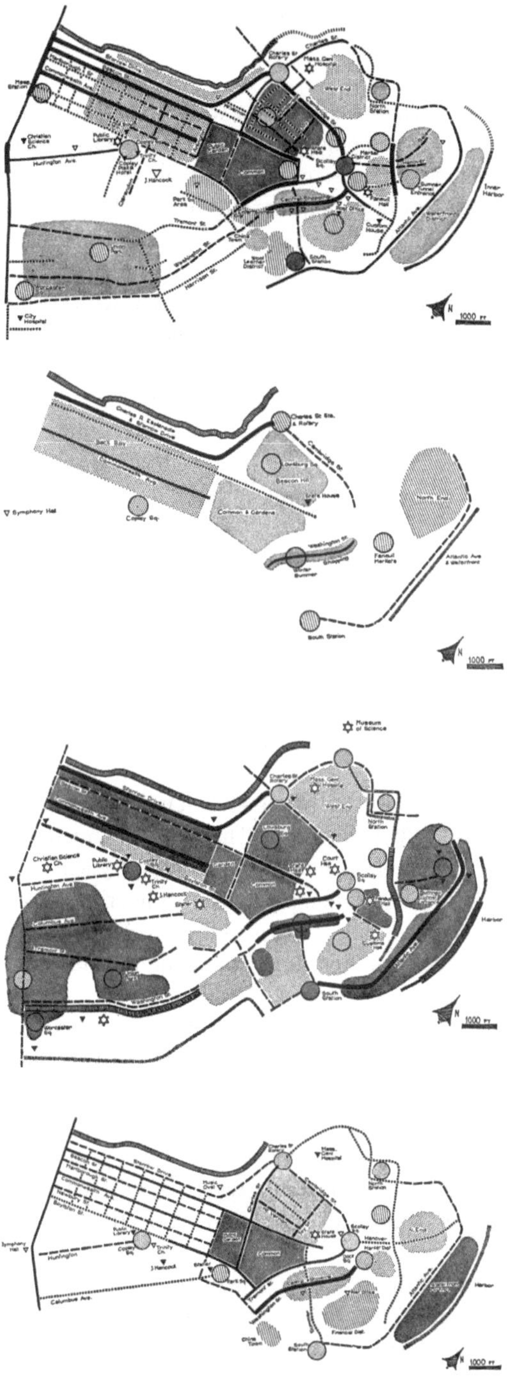

Fig. 8.1 Plan sketches of important elements in Boston, Kevin Lynch

The image of a given physical reality can change depending on the observer's perspective: a fast traffic route is a road for the driver and a boundary for the pedestrian.

Kevin Lynch further clarified that only a well-remembered image of the city, that is, an image that is clearly remembered by the inhabitants, can produce an identity that in turn implies an identification of the inhabitants. The image of the city contains three components that are important for the memorability of the image: identity, structure and meaning. All three elements are connected with each other. In Lynch's analyses, however, they are considered separately. Identity in this context means a clear recognition of the place. The place is detached from the context by the aspect of individuality and is thus identified. The viewer must be able to establish a spatial and structural relationship to the place, which in turn is linked to a sense of meaning. The existing structure, that is, the physical space, determines the identity of the place through its recognisability, meaning and sense. According to Lynch, however, meaning is very subjectively linked to the socialisation and history of the viewer and is therefore not a matter of planning for him.

> Perhaps the most important task of future work is (...): to understand the urban scene as a comprehensive and complex structure, to awaken an understanding of the interrelation between elements, of the overall structure and time sequences. The perception of the city is essentially a temporal experience that is directed towards a very large-scale object. If the environment is to be perceived as an organic whole, then capturing the individual parts in their direct context is only a first step. It will be extremely important to find ways to understand and manipulate the large connections, or at least to overcome the problems of the temporal sequence and the unfolding structure. (Kevin Lynch)

Understanding Urban Image by Michael Trieb

Michael Trieb's work "Urban form—Theory and practice (Stadtgestaltung—Theorie und Praxis)", published in 1977, builds on Lynch's ideas ten years later. In his research approach, Trieb attempts to take an interim stock of existing theories of spatial perception and in doing so deepens the psychological aspects of perception and the transferability of these findings to planning practice.

Various aspects appear to be more far-reaching in the context of his work: the differentiation of the levels of perception, the aspects of the image process and the possible influence on this process. Trieb addresses the fact that space—here referred to as "environment"—is not simply perceived, but that space is transformed over several levels in the process of perception until it is stored and remembered as an image of it. Trieb establishes the term "levels of environmental experience" and distinguishes between the "existing," the "effective" and the "experienced environment".

The field of action of the planners is first of all the existing environment; man perceives only a section of this existing environment and transforms this effective environment into an experienced environment. Individual parameters such as memories, associations and values play a formative role here. The experienced environment could be summarized as *"image"*, in accordance with the approaches of Kevin Lynch. It is based, on the one hand, on factors that depend on the observer's perceptive faculty and values, and, on the other hand, on factors that are shaped as patterns by the existing and effective environment.

> The experienced environment is the product of a process that takes place as an interrelation between the observer and his existing environment. This process is determined by the socio-economic structure of the existing environment and its spatial appearance, as well as by the meaning the observer gives it as an individual on the one hand, and as a member of various overlapping groups on the other. (Michael Trieb)

The existing urban environment thus describes all the calculable parameters. It is determined by physical and mathematical space with its topological and metric subgroups. Furthermore, the existing environment is only perceived by an omniscient observer. The effective urban environment integrates the aspect of limited perception. When a passer-by moves through urban space, he perceives the space from his own perspective, which shows only a section of the real existing environment.

The experienced urban environment implies the concept of individual evaluation and can be equated with the Lynch perception approach. It is the general mental image that a person forms of the external world. Trieb, however, emphasizes the aspect of power of synthesis and the importance of space for action and reaction.

When designing, the elements that make up a cityscape must be planned in terms of their use and significance as well as their appearance. Since the urban design elements form the basis of the experienced environment of the city dweller, they are one of the most important stages in the entire urban design process. The conscious planning of urban design elements requires a careful analysis of the situation at each location before any appropriate intervention is made.

2 Dimensions of the City

2.1 Design Qualities

An urban development quality cannot be easily determined on the basis of individual design features. In his books "Basics of architecture (Grundlagen der Baukunst)", Vitruvi already formulated the demand for quality and meant *an appropriate arrangement of the elements into a whole according to their intended purpose.* Although the demands on urban planning have remained the same in terms of their basic understanding, individual aspects have also changed over time.

Quality urban design brings together different aspects and requirements: functional utility value, social responsibility, ecological and economic appropriateness, cultural understanding and aesthetic design. Depending on the design and planning task, the weighting of the requirements varies. Objectives such as the promotion of an unmistakable character of a city, a lively atmosphere or the beauty of a neighbourhood ultimately serve to strengthen quality of life for people in the respective location. Accordingly, urban design quality must be discussed at various levels and striven for in a design process:

- Object quality (building, architecture)
- Urban structural quality (dialogue between building and spatial context)
- Functional quality (use and functioning)
- Open space quality (public space and landscape)
- Ecological quality (sustainability, dealing with nature)
- Quality of development (traffic routes and other development)
- Socio-cultural quality (social appropriation of the place)
- Economic quality (profitability and follow-up costs)
- Process quality (process and process culture)

The levels shown are interwoven in different ways depending on the planning task, so that they must be considered in context. A dissection into these different levels makes it possible to recognise qualities, to introduce them into a design process and to check the result later.

1st dimension
2nd dimension
3rd dimension
4th dimension

2.2 Dimensions of the City

In contrast to urban planning, urban design introduces concrete spatial dimensions. However, urban design goes beyond the third dimension of space. The following dimensions have a decisive influence on the shape and image of the city and interlock with each other during the design process:

1. **Dimension: structural level**
   The first dimension of urban design initially means structures in the sense of basic patterns. These can be grids, but also more differentiated geometric and organic patterns, which serve as a basis for further urban design.
2. **Dimension: length and width expansion**
   The second dimension of urban design introduces a length and width extension of the structures. To the structural statement are added area-based allocations of the built over and free areas. It can consist of an irregular or even structure; caesura and breaks can create corresponding tensions, as can be seen in the figure-ground plan of Manhattan.
3. **Dimension: altitude development**
   The third dimension gives the two-dimensional structure a vertical component. The pattern with its depth and length expansion experiences a height development, which in the close-up effect has an impact on the urban space that can be

experienced and in the long-distance effect shapes the city silhouette. The third dimension can be in harmony with the second dimension of the structure, or each can have its own statement, as the skyline of Manhattan makes clear.

4. **Dimension: shape and atmosphere**
   The shape is expressed in the design of the facades, their proportions, their materiality, but also in the roof landscape, the fifth facade of the city. The designed form can never be neutral in its effect. It always exerts a subjective effect on the person who looks at it or experiences it. The fourth dimension of urban design is the built form and its charisma, which is subjectively perceived by the observer. The built form and the activity that emanates from it spreads an atmospheric effect that influences the observer depending on the individual receptiveness and perception, both in a positive and negative sense.

   These four dimensions stand in a complex order to each other, depending on the location and task for an urban design.

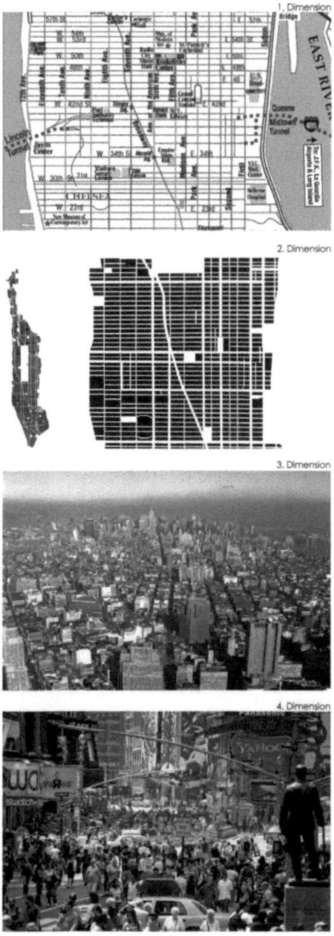

Fig. 8.2 The dimensions of the city using the example of Manhattan, New York

3 The Hierarchy and Collective Memory of the City

The city as such cannot be traced back to a single basic structure or basic idea, but consists of numerous parts, each of which has its own character.

The Italian architect and urban planner Aldo Rossi distinguishes between the "primary elements" and the "secondary elements" in the structure of the city and assigns different roles to both: the primary elements, such as churches, town halls, universities, etc., which play a central role in the sense of identity-forming elements and can be particularly accentuated, and the secondary elements, the residential buildings, the everyday buildings, which should fit in with their form.

> From a purely functional point of view, these primary elements can be seen as fixed activities of the community for the community or as precisely defined urban design facts that characterize the city as a whole as the scene of an event or as a building. As such they are an expression of the history and the idea of a city.... (Aldo Rossi)

Aldo Rossi introduces another important concept into the value of the city, that of the so-called "collective memory".

> Just as "memory is tied to facts and places, so the city is the site of collective memory, whose expressions are architecture and landscape ... In this sense, memory becomes a guide through the entire complex urban structure." (Aldo Rossi)

Fig. 8.3 Building hierarchies in Dortmund. Primary elements stand out against the secondary elements in the city silhouette

The hierarchy of the elements of the city is closely linked to "collective memory". Emphasized elements of the city, which due to their size, elevation, but also their function, have a special place in the memory of people, both residents and visitors. A church tower protruding from the city silhouette marks a special place in the city, but at the same time emphasizes a specific use.

Churches, as primary elements, shape the image of our cities to a large extent, but at the same time they are a central component of memory, that is, of what we remember. In contrast, the high-rise building with service uses at the entrance to the city emphasises the entrance to the city and may therefore have an important spatial significance, but it does not include any special use.

From these analytical thoughts on urban structure, the following insights can be derived for urban design:

- A differentiation of buildings and structures according to the spatial hierarchy and the hierarchy of meaning is important for the significance that can be assigned to an individual element in the context of the overall structure (see Urban Components of the City).
- The exposed location and the height development derive their justification from the hierarchy of the building.
- In the examination of the built city, special attention is paid to urban elements that have been removed from the context of the city, because they have significantly determined the memory of the city in the past.

Fig. 8.4 Hierarchy of the elements of the city of Lindau, Lake Constance. Church towers dominate the hierarchy of the city against the mountain panorama

4 Attitude towards the Built City

Cities, as engines of social and economic development, are places of modernisation and at the same time guardians of tradition. The search for spatial answers for growing, stagnating and shrinking cities and neighbourhoods also moves within this field of tension. Which parameters and laws play a role in urban design for further development of urban structures? Which structures are worth preserving, which ones can be redesigned and replaced by new structures in favour of a more convincing solution? What is the designer's position in the respective situation?

4.1 Constants and Variables

The processes of change in urban structures are sometimes more and sometimes less rapid. There are areas that have remained more or less constant and others that are characterized by extreme dynamics.

This juxtaposition of different ways of dealing with the built city, even within one and the same city, that is, a simultaneity of preservation, further development and transformation through new things, has existed at all times of urban development.

In dealing with the built city, a distinction must be made between two different categories

- the constants (historical city as a carrier of identity, striking buildings, etc.) which must be preserved and preserved. A well-founded analysis of the existing structure is the prerequisite for possible interventions, which may be necessary for the sustainability of the structure at certain points.
- the variables (abandoned sites, brownfields, "no man's land", etc.) that need to be developed and replanned. In these areas a new spatial quality can be developed that is more self-referential. Attention should be paid to the connections or interfaces with the surrounding context.

The interaction of constants and variables does not follow a predefined recipe, but has to be redefined each time. Their interplay is what makes urban structures so diverse and vibrant.

> The 'continued construction'' of architecture and cities is one of the essential tasks of the future: our cities must be concentrated, converted and improved so that they can cope with the demographic, climatic and mental changes that our society is undergoing... The approaches range from the adaptation of historic monuments to the open contradiction against the "old". (Andreas Denk)

Phases of Dominance of the "Variable" over the "Constant"

A concise example of the primacy of the variable over the constant is the phenomenon of "automobility and traffic" in different phases of urban development. The car has determined the shape and structure of the city to an extreme degree. Particularly in the field of mobility, technical achievements have radically changed the world we live in—not always in its favour. Initially, technical progress raised great hopes, as historical scenarios on the effects of motorised transport from 1885 show:

> The disappearance of the horses from the roads will be beneficial. Noise on the roads will be greatly reduced. With the speed of motor vehicles, there is no longer any need to worry about traffic congestion. Everyone will be able to move forward quickly and without stopping. Above all, the number of traffic accidents will decrease decisively, because a car is much easier to handle than a horse-drawn carriage.
>
> (English brochure "Horseless Age" —Age without horses!)

Some 40 years later, this assessment of Le Corbusier's technical achievements is translated into statistical studies and forecasts on the role of the automobile in urban development. Le Corbusier predicted that cities would "suffocate" if they were not developed and rebuilt in a way that was suitable for cars. They form the background for his "Plan Voisin", financed by the car manufacturer of the same name, with which he aimed to replace the old Paris centre with a car-friendly New City.

The urban design model of the car-friendly city from the 1950s ultimately had serious consequences for the way the existing urban fabric was handled, especially in numerous city centres where new traffic routes were planned and realised without any significant consideration for urban structures worthy of preservation.

4.2 Superordination: Subordination—Classification

Urban design requires context-related solutions, which represent a great challenge, especially with regard to the handling of existing buildings. There are no ready-made solutions, each situation requires an individual response. In order for a solution to be convincing, an intensive examination of the existing situation is essential. This leads—depending on the result—to different attitudes, which are reflected in urban design. A design can subordinate the new to the existing, but can also let the new become the determining element. However, urban design can grant the new its independence without impairing the existing structure and qualities.

Superiority

The principle of superordination brings the new building to the fore and thus ignores existing design criteria. The focus is on the independence of the new building, which can only be judged uncritical if the existing situation has no qualities and only the new building can create such qualities.

Subordination

With a subordinate attitude, the new takes a back seat to the existing structure and architecture. The principle of subordination becomes critical when something new is no longer recognizable as such, or when it is a recent reconstruction of a building of the last century, which is still practiced in this way today.

Classification
An important prerequisite for the integration of the new into an existing situation is first and foremost respect for the existing. Classification does not exclude the use of new materials or a new interpretation of existing buildings and typologies. Especially the sensitive visualization of old and new can be a valuable development of the existing qualities and structures.

Superordination, subordination, classification—it is not always possible to consistently maintain such fundamental attitudes, or the concrete design task only allows a mix of such attitudes with regard to the dimensions of structures, space and form. Decisive for the respective assessment is the concrete situation and the recognizable examination of the existing and its history.

In a city, in a place there exist—more or less distinctly—recognizable and datable sequences of historical layers. "History" is therefore above all a snapshot that can be changed and enriched with new layers at any time. Traditions, too, are always also a setting, an invention at any time of their emergence. This does not mean that they do not exist. But it does make it clear that traditions are cultural achievements that have their own value.

Fig. 8.5 Imperial Bath Aachen, Architects Kasper + Klever

In the third dimension of the city (height development), the building complex on the Büchel with a V-shaped silhouette, which opens up the view of the cathedral, behaves in the sense of subordination, and in the courtyard with the inclusion of the gable structure in the sense of classification.

In the fourth dimension of the city (shape and atmosphere), the building complex takes on the attitude of a superordination to the different sides. The materials and their embossing in detail consciously stand out from the existing.

The planning approach to the "Kaiserbad" in Aachen shows that the attitude towards the place is complex and can change in the course of history.

The location in the historical context of the city is characterized by its triangular squares, which were created by overlapping Roman and medieval ones. This second dimension of the city (length and width), which has been shaped over the centuries, is ignored by the project "Imperial Baths" in the sense of a superimposition, because the building complex slides into the triangular courtyard.

Literature

1. Urban Form and Urban Image

Boehm, Gottfried (Hrsg.) 1994: Was ist ein Bild? München: Fink.
El Khafif, Mona 2009: Inszenierter Urbanismus. Urbaner Raum für Kunst, Kultur und Konsum im Zeitalter der Erlebnisgesellschaft. Saarbrücken: VDM
Lynch, Kevin (1965) Das Bild der Stadt. Bauwelt Fundamente 16. Braunschweig.
Schirmacher, Ernst 1988: Stadtvorstellungen. Die Gestalt der mittelalterlichen Städte–Erhaltung und planendes Handeln. Zürich, München: Artemis.
Trieb, Michael 1977: Stadtgestaltung–Theorie und Praxis. Braunschweig: Vieweg.

2. Dimensions of the City

Krause, Karl-Jürgen 2006: Enzyklopädie der Stadtbaukultur. Dortmund. (unveröffentlicht)
von Meiss, Pierre 1994: Vom Objekt zum Raum zum Ort. Dimensionen der Architektur. Basel, Berlin, Boston: Birkhäuser

3. The Hierarchy and Collective Memory of the City

Rossi, Aldo 1975: Die Architektur der Stadt–Skizze zu einer grundlegenden Theorie des Urbanen. Düsseldorf: Bertelsmann Fachverlag.

4. Attitude towards the Built City

Urban Development as a Process 9

1 Process-Oriented Approach

In times of ever more complex transformation processes of cities and neighbourhoods, urban design means that, in addition to dealing with space as a product, space as a process is playing an increasingly important role.

The city consists of various components, not only of the built structures and open spaces, but it is also largely determined by social and communicative structures, by the users and ultimately by the people who, as actors, develop, understand and implement concepts or who themselves shape spaces through their activities. Therefore, urban development concepts must

- provide a **robust framework** that is also valid in changing circumstances;
- ensure a **scope** for participation processes without losing the basic ideas;
- create a **sound basis** for subsequent communication strategies, so that ideas are anchored in a broad range of stakeholders

> Every form is the frozen snapshot of a process. So the work is a stop of becoming and not a frozen goal. (El Lissitzky)

Urban design is increasingly linked to planning and conception for specific periods of time: for stages, for phases of realisation, also for "development pauses", in which urban space may be available for other functions and uses.

1.1 Temporary Uses

Temporary uses are not yet planned in the official planning cycle. At the same time, they—whether as open space or building use, as temporary structures or space activation—have long since become a highly regarded phenomenon not only in eastern Germany, but also in western Germany. The decreasing economic pressure

C. Reicher, *Urban Design*, https://doi.org/10.1007/978-3-658-34370-5_9

on the use of inner city brownfields is leading to the situation that areas and empty buildings are becoming undesirable "gaps", unused and underused urban spaces in the urban fabric. An immediate re-use is often not possible due to a lack of private demand, unclear future development directions or weak municipal capacity to act. Such phases of waiting and seeing tend to establish a process of trial and error without having long-term perspectives in mind.

Temporary uses are defined as uses of a derelict area or an empty building that are not permanently established. The decisive factor for classification and understanding of temporary uses is that two essential basic conditions are fulfilled:

1. The temporary use is **limited in** time There are no specifications for the type and duration of the limitation, where it is contractually regulated or merely tolerated. The use can even have an illegal status.
2. The temporary use does not meet the owner's long-term **economic expectations.** This condition relates above all to the actual yield expectations, because economic revenues are usually foregone during the temporary use.

1.2 Staggering and Phasing

Urban design concepts and projects are rarely implemented at short notice and in their entirety. Usually the implementation takes place step by step, in staggered stages or phases. This implementation process requires that urban design concepts must be planned and structured in such a way that they can be divided into clearly readable fields. These can then be used as basis for a step-by-step implementation. Each individual realisation step should be planned in such a way that no fragmentary structures are created after a single phase, but rather an intermediate state with the quality of a possible final state is created, a so-called "temporary final state".

Choreography of Hardware: Software—Brandware—Orgware

A desired urban state is created by bringing together and superimposing the different spatial levels of hardware (physically built space), software (uses, activities), orgware (sets of rules) and brandware (image production, image construction). This understanding of choreography is based on the concept of space as a matrix of these four spatial levels. Mona El Khafif speaks of an "urban script", which demands the inclusion of the factor time in the process and rhythm of space formation (see El Khafif 2009). The urban script controls the interplay of the spatial dimensions, while at the same time brandware can motivate the appropriation of space as software and orgware can animate networking of the actors. With the process dimensions described, the urban script thus goes beyond classical urban design instruments such as the master plan and expands the understanding of urban design as spatial production and staging, which at the same time must always be seen in relation to the specific conditions of the location.

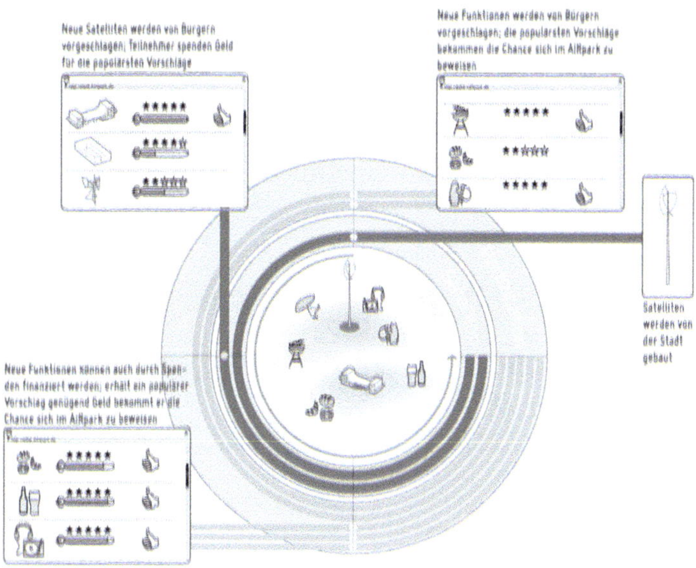

Fig. 9.1 Tempelhof trial, competition Tempelhofer Feld, Berlin

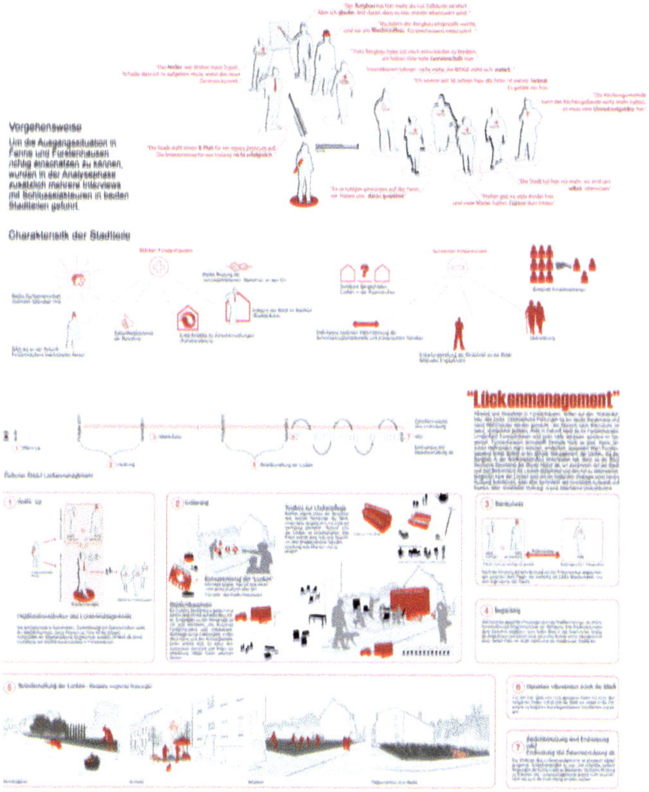

Fig. 9.2 Participation and "gap management", LEG Prize 2010 competition

> The last word is spoken by the citizens who fill the city with private and public life. They are the ones who give the city its face. Humane urban planning is characterized by self-determination. It gives the citizens the chance to contribute to the shape of the city themselves. (Hans Paul Bahrdt)

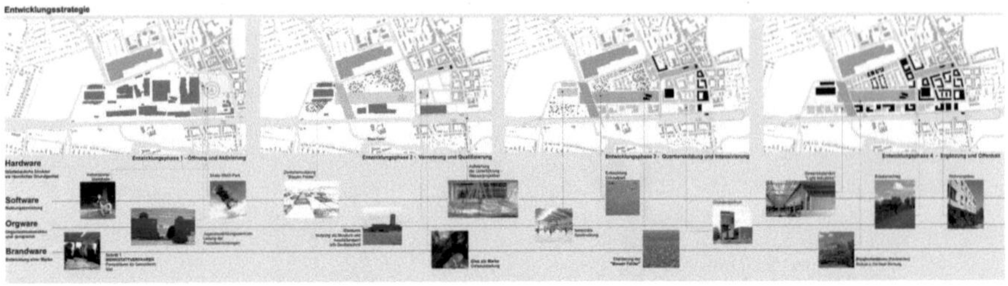

Fig. 9.3 Process design "Düsseldorf-Gerresheim", RHA Architects and urban planners

2 Participation

2.1 Participation Procedures and Actors

In planning practice, the involvement of stakeholders is playing an increasingly important role. A prerequisite for the high-quality implementation of urban development plans and projects is the persuasiveness of the urban design vision as such, but also its anchoring with decision-makers and the local population. This makes procedures for the involvement of various groups of actors in various planning phases indispensable.

Formal and Informal Participation Procedures

In principle, we differentiate between the formal and informal procedures of participation in planning. In urban land-use planning, there is a formal, legally required participation—as a contrast to the other procedures—in which citizens, associations and other authorities can voice their objections and concerns and include them in the further planning process.

Informal participation without prescribed procedural steps often serves to involve citizens in a planning process when an intensive discussion of topics and upcoming decisions seems to be useful. Open workshops, planning workshops or citizens' forums usually serve less to produce concrete plans than to adjust topics, focus on problem points and ultimately to communicate planning.

Charrette Method

For some years now, the "charrette method" has been drawing attention to itself. In the USA, the procedure has been used since the 1990s; in Germany, the procedure is still new, but it is already being used successfully and further developed for local conditions.

"Charrette" is a public planning method that directly involves citizens in design processes in the context of complex issues of urban development and urban design. Municipal decision-makers, housing associations, urban planners, entrepreneurs, landscape architects, architects, traffic planners, etc., work together with interested citizens in an interdisciplinary process to develop solutions for the future of the city or concrete planning projects within the city.

Urban design is a negotiation process. (Kees Christiaanse)

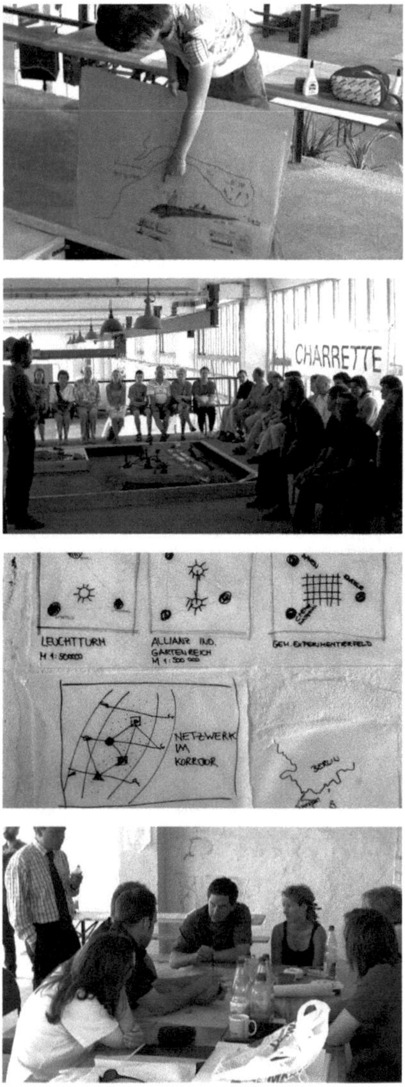

Fig. 9.4 Student Charrette Ferropolis. Task: Development of an urban design code as well as the use of water as a location for new buildings in the iron city

2.2 Open Source Planning

In software development and the exchange of knowledge and information, it is now impossible to imagine product development processes that are completely open and freely accessible to everyone—Linux, Mozilla, Wikipedia are prominent examples of this. Open source means the disclosure of the source code of a program or operating system and thus the possibility to intervene in the programming language used to create the program or operating system. The disclosure of the source code thus means a radical opening of the system previously controlled only by the manufacturer.

Beyond the field of information technology, the hope is spreading that the model of open source process can also be transferred to other specialist and planning areas—including urban planning. First of all, this would mean relinquishing the sovereignty of the planning authorities—municipality, city, region—in relation to fixed definitions and opening up to interventions by each participant in the urban process. The urban space is no longer defined in advance, but is treated as a product that is constantly being thought out further through participation of various interest groups. An open platform allows different opinions and models to be used productively. The establishment of a process-oriented, constantly updated planning of the city should not aim at deregulating the planning as a whole, but rather at a shift in which tasks are perceived and managed differently.

Fig. 9.5 Open source workshop Cologne-Ehrenfeld. Task: Development of new ideas for Ehrenfeld and experiment, what open source opens up in terms of possibilities in planning and where limits are set to this kind of action in real space

Königs, Heinemann and Schmidt already presented their **Divercity Strategy** in 2000, which is based on the open source principle. This strategy understands the city as an open system and includes a planning instrument that is intended to meet the requirements of the city and urban planning through general accessibility and constant updating. Essentially, Divercity describes urban planning as an open source project, the basis of which is a networked and freely accessible planning platform, in which urban development processes are not regulated by model-driven, time-spatially predefined specifications, but by direct and up-to-date input. The urban space is treated as a product that is continually improved through the participation of various competent interest groups. The reality of the city (dynamics, specialisation) can be actively incorporated into the planning process. Crises and convergences are productively built in, through which the planning process is constantly questioned and regulated. The potentials of deficient contexts can thus be used directly. The planner becomes the organizer of a process instead of being the author of a draft, as was previously the case.

The openly designed process begins with the collection of information, which is divided into **cold and hot information.** The former are statistical data, the evaluation of map material and literature, hot information always includes a discussion function, such as photos, impressions, opinions or interviews. Then follows the initiation of a project. Planning proposals for a defined space are presented. Condition for this step is the relinquishment of any rights to this first input, so the disclosure follows the principle of "copyleft". The relinquishment of planning sovereignty releases forces and at the same time creates the motivation to participate in the process of planning with others and with their contributions and help. After the initiation of the project, the process of copyleft begins. By modifying, copying, referencing, rearranging, extending or eliminating existing ideas, the original idea is further developed. Anyone can intervene at any time in further development of the temporary project states.

The testing of the open source principle in urban planning was carried out using an interactive urban planning and participation model, designed for the city of Schwabach, whose former barracks area described the planning field. The model consisted of two interlinked planning tools: a table model set up on site with freely combinable city modules and an interactive website. Local and global ideas and suggestions for the development of the area could be introduced on both levels of reality.

The unfolding design process was initially deliberately left unattended; the starting point was complete openness and free access to the planning model. The aim was to investigate whether stable and cooperative structures could be established even in the event of a radical withdrawal from overriding guidelines and mission statements. After initial fluctuations, a relatively orderly structure emerged, which then even tended towards uniformity after additional moderation. The transparency of the model was able to enormously increase the willingness to cooperate among the various participants. However, an excessive number of compromises also highlighted the danger of mediocrity. It should be noted that the freely negotiated basic framework proved to be very robust in further (unmoderated) phases.

With an operating system for computer users, the consequences of an open source system can be described relatively precisely. It is a homogeneous interest group, in which users and developers are identical in the beginning. A transfer of the principle to urban planning methods encounters much more complex action contexts, since, in contrast to the optimisation of an operating system and in the common understanding of the city, it is more a matter of dealing with diverging objectives of urban development. Nevertheless, the open source method can serve as a fixed point because it uses networked communication very successfully, demands transparency in a fundamental way and allows the involvement of a broad range of authors.

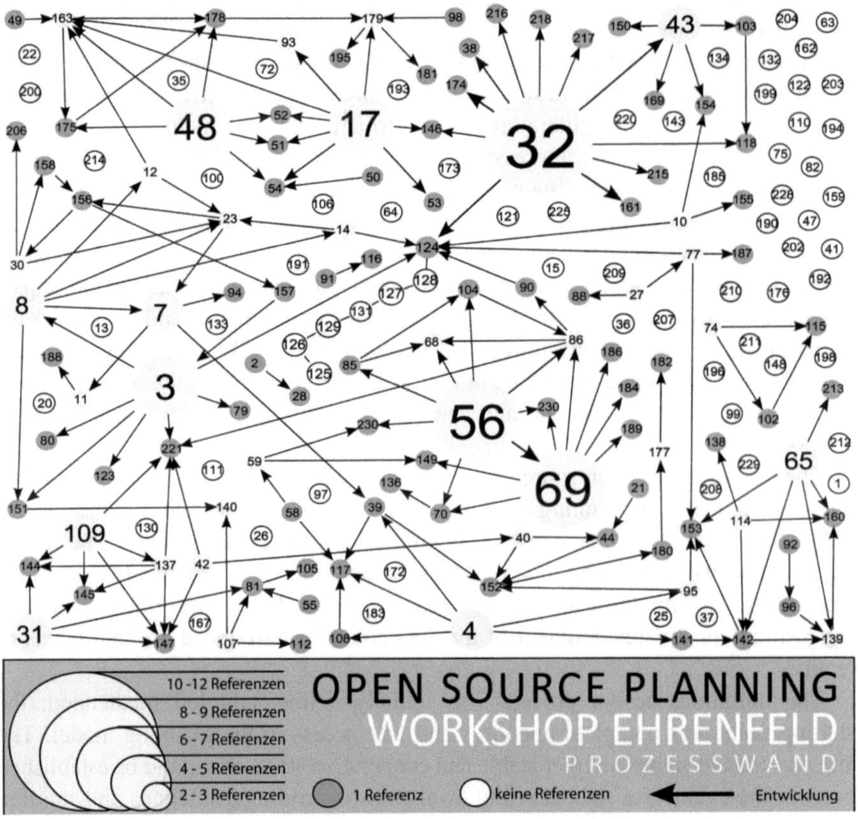

Fig. 9.6 Process wall. The wall shows the development strands and follow-up of the individual planning ideas as a result of the workshop in Cologne-Ehrenfeld

3 Image and Branding Strategies

Frank Roost

In view of the increasing competition between cities for businesses, tourists and residents, it is becoming more and more important for municipalities to distinguish themselves with strategies that involve a systematic further development of the city image or even the development of a kind of "brand" for the city or individual neighbourhoods. This approach is linked to the somewhat contradictory goal of developing the inner cities into modern service locations that both meet the same demands of potential visitors or investors everywhere and at the same time are intended to be unique and unmistakable. The approach taken by the municipalities is partly shaped by an entrepreneurial self-image: cities and regions are often marketed like a product, although they are of course much more complex than a consumer item. Nevertheless, improving the image of the city and developing a brand-like profile for cities as an inward-looking identity-building and legitimation strategy and, at the same time, as an outward-looking marketing campaign for the city, is an important municipal task today, and one that shapes planning action.

Definition of Terms

The English term "*image*" is derived from the Latin word for picture, "*imago*", and refers not only to the image in the narrower sense, but also to the inner imagination, that is, the picture that someone makes of something. However, these two levels are closely connected, also and especially in relation to the cityscape. Kevin Lynch therefore used the term "Image of the city" to illustrate the great importance of the legibility of building structures and derived approaches for a design strategy that offers users orientation and identification possibilities.

However, the frequent use of the term image in urban planning has a second root, namely in the context of city marketing. These approaches, which have been widespread in Germany since the 1980s and use targeted marketing strategies to convey the qualities of city centres, tourist destinations or business locations with the help of image campaigns, have their roots in British and US-American planning and are therefore characterised by a vocabulary and approach that was first used by the advertising experts there to market consumer goods.

City marketing and images serve as an outwardly directed marketing strategy for the city (see Helbrecht 2006), but also aim to create identity internally. Successful city images in this sense are therefore not an artificial product, but are developed from endogenous potential and thus illustrate the chances of a structural change that is taking place.

A more recent development of such approaches is branding, which also has its roots in British and US marketing. The term is derived from the English word "*brand*". Branding, which is particularly common in the private sector, describes the efforts of a company to communicate its unique selling proposition in comparison to its competitors and to ensure that potential customers associate positive associations with the company or brand name, thus creating a competitive advantage. Similarly, the systematic further development of the image of a location is also known as *place branding* (see Stöber 2006), which can refer to a shopping street, a neighbourhood or an entire city.

3.1 The New Role of Images in Urban Production

The increasing importance of image and branding strategies in planning must be seen in the context of the so-called festivalisation of urban politics (see also Häußermann, Siebel 1993): Since the 1990s, large-scale projects in particular have been increasingly staged and often realised in connection with major events. Since the events are more media-compatible than conventional planning, complex planning goals can often be communicated more easily with them, so that an aesthetic, urban planning and thus ultimately also urban policy consensus can be created for the procedure. A typical form of this festivalisation is the linking of large-scale urban development measures with the organisation of Olympic Games and World Expositions. Although such large-scale events have traditionally served as an opportunity for national and urban self-expression, since the 1990s they have increasingly been used systematically as a prelude to the structural restructuring of the host city and, by linking them with infrastructure and housing projects such as those in Barcelona or Lisbon, have been specifically placed in the service of urban redevelopment.

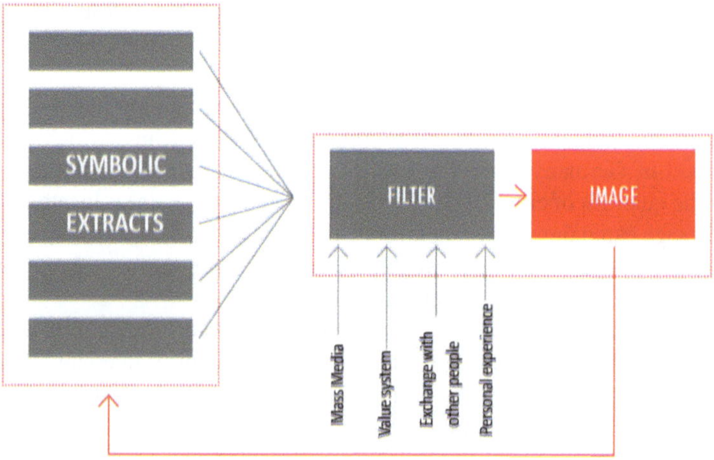

Fig. 9.7 Production of Image, Karl Ganser

Following on from the success of these strategies, many municipalities are now relying on sensational effects from the outset to popularise their large-scale projects. The buildings realized under these auspices are then no longer tied to an event, but function through their architecture as a public-effective and image-boosting attraction. They often have a solitary building of outstanding design, which as a so-called "flagship" has the function of carrying the banner, communicating the change and symbolising a new image or promoting the new or redesigned neighbourhood and the city as a whole as a tourist destination. A characteristic feature of this is a combination of old and new, which serves as a staging of innovation and tradition to further develop the city image in the broader sense—that is, beyond tourism marketing, also to profile the business location and represent the identity of the inhabitants of the entire region (see Bodenschatz 2005).

In addition to such public buildings or buildings erected in public-private partnerships, privately financed and used building complexes can also function as flagship projects. In this context, globally active branded companies are taking advantage of the booming city tourism and are setting up so-called branding centres in the centres of metropolitan areas, which serve as a stage for them to present their products in an effective advertising manner. Such facilities, such as the Sony Centre Berlin, serve the corporations as three-dimensional advertising. In the same way that major projects act as image-building flagships for the cities, the branding centres are also intended to serve the investing groups as advertising flagships for the brand image. The marketing strategy of the investor and the urban development strategy of the municipality thus merge into a system of urban production in which mutual image enhancement is the main focus (see Roost 2008).

In view of these potentials for further development of images, there are corresponding approaches not only for entire cities or outstanding city centres, but comprehensive branding and image strategies are also developed for individual residential neighbourhoods. Branding techniques can even provide approaches for profiling disadvantaged urban districts or large housing estates. In Rotterdam, for example, the so-called **neighbourhood branding** is being used to re-understand and reposition neighbourhoods with a great need for renewal and a difficult image. For example, the large municipal housing companies use market research findings to better respond to the needs of their customers, who are predominantly from socially disadvantaged population groups. This is not least a matter of differentiating people not only according to income groups and social status, but also according to lifestyle, consumer wishes and housing needs, in order to better profile the settlements in the long term and tailor them to the wishes of the residents.

3.2 Image Carriers and Built Images

A special role in the image-oriented forms of planning is played by joint coordination of further development of image and structural and spatial design, because visual elements are a central element in the communication of images. Particularly symbolically charged individual buildings are of great importance as recognisable image carriers and serve to visually condense a regional or urban image. This includes, for example, urban redevelopment projects that focus on the (new) staging of the architectural heritage. Such measures range from the upgrading of existing historical urban spaces to the reconstruction of lost key buildings (as in the case of the new palace buildings in Berlin or Braunschweig) and entire urban districts (such as Dresden's New Market). In the old-industrial regions such as the Ruhr Area, it is primarily renewal strategies for restructuring and reinterpreting post-industrial landscapes that include measures ranging from aesthetic redesign to museumisation.

In addition to such approaches, which often make use of potentials of the traditional European city, forms of urban production that aim at the symbolic power of explicitly modern buildings also play a role. Thus, the "skyline" of a city serves as a form of expression for economic power and international control functions. Similarly, ambitious are the so-called flagship projects, which are realized as sensational solitary buildings by globally active star architects and on which high expectations are placed as impulses for urban development—such as increasing the attractiveness of the city for tourists, the advertising function for a new neighbourhood or even the support of a metropolitan city image, which is intended to promote structural change of a municipality into a service location.

Fig. 9.8 Guggenheim Museum Bilbao, Spain, Frank O. Gehry

Different characteristics can be distinguished: The most widespread form of such image-oriented projects are **representative cultural buildings**, which have been developed in Great Britain since the mid-1980s in the face of deindustrialisation and the reduction of state funding and which were closely linked to the development of new city marketing concepts. These early examples, mostly realized in public-private partnerships, include above all museum projects aimed at broad impact of popular culture, such as the Film Museum in Bradford or the Music Museum in Sheffield. Of note here is Glasgow's "City of Architecture and Design" campaign, which is an early example of how architecture and design can raise the profile of cities. At about the same time, the "Grand Projets" began in France—initially in Paris, Lyon and Lille—which, although much more clearly characterized by the active participation of state actors up to and including the central government, were no less aimed at changing the image of local authorities and also reflected even more strongly the social function of the projects, namely to create a consensus for the approach in terms of urban policy, planning and aesthetics. At the same time, as large-scale and outstanding architectural landmarks, they set the architectural and urban design standards for the following phase of large-scale projects designed by star architects.

Since then, numerous cities have created or re-staged cultural institutions of international standing, which are intended to increase worldwide appeal of the city as a tourist destination. Entire city districts are being geared to this function and serve to raise the city's cultural profile as a soft location factor (as in the new construction of the Tate Modern in London and the Museumsufer in Frankfurt) or to modernise the city's image (Museumsquartier Wien). The most popular variants are cultural facilities in waterfront locations designed by star architects in many places. They often serve as image bearers of converted old industrial port neighbourhoods into new work and residential locations, which are explicitly designed to promote transformation of a community from an industrial to a service location, such as the Imperial War Museum North in Manchester as a lighthouse on Salford Quays or the Sfera in Genoa as a prelude to the inner city port redevelopment of Ponte Parodi.

The pioneer and most famous example is the Guggenheim Museum in Bilbao. The secondary effects expected from such image-oriented projects are therefore often referred to as the "Bilbao effect" and as such have found their way into the German-speaking professional community. Although Frank Gehry's spectacular museum building naturally played the central role in this strategy, it is also important to pay tribute to the urban development and planning framework that made this success possible in the first place: The Guggenheim Museum is part of a much more comprehensive project, the new Abandoibarra neighbourhood. The Guggenheim Museum is part of a much larger project, the new Abandoibarra neighbourhood, a former shipyard site that has been converted into a prestigious residential and service area along the course of the Nervión River. The architect César Pelli developed the urban master plan for the neighbourhood, which included several high-rise buildings, a new riverside park and the construction of a museum directly on the waterfront. In order for the Guggenheim building to achieve its great urbanistic effect on the riverbank, extensive preparatory work was necessary, including the laying of a railway line and the construction of new bridges. The Bilbao Effect is thus not only based on a masterfully successful solitary building by a single architect, but is the result of a comprehensive planning strategy that decidedly favours the effect of such a project and in the first place creates the creative and economic scope for such elaborate buildings.

Fig. 9.9 Skyline of Chicago, by land and by sea

This combination of comprehensively conceived urban planning action and the growing importance of the symbolic dimension of design therefore gives rise to new tasks for practice that go beyond conventional questions of interaction between architecture and urban design. For example, it is necessary to consider which addressees are addressed in the planning of symbolic projects—both in terms of social composition of users and the area of origin (local—regional—supra-regional—international) of different target groups (consumers, tourists, tenants, investors, etc.). This includes the question of the extent to which the projects act as a means of perception in global competition with the outside world, or to what extent, against the background of ongoing suburbanisation, they also serve to create identity within a fragmented metropolitan region. And at the same time, the mechanisms of media dissemination must be taken into account: This concerns images of new urbanity and their anticipation in the creation of outstanding solitary buildings, urban spaces or pathways and viewpoints as "constructed images". This also includes the question of the continuing significance of elements of traditional European urbanity, which are obviously always a point of reference in new staging approaches. And the influence of new media on cultural and architectural production of the city must also be taken into account when it is designed as a "built image" in terms of its mental recognition and media reproducibility.

Image and branding strategies thus complement urban development concepts and planning. Their significance for urban design and urban planning now goes far beyond simple marketing concepts in the sense of logo design and advertising campaigns. The future potentials of branding and image building therefore lie beyond the Bilbao effect hoped for an entire city—which is probably difficult to repeat. It is rather important to examine to what extent individual elements of techniques of image building can be applied to the most diverse planning approaches in the broadest sense—from the (re)staging of the architectural cultural heritage to the profiling of individual city districts.

Literature

1. Process-Oriented Approach

El Khafif, Mona 2009: Inszenierter Urbanismus. Urbaner Raum für Kunst, Kultur und Konsum im Zeitalter der Erlebnisgesellschaft. Saarbrücken: VDM

2. Participation

Königs, Ulrich; Heinemann, Christoph; Schmidt, Christoph: Divercity. In: IFG Ulm (Hrsg.) 2000: Strategischer Raum–Urbanität im 21. Jahrhundert. Frankfurt am Main

3. Image and Branding Strategies

Bodenschatz, Harald 2005: Renaissance der Mitte. Nachmoderner Zentrumsumbau in London und Berlin. Berlin: Verlagshaus Braun.

Häußermann, Hartmut; Siebel, Walter: Die Politik der Festivalisierung und die Festivalisierung der Politik. In: Dieselben (Hrsg.) 1993: Festivalisierung der Stadtpolitik. Stadtentwicklung durch große Projekte. Opladen.

Helbrecht, Ilse: Stadtmarketing und die Stadt als Ereignis. Zur strukturellen Bedeutung symbolischer Politik. In: Birk, Florian (Hrsg.) 2006: Stadtmarketing–Status Quo und Perspektiven. Berlin: Difu. S. 263–278.

Roost, Frank (2008): Branding Center–Über den Einfluss globaler Markenkonzerne auf die Innenstädte. Wiesbaden: VS.

Stöber, Birgit (2006): Von brandneuen Städten und Regionen–Place Branding und die Rolle der visuellen Medien. In: Social Geography Discussions, 2, S. 215–251.

Informal Urban Design

10

1 Understanding of Informality

Only a few years ago, the term "informal urban development" would probably have been understood to refer primarily to the so-called "informal instruments" of urban planning, to the urban development concepts and framework plans that are not fully regulated by law and that city administrations can design according to their own needs in terms of their preparation procedure and contents of plans (see p. 184). However, for about 10 years now, the terms informal urbanism or informal town planning have been associated with other aspects of current urban development, which are becoming increasingly important and are placing new demands on urban design and planning activities.

This new understanding of informality has long since ceased to be limited to the countries of the global South or to crisis regions, nor is it exclusively equated with a lack of formal regulatory structures and makeshift infrastructures or with illegality. It is rather a matter of a thoroughly desirable "informality" of planning and urban development as a whole: open urban development processes, a variety of civil society actors and temporary uses of space characterise the image of our cities today—possibly more than many a planning procedure. Conventional top-down planning approaches and their traditional constellations of actors are increasingly being questioned. At the same time, the call for more self-design by citizens beyond formalized planning procedures is growing louder. When new approaches to urban development and urban planning are considered today, it is no longer exclusively in categories of control and order. Nor is it a question of "a little more citizen participation", but of a real rethink. For urban development will in future move even further away from the development of building land towards urban development work.

C. Reicher, *Urban Design*, https://doi.org/10.1007/978-3-658-34370-5_10

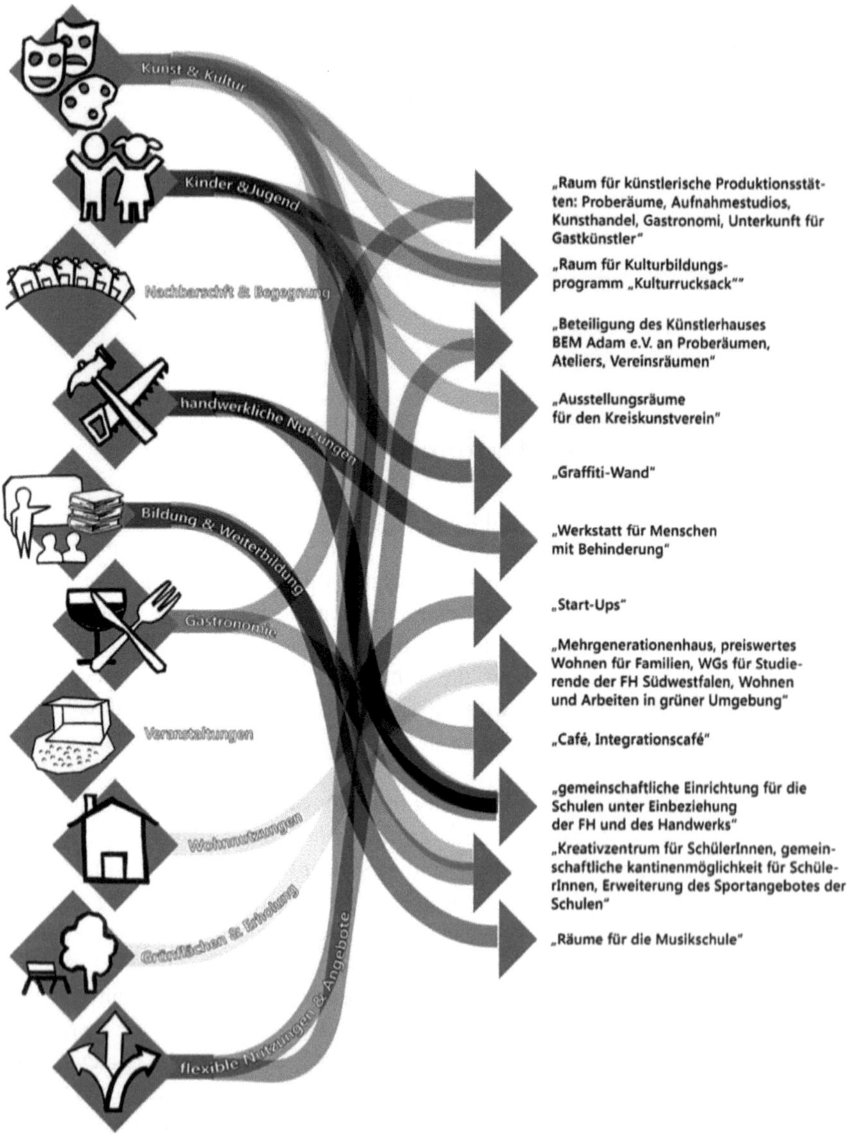

Fig. 10.1 Civil society as an actor, Master's draft 2016/2017 "Utilization concept Merkur light bulb factory Soest" Leitenberger/Witter

1.1 Civil Society as an Actor

The chances of this are good: the range of informal activities is becoming ever wider, and the spectrum of actors is becoming more diverse. These now include cultural projects, self-organised sports opportunities for young people, community gardens, village meetings and citizens' groups, shops to maintain local supplies, self-administered housing projects, citizens' buses, FabLabs and repair cafés, citizens' energy cooperatives and citizens' foundations, as well as citizens' initiatives and occupational groups. Everywhere, people are taking initiative to once again exert a more direct influence on design and use of their environment.

All these projects have manifold references to planning goals and measures, they enliven the city and should therefore be considered by planners at an early stage when thinking about the future of cities and neighbourhoods, suburbs and villages. This does not mean that the existing planning instruments are outdated, but rather that they will only develop their effect, or better, if they are understood more strongly than before as a space of opportunity for civil society actors. In the future, they will continue to form a basis for controlling urban development. However, it is necessary to do even more justice to the increasing differentiation and dynamization of uses, lifestyles and economies in urban development. Against this background, traditional planning tools must be supplemented by new models for action. Informal urban design should therefore be used alongside the strategies of urban planning and urban development described in the other chapters of this book.

1.2 Definition

In the following, the term "informal urban design" is to be understood as a specific form of urban design and planning that is rooted in everyday urban life and serves to satisfy practical needs. The term thus refers to a wide range of activities that contribute to urban development but have so far hardly been attributed to it or recognised as meaningful strategic building blocks. Some of these activities (urban gardening, temporary uses) are already closer to the instruments of urban planning, others (trend sports, open workshops) are less easy to integrate. At the same time, the term changes the view of the actors: alongside the planner and designer, the user appears as an equal active partner who should be involved as a co-producer in the conception and design of urban spaces.

In the course of this development, new planning methods become necessary which take into account complexity, non-linearity, uncertainties and self-organisation. Whereas traditional urban planning aimed at establishing a final stage of development and the most precise spatial design of a future urban neighbourhood, the central task of informal urban design is to actively shape the development itself. It is not the discussion about the formal urban design that is in the foreground, but the question of possibilities for urban activation and urban use.

Informal urban design is thus also a means of creating a better balance between experts and civil society in the interests of democratic spatial planning and of providing low-threshold access for groups that are more difficult to reach. Informal urban design can better promote the willingness to participate and the integration of citizens as actors than framework or urban land use planning that is reduced to building structures and design means.

Especially in the concept of informal urban design, the process regains an important significance. It must, however, be clearly distinguished from conventional planning practices and (in contrast to the early consolidation of planning of economic actors) accept the contingency of practice, even make it a principle. Concentrating on the endogenous potentials of the planning area and allowing room for negotiation to determine objectives breaks through linear administrative procedures and creates uncertainty. The outcome of such processes must be left partially open in order to leave room for creativity and creative power of civil society. In such processes, planning does not control the actions of local communities and civil society structures. It sets a framework for self-control and self-organisation to develop.

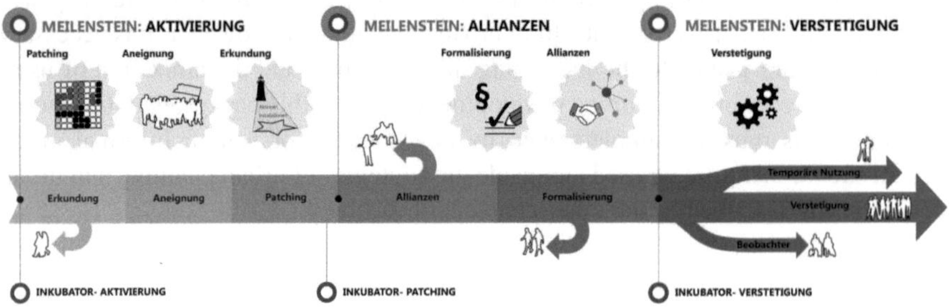

Fig. 10.2 Exemplary planning process in informal urban development, master design 2016/2017 "Utilization concept Merkur bulb factory Soest" Leitenberger/Witter

2 Plan and Design with Informality

Stephen Willinger

The following considerations on greater involvement of informal activities of civil society groups by planners are based on new ways of dealing with civil society actors. In the following, a few examples will be used to indicate which directions are conceivable here. In each case, the focus is on supporting informal actors and providing suitable open spaces.

2.1 Opening up Urban Space

Creating Information Bases

The database for green temporary uses in Leipzig, which is accessible on the Internet, contains about 400 brownfields in the entire city area, which could be used differently in the meantime. Each area is described in more detail by its address and details of its useful life (temporary, permanent). If citizens are interested in a specific area, they can contact the respective owner. The website shows plots of land where the owners are basically interested in a new use. The database was developed in 2007 as part of a commission from the City of Leipzig, Office for Urban Renewal and Housing Promotion, to enable the creative use of vacant areas and spaces in the city in transition. The homepage is sponsored by the foundation "Citizens for Leipzig".

The (civil society) makers of the vacancy reports already established in several major German cities have self-organised these platforms on the Internet and built them up as a critique of prevailing exploitation mechanisms in the real estate industry. Beyond their actual purpose, they serve as an occasion to network different scenes (temporary users, young planners, cultural workers) and can thus become contact persons for city administrations.

The platform openBerlin is a tool for participatory urban development, spatial transformation and idea production. Projects, project seekers and urban open spaces are networked on the internet and at events. Already successful projects share their knowledge and thus take responsibility for their city. The aim of the project, which was initiated by civil society, is to bring urban policy and administration back to a public-interest approach to public land.

Fig. 10.3 Screenshot openberlin

Mediate Temporary Uses

Another example of this different way of thinking are temporary uses (see p. 222), which can have a variety of positive effects on urban development. They are an important development component, particularly in the development of brownfield sites or large-scale areas. However, the users do not follow planning considerations, temporary uses cannot be planned. Integrating them into strategic, urban development projects, therefore, requires a special understanding of the interests and motives of users and an adapted planning concept. The task of the ZwischenZeitZentrale, which has been mediating between the needs of temporary users and landowners on behalf of the city administration in Bremen for several years, is to provide the necessary impetus. It

- detects vacant spaces and areas and checks them for their suitability for temporary use
- wins private owners to provide spaces and areas
- develops concepts for temporary uses together with potential temporary tenants
- accompanies the projects in the implementation
- combines the interests of the administrative departments involved and
- helps to shape the dialogue between authorities and temporary users.

Fig. 10.4 How the ZwischenZeitZentrale (ZZZ) in Bremen works

2.2 Enabling Appropriation

Developing New Networks

Beyond this spatial perspective, new forms of creative governance with a focus on the development of innovative networks are needed to give informal actors more scope in urban development. To this end, platforms must gradually be established where committed actors can network and then meet with planners to discover each other's worlds and exchange ideas about possible futures. In Madrid, a group of architects has voluntarily opened an office and an Internet platform that is to function as a "nursery for urban initiatives". After only a few years, there is already an extensive network of different agents, not architects but citizens of the city. "When the city administration is planning a project at a location, you can see through our mapping which people are already working there. For example, there is afternoon childcare for children and there is the bicycle club that has mapped the least travelled routes. The people have local knowledge . . . Building the city together with these initiatives is really urbanistic". It is imperative for administrations to support these networks and to talk to them. This can only be done in their own way, namely informally, unbound, with encounters, even chance encounters.

Support Users

Leipzig was the city in Germany that prepared the inclusion of the informal in the instruments of today's urban development with temporary open space installations and systematic support for the appropriation of empty buildings. The administration had recognised that unusual initiatives and experimental uses of space by civic groups can provide important impulses for stagnating urban districts. Thus, new communication structures were established with the local actors, who thus became key players in urban development. In 2012, Head of Office Karsten Gerkens stated: "We cannot prescribe the objects that need to be worked on. HausHalten e.V. must say which objects they can process. We from the administration do not know the whole complex interrelations, we cannot conclusively assess the pros and cons for a project. There we lack the background of experience, which the actors have partly built up over years. "The creativity of HausHälter does not manifest itself in social utopias, but rather in the skillful organisation of coordination processes, in building trust in the triangle of homeowners, informal users and administration. HausHalten advises the users intensively and then concludes contracts with the house owners. This relieves the users of legal difficulties and the association supported by the city acts as guarantor. This pragmatic approach strengthens users, accustoms owners to new approaches and gradually changes administrative routines.

The North Rhine-Westphalian state programme "Taking the Initiative" seeks contacts with civil society initiatives that convert vacant buildings, create cultural offerings and provide economic impetus. The users receive intensive coaching and become spatial developers.

Prefer Users when Allocating Space

In connection with the sale of public real estate, there is increasing criticism that purely economic criteria are decisive here. Aspects of a more broadly understood "urban return" on uses are left out. It is obvious that projects such as the Princess Gardens (Prinzessinnengärten) or the Bar25 in Berlin have sparked off controversy over such a property policy, because they are prototypical of the impulses that informal urbanism is capable of generating. Now practice is slowly changing, and procedures are being developed that also take social and cultural aspects into account. For example, in the allocation of plots around the Flower market hall in Berlin's southern Friedrichsstadt, preference was given to informal carriers. Based on the development model for an art and creative neighbourhood, the Berlin real estate fund, in a formal bidding process, selected investors who offered less money but more content relevant to urban development.

Informal Financing

Even though the involvement of local actors in urban development is becoming increasingly important, all too often ideas and projects get stuck at the idea stage because there are not enough funds to realise them. Crowdfunding means the financing of a project by a large number of people—the crowd. Crowdfunding is usually handled via an online platform on which the project ideas are presented. A pre-defined minimum sum must be reached within a defined period of time in order for the project to be realised. If the target sum is not reached, the supporters receive their money back. As urban crowdfunding, this type of informal financing is becoming increasingly popular and complements other new financing instruments such as the disposal funds. Crowdfunding can be organised by the project developers themselves. However, there are also already municipalities that are setting up their own local or regional platforms to support creative civic projects in this way.

2.3 Planning Differently

Planning with Users

When planners receive orders, the goal is often already set: to build. Informal urban planning calls for a brief pause, a review of the programme and a close examination and appreciation of the existing buildings, open spaces and—above all—uses. The Assemble group of architects was awarded the prestigious Turner Prize in 2015 for having successfully developed several sites and neighbourhoods using this principle of non-planning. In London's Clerkenwell neighbourhood, a disused gas station was converted into a temporary movie theatre in the summer of 2010. The self-initiated project was initiated by a group of young architects, designers and artists under the name of Assemble, who are committed to revitalizing neglected urban areas through so-called pop-up projects. With such non-profit temporary uses, the group aims to draw public attention to the great potential of inner city brownfield sites in the UK. Closed with an ornate curtain that could be lowered from the canopy of the gas station, the Cineroleum was furnished in the simplest way possible with reminiscences of the lavish interiors of historical film palaces. Familiar elements of cinematic iconography were reinvented with cheap, donated or re-used materials. Specially developed manuals for the installation invited the residents to participate in the construction process. Their manual skills were irrelevant. Over a hundred volunteers took part, learning together and developing the design in a trial-and-error process. A 5-week film programme attracted more than 2000 visitors—every screening was sold out. At the same time, a new experience of going to the cinema was created: separated only by curtain thickness from a busy street, the Cine-roleum offered not only a shared film experience, but also a visual spectacle for passers-by: With the credits the glittering curtain lifted, the space for the film dissolved and the cinema spread into the street space.

Fig. 10.5 Photo Cineroleum closed, Assemble/Morley von Sternberg

Fig. 10.6 Filling station, Assemble

Fig. 10.7 Cineroleum open, Assemble

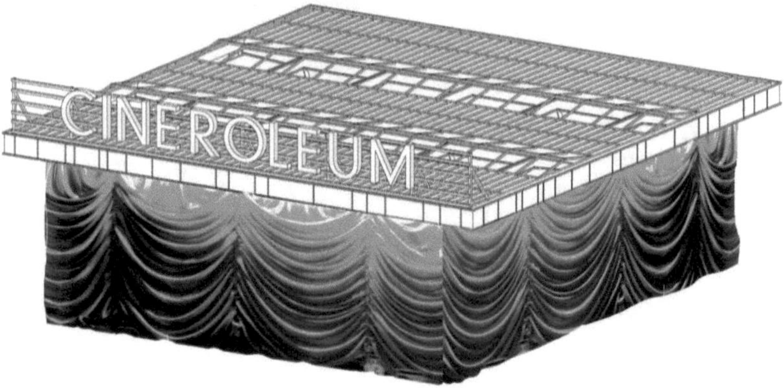

Fig. 10.8 Cineroleum closed, Assemble

Fig. 10.9 Photo Cineroleum open, Assemble/Morley von Sternberg

Empty Space as Potential

Planners have known for a long time that important impulses for a different urban development arise in spatial niches, in neglected and unintentionally created residual spaces. Such places would have to be established or secured as planning places. As early as 1987, in his contribution to the urban planning competition for the new city of Melun-Sénart near Paris, Rem Koolhaas postulated voids (gaps, empty spaces) as decisive urban components of any city. Since then, the term "voids" has been used to describe all the places in a city that are not conclusively defined in terms of their use, nor can they be because of certain obstacles: Reserve areas, fallow land, clearance areas, vacant buildings, polluted areas. They are constantly being created anew by structural changes in cities and offer niches for creative developments, potential areas beyond economic constraints, small freedoms in urban space otherwise determined by rules. Informal urban development should perceive these characteristics as qualities and not simply overplan and build on them. Rather, it is precisely these zones of lower density that can become development poles for civil society activities and projects. Fallow land or dilapidated buildings should be recognized as a silent reservoir for project developers and not closed off too early by planning specifications.

City on Trial

The Eichbaum subway station of the U 18 line from Essen to Mülheim/Ruhr is a location with deficiencies in terms of space and design that will have to be renovated and converted in the near future. The A40 motorway and the route of the U18 line cut through the district of Heißen in two parts. So far, young people are the only ones who identify with this place—but their use (hanging around, wild spraying) further aggravates the problems.

With a participation process, planners became active on site and tried to find out how young people use the site and what suggestions for the use they have. The wish of a youth group of Polish origin from the neighbourhood, who are intensively involved in boxing, was to convert the Eichbaum station into a temporary boxing arena. The young people accompanied the entire transformation process until the arena was completed. The boxing championship took place over two days: Boxing and rapping took place in the boxing ring on the disused platform section. Showboxing, open training for newcomers, cheerleading, concerts, rap battle and of course, official point fights thrilled a colourful audience.

Fig. 10.10 Photo Eichbaum underground station Mülheim/Ruhr, raumlabor/Andreas Krauth

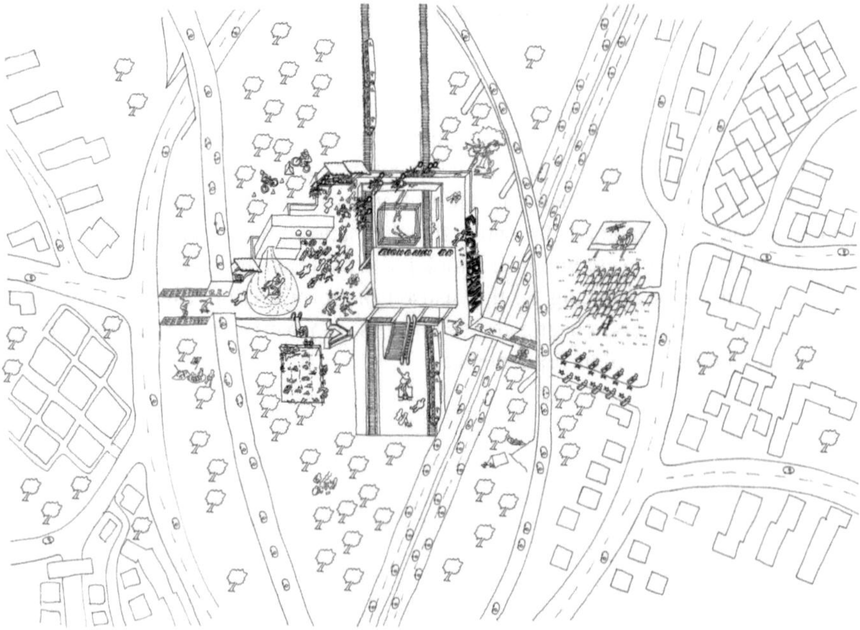

Fig. 10.11 Drawing Eichbaum underground station Mülheim/Ruhr, raumlabor

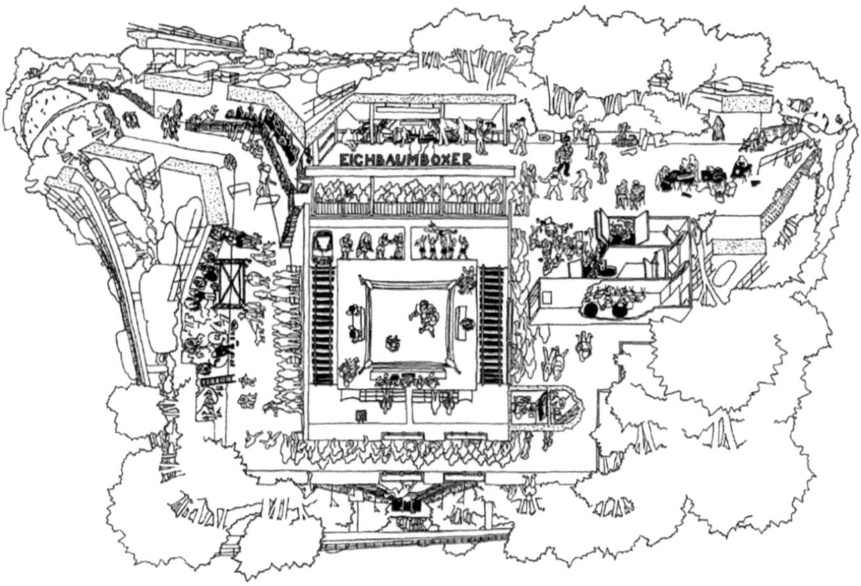

Fig. 10.12 Drawing Eichbaum underground station Mülheim/Ruhr, raumlabor

Planning for Users

Northwest of Munich's city centre is a 20 ha large, partially inaccessible area that will open up to the city for new development. In recent years, temporary users from the art and culture scene have settled in one section. This special profile will be pursued in the urban planning of the teleinternet cafe office and forms the starting point for a site-specific strategy. On the one hand, the design offers the chance to realise residential construction very quickly. This rapid densification relieves the areas already in use and thus acts as a valve for the development pressure that weighs on the entire neighbourhood. This gives the area with the existing temporary uses time for development, which is to be stabilised as an urban experimental space for creative uses in combination with special forms of housing, small-scale manufacturing industry and alternative cultural offerings. The speed of development of the building blocks is individually controlled with the aim of maintaining and carefully developing atmospheric and programmatic diversity.

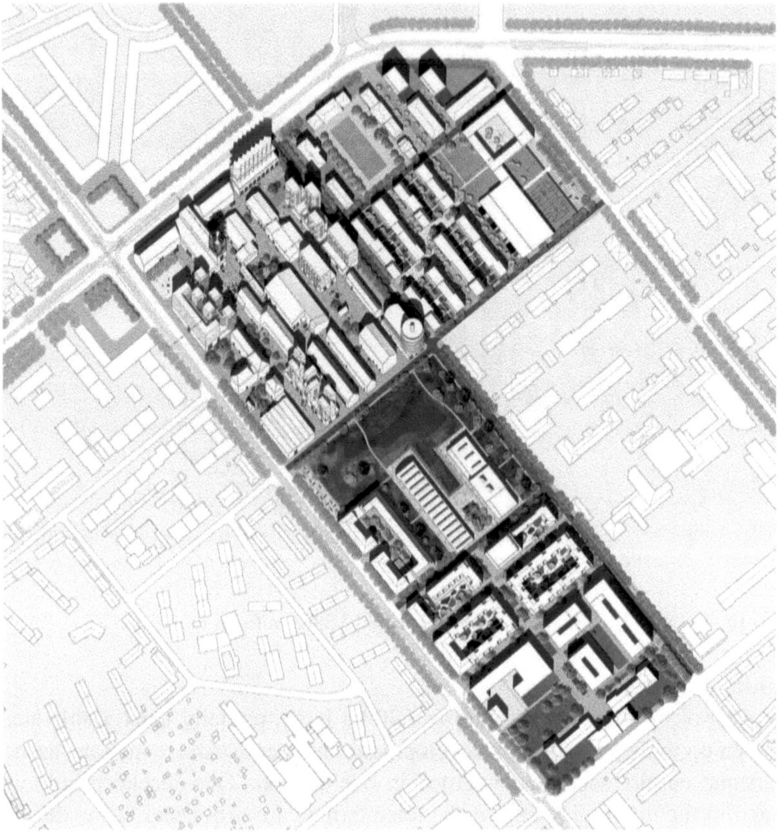

Fig. 10.13 Urban and spatial planning design for the creative neighbourhood Munich (M 1:1000), Teleinternet cafe and Treibhaus

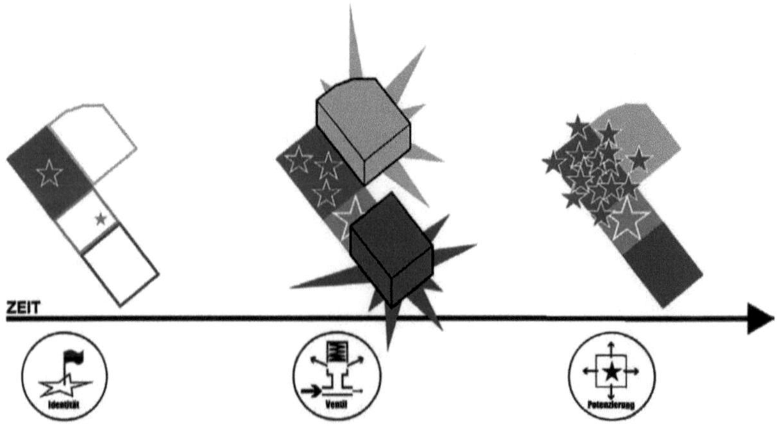

Fig. 10.14 Process development strategy, Teleinternet cafe and Treibhaus

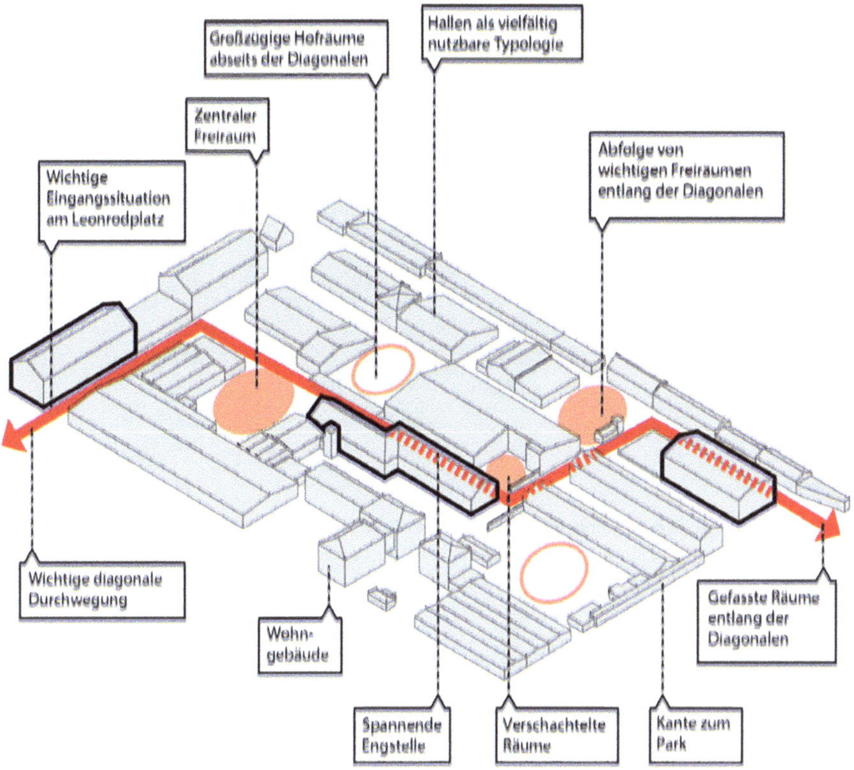

Fig. 10.15 Presentation of the existing situation and site-specific quality, which are to be maintained and further developed in the transformation process. Teleinternet cafe and Treibhaus

Fig. 10.16 The laboratory as an urban experiment space, Teleinternet cafe and Treibhaus

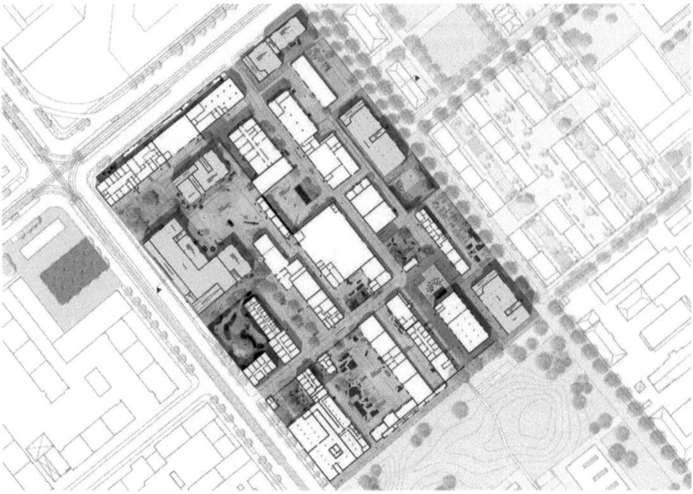

Fig. 10.17 Presentation of a possible development of the laboratory, 2020: Open spaces and vacant existing buildings are capitalised, dilapidated existing buildings are replaced by new buildings in certain cases. Teleinternetcafe and Treibhaus

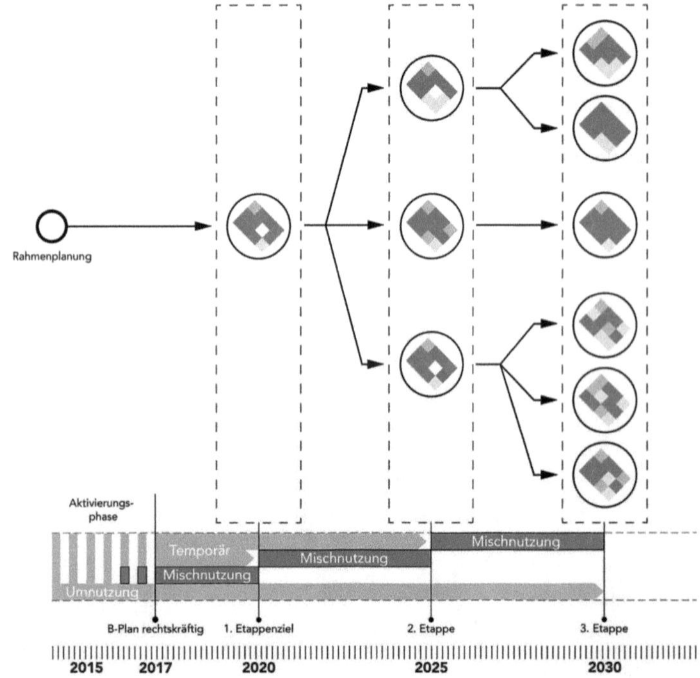

Fig. 10.18 The development of the laboratory leaves many different development possibilities open within a defined framework. After each stage of development, there should be an opportunity to evaluate the steps taken so far, to adapt strategies and to set new priorities. Teleinternet cafe and greenhouse

3 Implementation of Informal Urban Design Using the Example of Dortmund and Acharnes (GR)

Päivi Kataikko-Grigoleit.

3.1 (In)Formal Urbanism: Concepts for Dortmund's Nordstadt

Dortmund's Nordstadt has a metropolitan scale in many places, is characterised by buildings worth preserving and generous green spaces, but—because of its population structure—not only has a charming mixture of different cultures, but also has to deal with many problems that can be successfully solved in a sustainable city. More than half of the population in Nordstadt has an immigrant background and over 40% do not hold a German passport.

Despite a variety of aid programmes—such as "Socially Integrative City", "Urban redevelopment (Stadtumbau) West" or "URBAN II"—problems such as crime, drugs and prostitution could not be driven out of the district. Nonetheless, voices are also being raised that regard the district as a creative neighbourhood. Since rents there are low and there is enough space to rent, many artists and other creative people have settled here. Many students who do not want to be put off by the predominantly bad reputation of Nordstadt can also find beautiful and affordable Wilhelminian style apartments here.

In the following, four approaches to the implementation of (in)formal urbanism that have emerged within the framework of the Master's programme at the TU Dortmund University, specialisation in urban design, are presented.

One of the groups developed the project "Living Room". The proposal is to design the public space in Nordstadt in such a way that it supports self-organisation of skills and talents and the exchange of knowledge in the best possible way.

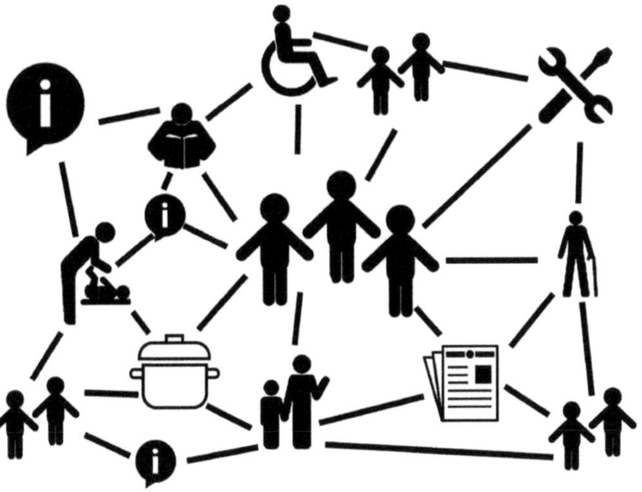

Fig. 10.19 Self-organisation of skills and talents

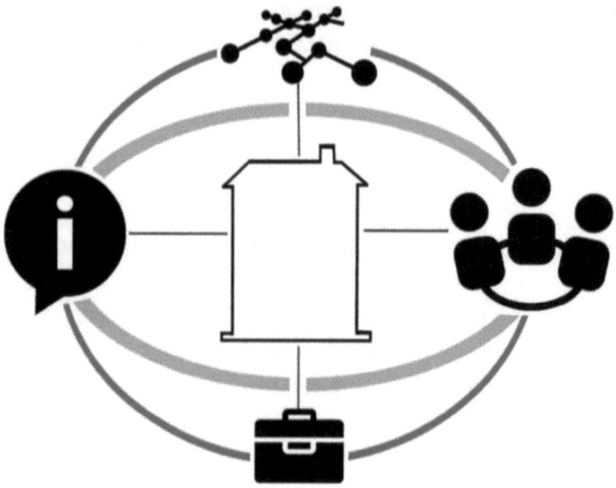

Fig. 10.20 Networking the neighbouring areas of interest

Another group designed the "Dortmund Network" project, which aims to network initiatives and activities and to create a "felt" sense of belonging. An important component of this is the upgrading of public space through user-friendly facilities, which, among other things, should encourage people to discover space.

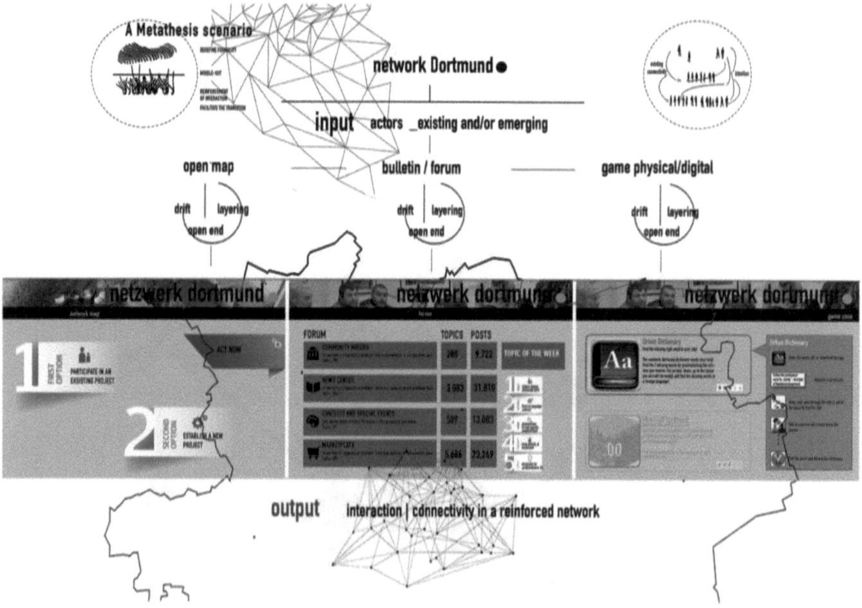

Fig. 10.21 Interactive platform

A similar approach is taken by the "container" proposal, which aims to make information available to Nordstadt residents who do not have access to the Internet or even read it. This involves working with overseas containers that are temporarily placed in the respective action space, thus providing a concrete location for the exchange of information in public space. From there, existing networks are strengthened, social and spatial potential is bundled, and citizens are sensitised to the appropriation of public space by the neighbourhood.

Fig. 10.22 Image campaign

The same basic idea of a low-threshold contact point was taken up by another working group. Here, a van within Nordstadt is to act as a contact, mediation and coordination point under the motto "Seeds to share". In this context, the concept focuses on the networking of different needs and competences that are exchanged there.

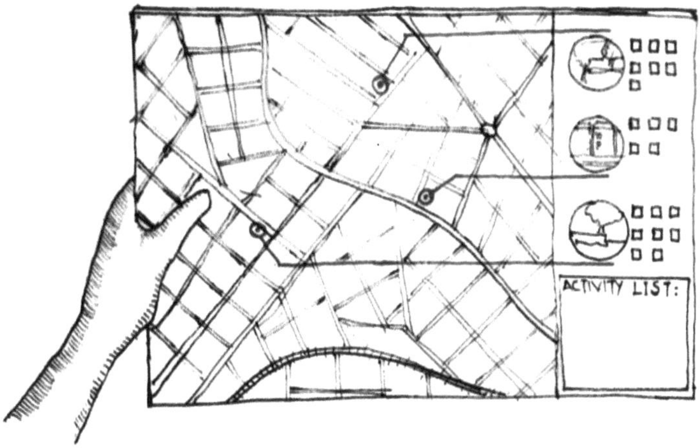

Fig. 10.23 Low-threshold medium for recording activities

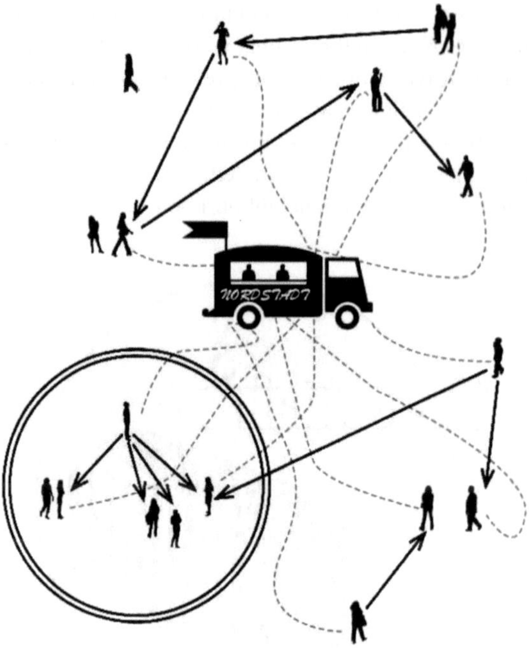

Fig. 10.24 Vans as initiators and contact points for activities

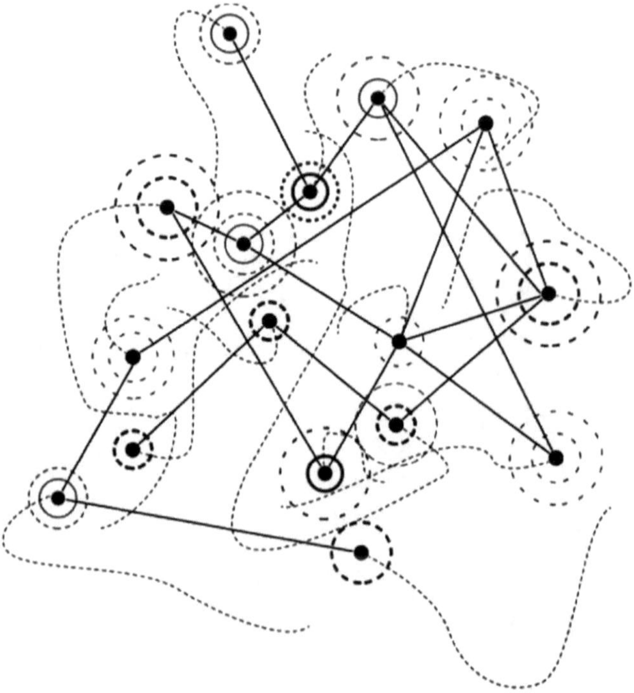

Fig. 10.25 Transmission of the various networks in the districts

3.2 (In)Formal Urbanism: Concepts for Acharnes (GR)

The municipality of Acharnes, north of Athens, which is characterised by both industry and agriculture, has a high degree of mixture of social classes; from the prosperous garden city of Tharakomacedones, to the former Olympic village, to illegally settled and subsequently legalized areas, such as Agios Ioannis Rossos and Platonas, a wide variety of urban compositions can be found. Acharnes is one of the communities in which many different ethnic groups have settled; in addition to the largest immigrant group, the Potos Greeks, the neighbourhoods of the Roma are particularly well organised. Both groups are strongly represented in local politics. In addition, several refugee camps have been established in Acharnes in recent years, although they do not have a good reputation.

In addition to the planned urban expansions, numerous illegal settlements have been built in Acharnes at the same time. Most of them have been legalized afterwards, or their legalization process is being prepared in the meantime. The difficulties are mainly such plots of land (and their buildings), which are either in the flood area or under the high-voltage transmission lines running through the area, as well as the lack of financial conditions and personal motivation.

The students' task was to develop visions for the future and development strategies for the Acharnes municipality, taking into account the many aspects of informal urbanism.

One of the four contributions, "Act in transition", puts forward the thesis that a successful transformation can only be achieved in close dialogue with the residents. Instead of the uncompromising implementation of a major master plan, the paper presents the legalisation process as a series of selective, cautious interventions, which on the one hand are intended to have a positive impact on the entire district, and on the other hand to act as a driving force for further implementation measures.

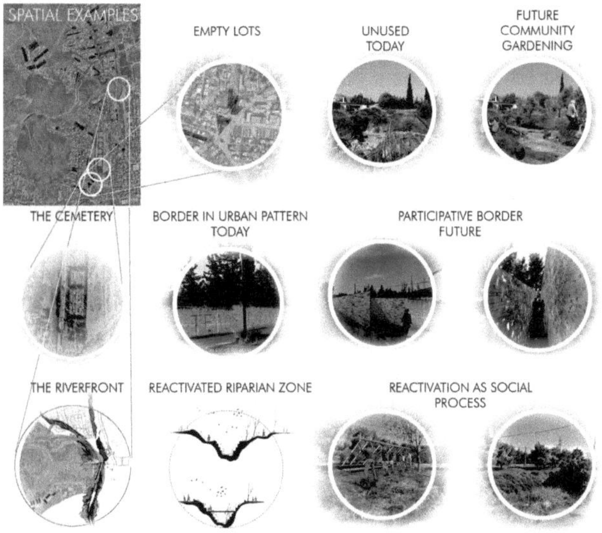

Fig. 10.26 Selective intervention in the urban district

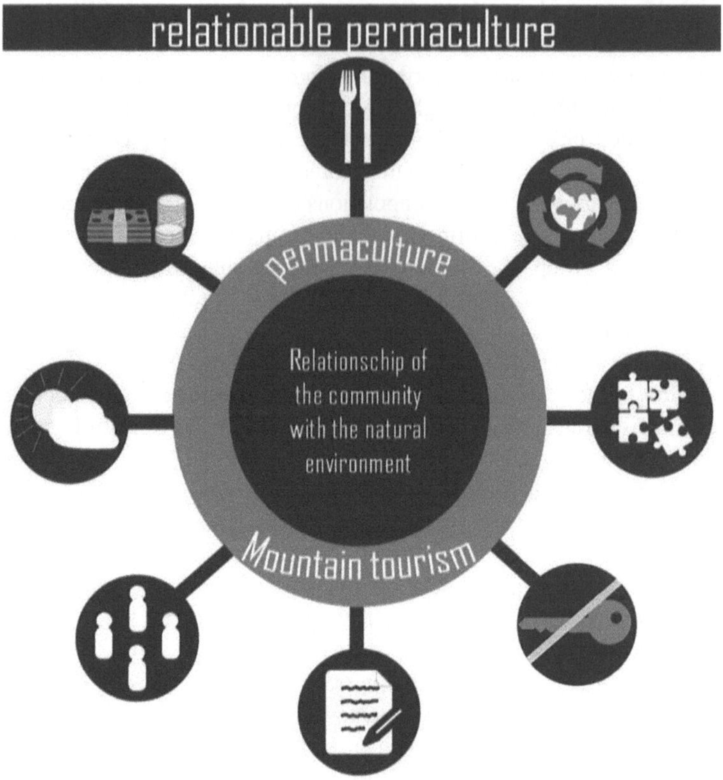

Fig. 10.27 Careful cultivation of existing resources

A further work, "Relational permaculture", presupposes that a common, supporting spatial framework must be created for the cohesion of the urban districts with their different, sometimes conflict-ridden ethnic groups. The nearby nature reserve is considered suitable as a basis for this. This should not only be used for recreation, but it is proposed to carefully cultivate the resources that are already available there (medicinal plants, etc.). The problem of flooding in the riverbank elk is solved by innovative rainwater harvesting.

Since some of the buildings in the two informal settlements of Agios Ioannis Rossos and Platonas are located directly under high-voltage power lines or on the banks of the river that are at risk of flooding, the work "Connecting Living Space" initially states that a successful legalization of the two settlements can only be planned and carried out with the close involvement of those affected. The diversity of the population structure represents both a potential and a challenge. The "Connecting Living Space" is being developed as a mobile information, health and education service that connects not only the two disadvantaged neighbourhoods, but also the city districts beyond.

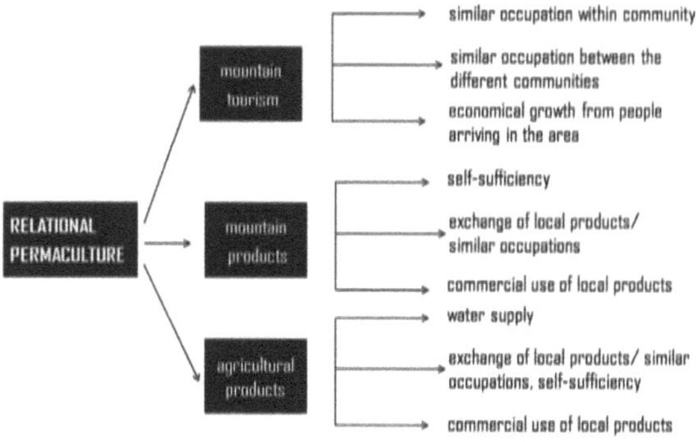

Fig. 10.28 Careful cultivation of existing resources

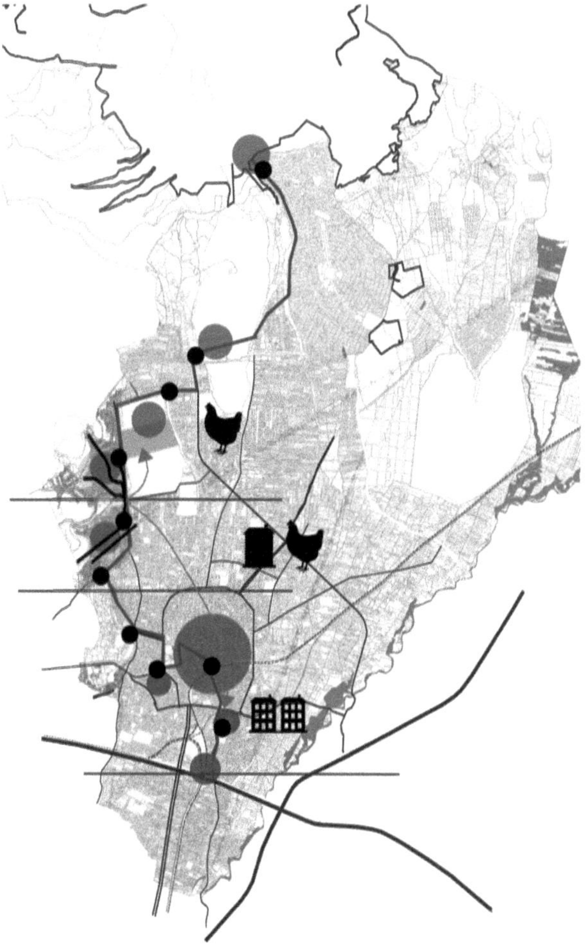

Fig. 10.29 Mobile information, health and education services

The project “AGROSCAPE/WATERSCAPE—envisioning landscapes” also offers a multi-layered future vision for urban development in Acharnes and proposes an implementation of an innovative combination of agricultural production/research with water management. By combining these two components, green and blue infrastructure, knowledge-based economic uses as well as a distinctive identity should be created on the site.

First of all, the proposal provides for the establishment of a department of the Agricultural University in Acharnes. Care should be taken to ensure that the university does not emerge as an exclusive institution, but that the local population is intensively involved in the everyday activities of the university. For example, experimental workshops could be offered in the field of agriculture.

Through innovative water management, rainwater should not only be developed as a resource in agriculture, but grey water should also be used in local households.

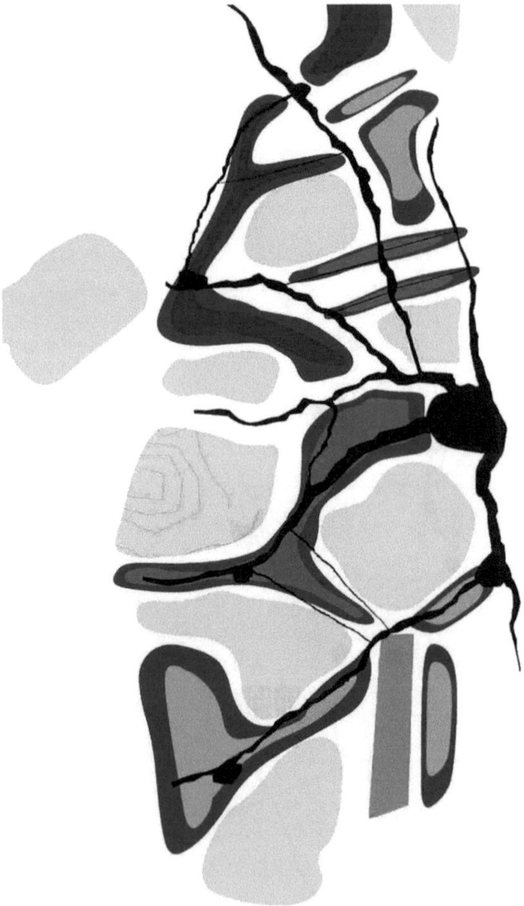

Fig. 10.30 Urban planning vision

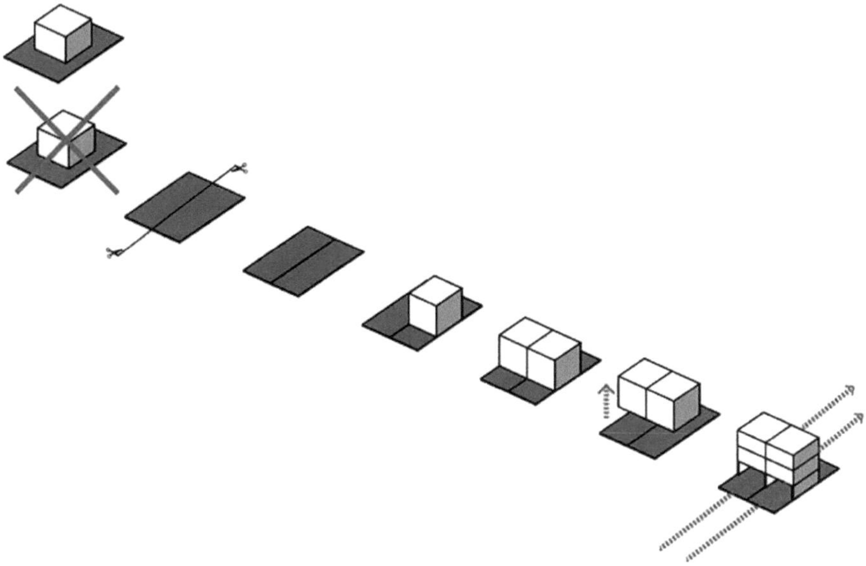

Fig. 10.31 Acharnes as a trial field for new house typologies

Literature

1. Understanding of Informality

Willinger, Stephan (2014): Governance des Informellen.In: Informationen zur Raumentwicklung, Heft 2.2014

2. Plan and Design with Informality

Oswalt, P., Overmeyer, K., Misselwitz, P.: urban catalyst (2013): Mit Zwischennutzungen Stadt entwickeln

Stadtstattstrand (Verf.), BBSR (Hrsg.) (2016): Die Freiraum-Fibel. Meine Stadt–mein Block–mein Garten

Urban Catalyst studio (Verf.), BBSR (Hrsg.) (2016): Jugend. Stadt.Labor. Offene Räume für die Stadtentwicklung

Michael Ziehl, Sarah Oßwald, Oliver Hasemann, Daniel Schnier (Hrsg.) (2012): Second Hand Spaces. Über das Recyceln von Orten im städtischen Wandel

3. Implementation of Informal Urban Design Using the Example of Dortmund and Acharnes (GR)

Päivi Kataikko-Grigoleit, Panagiota Koutrolikou, Konstantinos Moraitis, Christa Reicher, Juliane Stegner (Hrsg.) (2017): (In)formeller Urbanismus

Regional Urban Design 11

1 Urban Networks and Agglomerations

When we talk about the twenty-first century as the urban age—the century in which the majority of humanity lives in cities for the first time—we usually focus on the rapidly growing mega-metropolises: Mumbai, Cairo, Sao Paulo, Shanghai, Istanbul, or the traditional global cities such as New York, London, Tokyo, Moscow. However, such multi-million cities are only one aspect of the global urbanisation process, because the majority of city dwellers do not live in the mega-cities, but in small, medium-sized or large cities below the five million mark, which however—and this is the characteristic of this urban age, especially in the European context—are increasingly organizing themselves into large agglomerations. This has given rise to spacious yet densely populated urban regions, some of which can do without megacities altogether: the Dutch Randstad, the Limmat valley in the Zurich area, the Veneto (with Padua, Venice, Verona, etc.), the Rhine-Main area and the Ruhr area, which is almost paradigmatic for the polycentric agglomeration type in the German-speaking world, are well-known and often described examples.

Regions consisting of a large number of individual cities are therefore not a special case. City networks are even considered a promising model for present and future urbanisation processes. What may be convincing empirically and theoretically, however, raises some practical questions: How to design such large-scale urban landscapes, according to which principles and by which means should urban development take place on a regional scale?

C. Reicher, *Urban Design*, https://doi.org/10.1007/978-3-658-34370-5_11

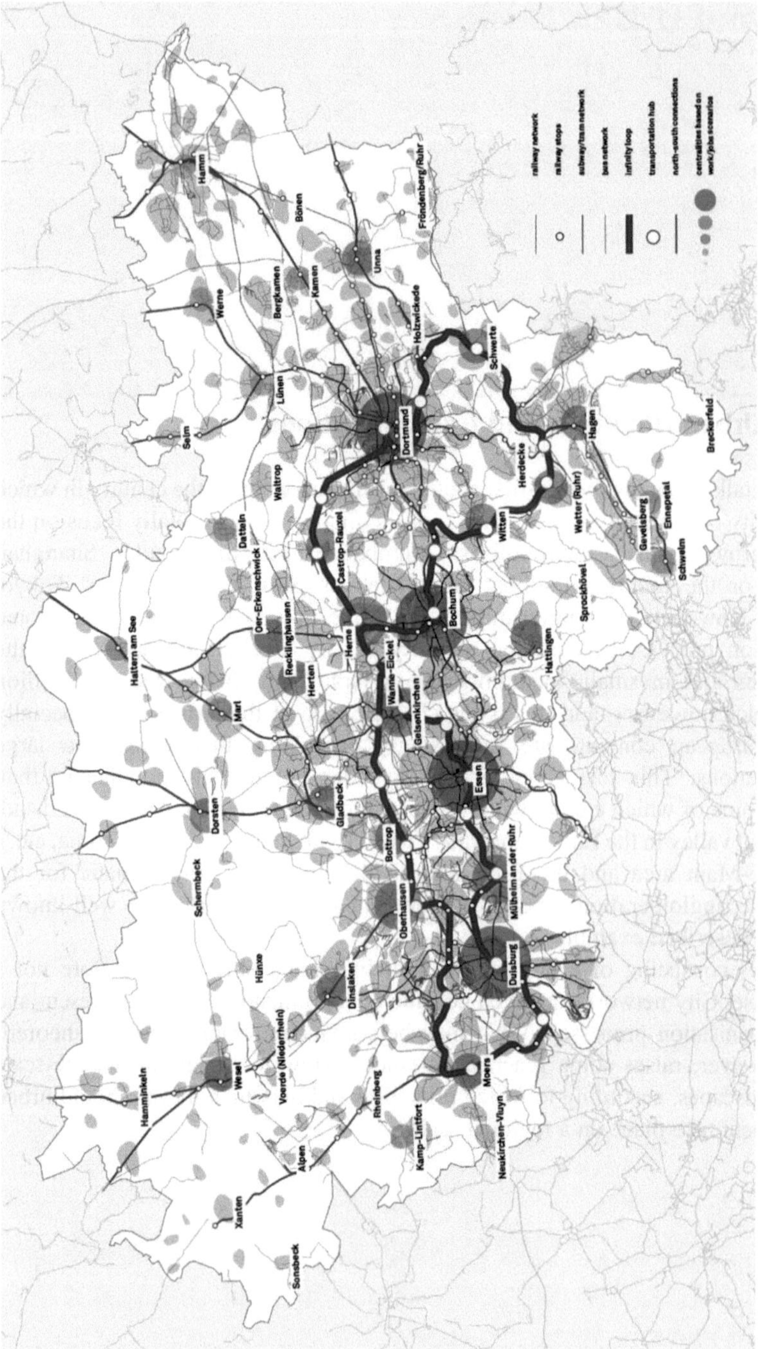

Fig. 11.1 Polycentric spatial structure of the Ruhr Area, RHA architects and planners, Aachen/Dortmund

2 Regional Urban Design Taking the Agglomeration Ruhr as Example

The example of the Ruhr area is a good illustration of what a systematic analysis of an agglomeration looks like and how regional concepts can be developed based on specific laws. Even if this approach cannot be transferred to every region, it can provide orientation for other regional urban design concepts.

Urban design visions for the Ruhr Area necessarily refer to the regional scale, and are therefore not the same as urban planning visions for this or that city within the region. The focus is on the "architecture" of this region as a whole, that is what it supports and represents.

2.1 Analysis of Spatial Characteristics

Polycentricity

A characteristic feature of the Ruhr region is its polycentric spatial structure. Although the coexistence of cities within a region is not a unique selling point of the Ruhr Area, this enormous density of sovereign large, medium-sized and small cities is without precedent in Germany and possibly also in Europe. The Ruhr Area is characterised as a polycentric region, as it is formed by several city centres of different configuration and importance. Across all spatial levels, a heterogeneous network of open and built-up spaces extends, forming a network-like structure without dominant centres. Polycentricity is now considered a good prerequisite for sustainable urban and regional development.

Breaks and Caesuras

A dense network of traffic and industrial infrastructures runs through the entire Ruhr area and forms breaks in the settlement and landscape area. Due to the industrial past, which has led not only to large industrial areas but also to a massive expansion of the one-way network, the separating effects in the settlement structure can still be seen today. In the second half of the twentieth century they were supplemented by new breaks, a dense network of motorways and wide urban roads. Today, many railway lines have lost their importance for freight transport and represent potential for a new leisure route system or a regional cycle path. The same applies to the derelict industrial sites, which in the dense settlement structure are important potential areas for new uses.

Fig. 11.2 Broken settlement area. Layers of a region, 2011

Inner Edges

The region as a whole has a rather flat density (a lot of housing development, few decidedly metropolitan neighbourhoods) and is therefore to be classified more as a post-suburban than a classic metropolitan conurbation. Characteristic is the extreme interlocking of settlement and open space, not only on the edges of the conurbation, but already in the actual urban fabric. This multitude of inner city edges—lines of contact between built-up and undeveloped space—is a fundamental principle of regional spatial structure. Because such landscape borders have always been attractive to people—settling on the edge is possibly a kind of anthropological constant—this represents enormous potential for the future development of the Ruhr region.

Fig. 11.3 Lines of movement of residential areas with green spaces; Gelsenkirchen-Herne area. Layers of a region, 2011

Ethnic and Social Mosaic

The analysis of the population structure is an important basis for every urban design concept. The population of the Ruhr area is strongly influenced by people with a migration background, due to the influx of workers in the last century into the largest industrial region in Europe. Today, a clear overlap of ethnic and social segregation is discernible, that is, areas in which a high proportion of migrants overlap with a high percentage of socially disadvantaged people. Overall, a small-scale pattern and a colourful coexistence of different locations and groups of inhabitants can be identified. The task of urban design is to provide assistance in order to prevent segregation tendencies from increasing.

2.2 Specific Urbanity

Ways into the spatial future of the Ruhr Area must be derived from an analysis of the development conditions and potentials. A number of specific peculiarities have crystallised in the overall picture, which can be described as "Ruhrbanity", that is, the specific urbanity of the Ruhr Area.

Future urban development strategies for the region must be developed with the knowledge of such special features and characteristics. Visions, strategies and urban design around the "inherent logic of the cities", as the urban sociologists Helmuth Berking and Martina Löw call it, should not only focus on the existing character of the Ruhr Area ("Strengthening strengths"), but also include new, distinctive qualities based on new urban challenges.

2.3 Regional Guard Rails

On a regional scale, there are two main strands of content that can be used to develop the regional structure and make it sustainable: landscape and mobility.

Landscape Guard Rail

The Ruhr area has a wide variety of landscapes. This diversity can currently only be experienced to a limited extent, although many topographical, hydrological and geological features are important for the appearance of the Ruhr area. However, such features are often overshaped or superimposed.

On a regional scale, in addition to the remaining extensive open spaces and the newly created post-industrial landscapes, the regional green corridors, which run predominantly in a north-south direction, are of particular importance. Together with the open space belts running in an east-west direction along the Ruhr, Emscher and Lippe rivers, they form the primary open space framework of the Ruhr area. Originally, the green corridors were created at the beginning of the twentieth century to prevent the Ruhr from growing together completely. In the course of the International Building Exhibition (IBA) Emscher Park they were further developed and, in many places, provided with new programmatic tasks that go beyond pure open space protection. However, their development potential is far from exhausted today, both in the programmatic sense (with new tasks, new designs, new forms of use, etc.) and in the territorial sense. So far, the green corridors have only structured the core zone of the Ruhr area, while its outer edges are insufficiently integrated or not integrated at all.

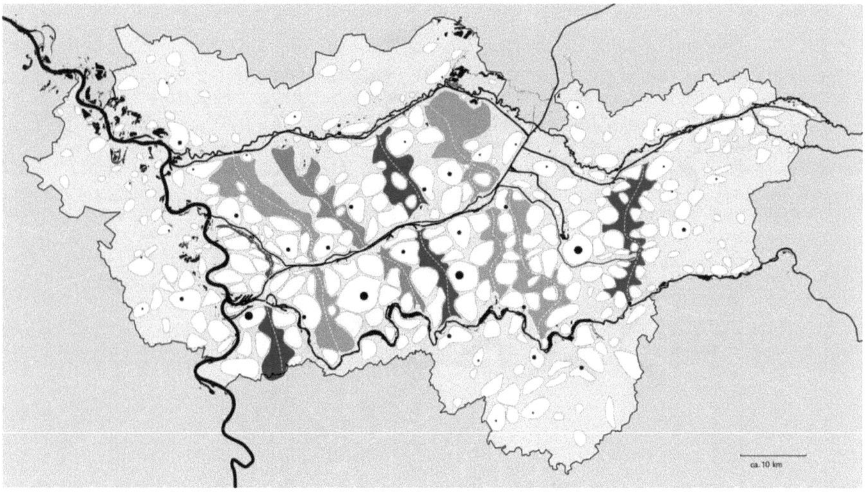

Fig. 11.4 Nameless green corridors/green corridors thinking ahead, RHA architects and planners

Extending the green corridors to the outer edges of the region and integrating open spaces there into the regional system of networked open spaces can thus be an important contribution of regional urban design.

The hitherto nameless green corridors (they only have abbreviations such as A, B or C) are also to be given names that create identity, so that they can actually be understood and experienced as landscape spaces.

The green corridors must be more closely networked and qualified with the adjoining landscape areas. Four different zones are defined in the open space: Core zone, landscape zone, hem/inner edge and natural taboo zone/natural space.

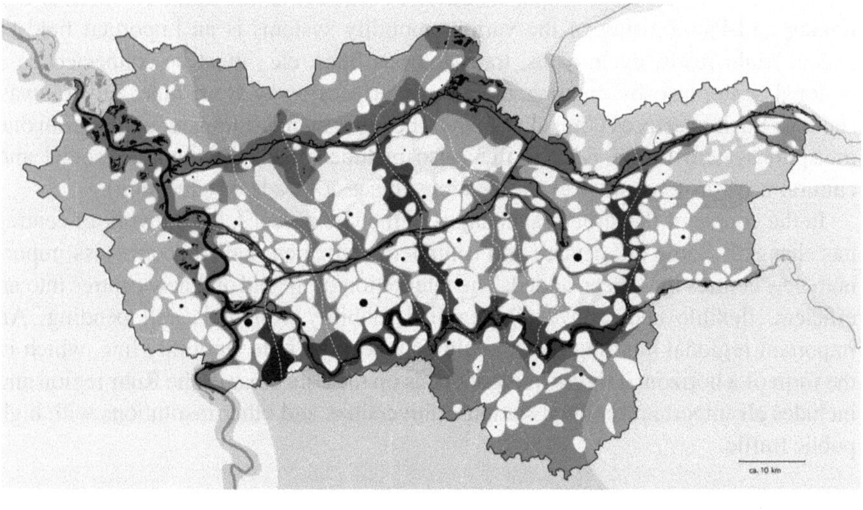

Fig. 11.5 Green corridors networked with adjacent landscape spaces/green spaces

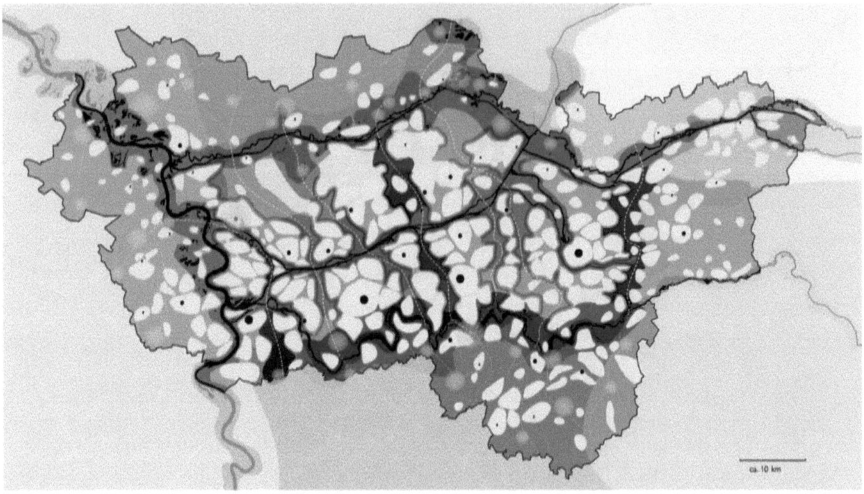

Fig. 11.6 Green corridors with adjoining landscape spaces/green spaces qualified, RHA architects and planners, Aachen/Dortmund

Mobility Guard Rail

The extensive spatial structure of the region has created a conglomerate of different transport infrastructures. A dense network of motorways and motorway-like traffic routes, a hardly less dense network of railways, stops, stations with numerous local and long-distance connections for freight and passenger transport, a large number of municipal public transport companies, each of which operates its own underground, tram and bus network.

The improvement of transport infrastructure, but above all its adaptation to new mobility needs and changing spatial structures is a central task in the Ruhr Area. The linking and harmonising of the various mobility systems is an important field of action: Main roads, cycle paths, tram and bus lines, etc., must be connected on a regional scale, also by creating mobility interfaces away from inner city railway stations. Such interfaces ("hubs") can even be more than just transfer points from one transport system to the next, if they also include other, for example, social and cultural uses (as temporary event locations, places of trade or education).

In the course of the structural changes of the last decades, the network of centres has changed: Some old centres and centres of settlement have become less important, new centres have been added. The integration of the old and new centres into an efficient, flexible and, above all, regional mobility system is still pending. An important regional building block could be a new "infinite" mobility line, which in the form of a horizontal figure of eight opens up the core areas of the Ruhr region and includes all important regional stations, universities and other institutions with high public traffic.

In principle, various models are conceivable for this new mobility line ("Infinity Line"): a rail-bound express railway on existing tracks and routes or a flexible express bus line which uses the existing road network but can react more easily to changes in the spatial structure, or a combined mobility system of bus and bicycle. For example, the "Infinity Line" as a new mobility guard rail could improve regional accessibility in the long term and make its catchment area a priority area for development projects.

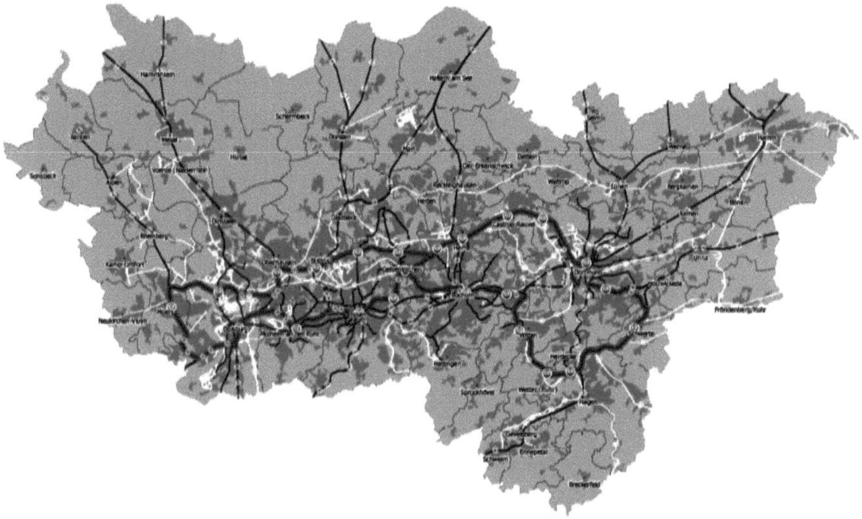

Fig. 11.7 Mobility line "Infinity Line"

3 Urban Challenges Between Neighbourhood and Region

The future urban development objectives will not only highlight the special features of a region, but will have to provide answers to the general challenges that more or less apply to all cities and urban regions—regardless of whether cities are "growing" or "shrinking", regardless of whether cities want to become "metropolises" or explicit "green cities". Often such simplistic dualisms do not face each other anyway, but are part of a comprehensive and contradictory transformation process that cities undergo throughout their history.

In many European cities and agglomerations, two processes can currently be observed which at first glance appear contradictory: reurbanisation and regionalisation. Inner cities and city neighbourhoods close to the city centre are becoming more attractive, especially for people who are (possibly rediscovering) the advantages of urban life. There are many reasons for this return to the core cities as a place to live; however, it is linked in particular to changes in the world of work or the trend towards having to combine living and working much more effectively than living in the suburban fringe areas of a conurbation allows in most cases. Here, well-organised districts with a wide range of functions have a clear advantage. At the same time, a further regionalisation of the urban environment is taking place: This does not only mean the extended scope of action of the inhabitants living in the different cities of a region, but also the development of new, mixed-use locations away from the old city centres. These can be university locations that are being enriched with new uses (living, working, leisure), old industrial sites that are developing into cultural and creative neighbourhoods, or even the first generation of shopping centres outside the core cities that—now easily accessible by public transport—are now being "urbanised", for example with sports, cultural or educational facilities. The spatial-structural consequences of both trends, reurbanisation and regionalisation, are foreseeable, should they prevail: a stronger focus on old and new cores within the urban regions, combined with the task of linking these cores as efficiently as possible. Urbanised regions or urban landscapes have clearly different characteristics and their very own logics and thus demand new forms of regional urban planning, beyond familiar patterns and procedures.

Literature

1. Urban Networks and Agglomerations

Reicher, C. (2010): Leitlinien für einen regionalen Städtebau RUHR. In: Charta Ruhr: Denkanstöße und Empfehlungen für polyzentrale Metropolen. Klartextverlag Essen

2. Regional Urban Design Using the Example of the Agglomeration Ruhr

Reicher, C; Kunzmann, K.; Polivka, J. U.A. (Hrsg.) (2011): Schichten einer Region. Kartenstücke zur räumlichen Struktur des Ruhrgebiets. Jovis Verlag, Berlin

Leggewie, C.; Reicher C.; Schmitt, L. (Hrsg.) (2016): Geschichten einer Region. AgentInnen des Wandels für ein nachhaltiges Ruhrgebiet, Kettler Verlag, Dortmund

Polivka, J.; Reicher, C.; Zöpel, C. (Hrsg.) (2017): Raumstrategien Ruhr 2035+. Konzepte zur Entwicklung der Agglomeration Ruhr, Kettler Verlag, Dortmund

RHA Reicher Haase Assoziierte (Hrsg.) (2017): Next Ruhr. Entwicklungsperspektiven für das Ruhrgebiet, Dortmund

3. Urban Challenges Between Neighbourhood and Region

12 Selected Task Fields

1 Solar and Energy Efficient Urban Development

General Conditions

Experts predict a further warming of 1.4–5.8 degrees Celsius over the next 100 years, depending on population growth and energy consumption. Even with an assumed moderate temperature increase of 2 degrees Celsius over the next few decades, increasing heat waves, longer periods of drought as well as extreme precipitation, seasonal storms and flooding must be expected in Europe due to the expected rise in temperatures. These threatening climatic extremes require answers and new solutions in urban design: increased use of renewable energies and adaptation of urban structures to the changed framework conditions, that is, resource efficiency: this also means avoiding negative environmental impacts and activating all existing natural resources with serious consequences. Two approaches are being pursued in urban design

- CO_2 reduction strategies and
- Adaptation strategies, in the sense of adapting urban structures to climate change.

Review

The relationship between man-made and the natural environment can be traced far back in all cultures. Vitruvius, for example, recommends that when choosing a location, attention should be paid to favourable orientation to the cardinal points and natural lighting.

C. Reicher, *Urban Design*, https://doi.org/10.1007/978-3-658-34370-5_12

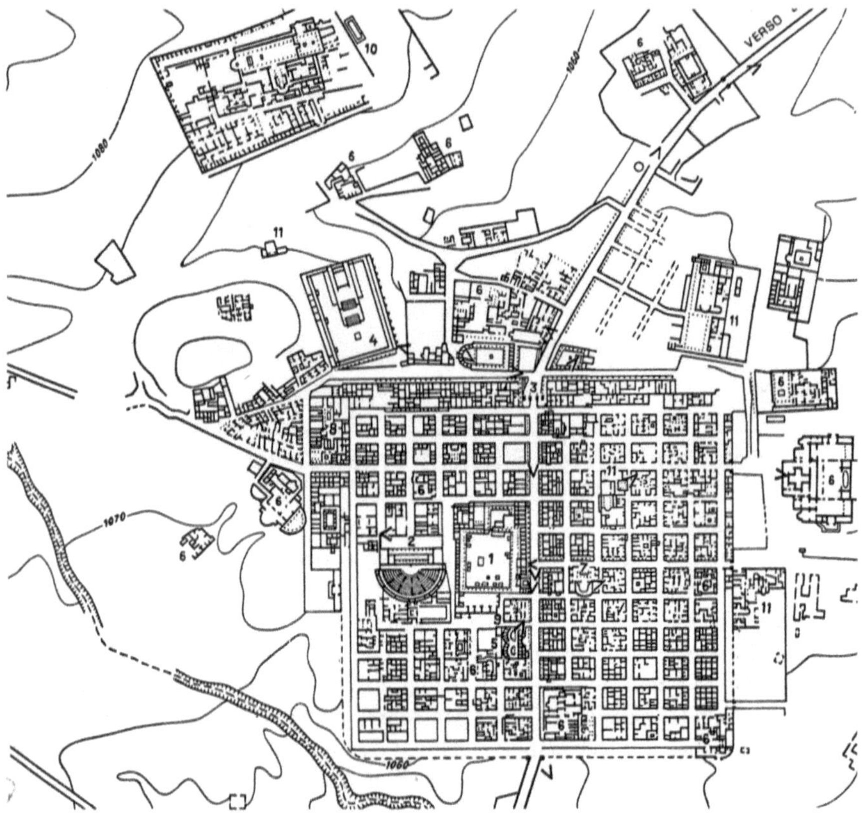

Fig. 12.1 Model of the Algerian city of Timgad

In the nineteenth century—caused by unhealthy living conditions in densely industrialised cities—the call for "healthy green" was heard. One example of this is the garden cities created by Ebenezer Howard. Urban expansion concepts such as Ildefonso Cerdá's plan for the urban expansion of Barcelona also take up the idea of more green in the city (see Chap. II Ildefonso Cerdá). Initially, the focus is on hygienic considerations, but these also lead to a new approach to dealing with nature. In this context, nature is seen as a capital that people can acquire for their physical and mental health, provided that they preserve it and use it sparingly. In this sense, the allotment garden movement was born, which was propagated and promoted by the doctor D.G.M. Schreber in the middle of the nineteenth century next to public playgrounds, at that time a revolutionary idea.

The housing reform movements of the 1920s attempted to address the problems of mass housing construction by means of a strict division of functions and rationalisation. Here, too, the starting point was the demand for light, air and sun for the entire urban population. Sunlight, natural ventilation and greenery were seen as fundamental factors for healthy human living conditions. This desire is most consistently expressed in the southwest-facing row development.

The Charter of Athens (1933) continues these thoughts and formulates a more far-reaching idea of the new city that includes all areas of life, living, working, education, leisure and recreation. Starting from the needs of working people, it calls for a functionally structured city. This includes, for example, high-rise residential buildings to allow more freely accessible green spaces.

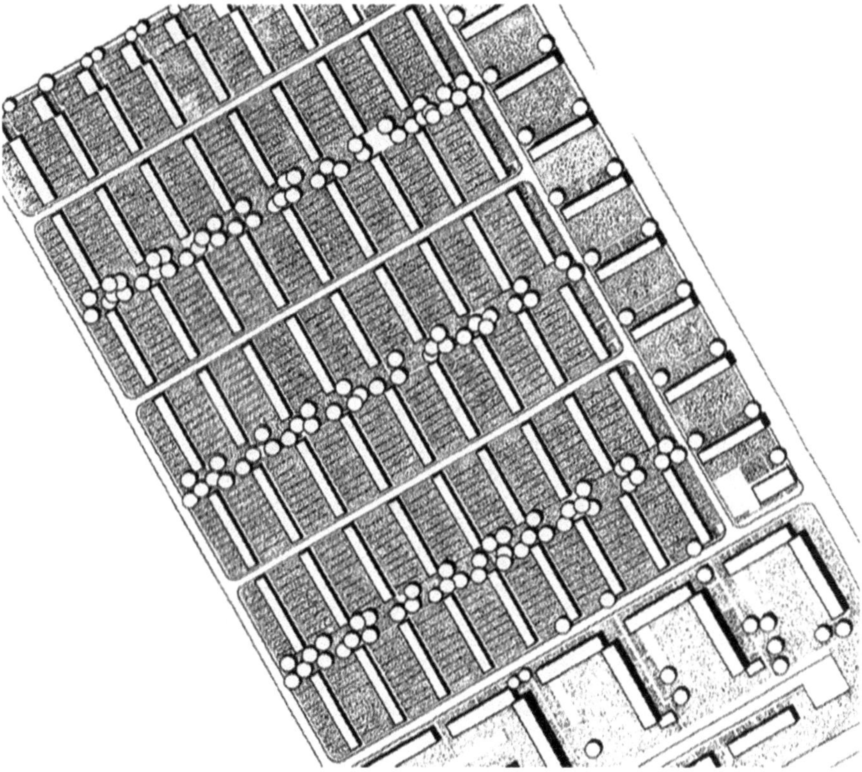

Fig. 12.2 Frankfurt Westhausen by Ernst May, 1929–1931

It was not until after the Second World War that another topic gradually gained importance in architecture and urban planning: securing energy supplies. In the 1950s and early 1960s—in addition to the beginning peaceful use of nuclear energy—serious research was conducted for the first time into energy sources that could one day replace the finite fossil energy reserves. Modern technology and science are being vigorously applied to explore solar, wind, geothermal and other forms of renewable energy. It is a time of technological optimism, when there is a general belief that any problem can be solved with the help of modern science, such as Fuller's idea of building a dome over Manhattan to create a pleasant, balanced climate in the city. At the same time, the first critical report by the Club of Rome, a grouping of eminent scientists, politicians and business leaders, has received a worldwide response. The report "The Limits to Growth", a comprehensive review of the state of the world, is published in German in 1972.

The oil crisis in the 1970s became the impetus for a further wave of research into non-fossil and renewable energy sources. The motivation is first of all to find a way out of the dependence on oil sources in distant countries and again to worry about the end of oil reserves. There is no direct impact on urban planning, but a social awareness of the earth's vulnerability is growing. Around the same time, Paolo Soleri coined the term "Arcology" (architecture+ecology) and initiated the construction of the car-free, solar-powered town of Arcosanti in Arizona.

The reactor accident in Chernobyl in 1986 and renewed negative ecological forecasts, especially with regard to global warming in the early 1990s, are the reason for the first World Environment Summit in Rio de Janeiro in 1992. There the term "sustainability" is coined. Sustainable development preserves the general quality of life, ensures continued access to natural resources and prevents permanent environmental damage. The danger of CO_2 for the Earth's climate is becoming a focus of attention. A new dimension is being reached which forces us to act.

In the current discussion on sustainable urban design models, complex concepts are being sought that harmonise technical achievements with ecological, social and spatial aspects. This requires integrative, interdisciplinary solutions. Basically, strategies for dealing with existing buildings, new buildings or the development of new neighbourhoods and urban districts must be differentiated.

An overall energy concept must take into account urban design aspects and structural organisational principles as well as energy requirements and bring them into balance.

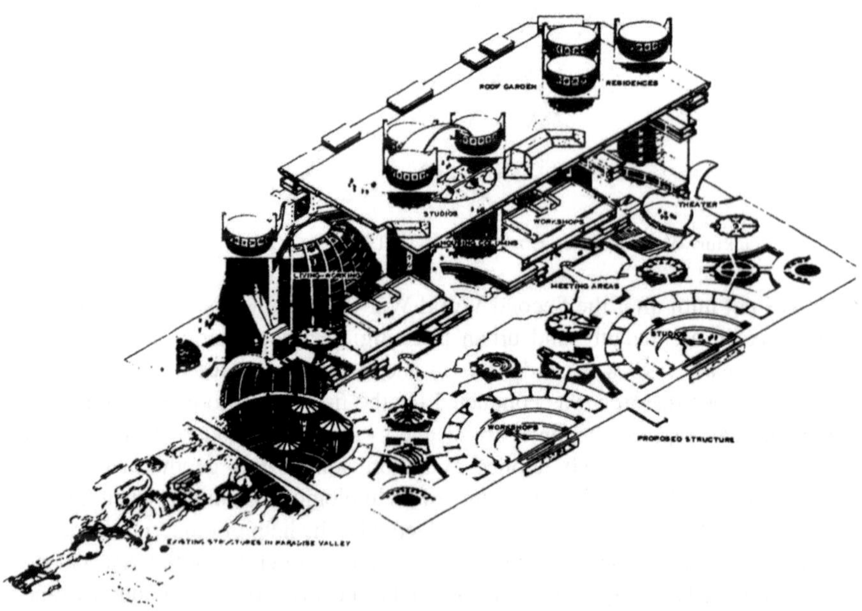

Fig. 12.3 Car-free town of Arconsanti, Arizona, USA, Architects Paolo Soleri

Fig. 12.4 "Help through green". Cover pages of early issues of the Working community for garden and landscape culture, July 1952, 1956

Fig. 12.5 Geodetic dome by Fuller, Expo Montreal 1967

Fig. 12.6 Example conservation of the stock, church tower, Rostock

Fig. 12.7 Example Technical modernisation, Vitus solar house, Rostock

Fig. 12.8 Example of revitalisation, Aidhausen

Solar Urban Design in Existing Buildings

Existing urban structures bring—due to their respective context of origin—different prerequisites for the refurbishment including solar technology. In principle, solar technology retrofitting in the existing structures should be carried out in as compatible a manner as possible. The possible scope of intervention is determined by the design quality of the existing building stock:

The more valuable the historical, architectural and/or urbanistic quality of the existing building, the more carefully the intervention in the form of solar energy upgrading should be carried out. The use of solar energy and the interests of monument protection are not in principle mutually exclusive. The lower the architectural and urban design quality of the existing building, the greater the scope for refurbishment and solar energy conversion measures.

The following graduated strategies can be distinguished:

Conservation of the Built

If the existing buildings are particularly worthy of preservation as part of the architectural heritage or are even protected as historical monuments, energy measures should have as little effect as possible on the appearance of the building or the neighbourhood. Internal insulation and integrated photovoltaic systems or alternatively other regenerative heat sources such as geothermal energy can be used.

A prerequisite for this cautious intervention is an intensive analysis of the existing building and a careful and balanced design of possible energy-related measures.

Technical Modernisation
Existing older buildings will be modernised, refurbished and supplemented with new building elements such as balconies, new windows, insulating facade cladding and solar energy elements. This changes the external appearance. The degree of change depends on the type and the way the respective measure is handled. Care must be taken to ensure that in this case, too, good architecture is created in the urban context.

Upgrading Renovation
If the existing building stock is problematic from an urban design point of view or due to the condition of the building fabric, a more intensive intervention can be carried out. This means demolition and replanning, if necessary, not only of the buildings but also of the urban situation. An upgrading redevelopment can consciously lead to a changed or new appearance of the urban space conditions as well as the building character.

Which path is taken for energy upgrading and refurbishment depends on the existing urban development situation and the building substance. An intensive analysis of the sensitivity to intervention in terms of design and urban planning is an important prerequisite for defining a sensible renovation strategy.

Solar Urban Development in New Buildings
In order to be able to make optimum use of solar energy, the location of the buildings in relation to the points of the compass and thus to the sun is also important. Since the optimization of housing construction in the 1920s, the orientation and opening of living rooms and balconies to the south, southwest or west has been required. After-work sun is the keyword. Buildings are aligned and floor plans are zoned accordingly. Bedrooms and often also children's rooms are then usually located on the east or even north side. The orientation of the buildings therefore has a significant effect on the organisation of the floor plan.

Orientation of Buildings in Solar Urban Design
The inner zoning of the building also supports a positive energy balance. The south orientation of the buildings allows an opening to the south and an extensive closing to the north, which means a great gain in passive solar energy use.

Energy buffer zones such as conservatories—as non-heatable zones between outside and inside—also contribute to a good energy balance. There are excellent examples, especially in multi-storey residential buildings, which combine a purely south-facing orientation with a zoned floor plan in a sensible way.

For a southward orientation, the urban arrangement typology of the line is most suitable, which—often independent of the surrounding context—can be aligned according to the desired compass direction. The advantage of this arrangement pattern is offset by a number of disadvantages:

- the one-sided orientation/north room or less deep and therefore less economical buildings and larger plots of land,
- a form of arrangement that allows for little urban quality, is rather anti-townish and is hardly oriented towards communication and community.

However, buildings and floor plans oriented to the west and east are not unfavourable either: all rooms of an apartment are directly exposed to the sun during the course of the day. The sum of the irradiation time from early morning to late evening is, apart from the winter quarter, more favourable and compensates for the lower intensity. By adding solar technology, the energy balance of both orientations can be further improved.

The combination of both building orientations, also taking into account solar energy issues, provides a sufficient instrument to develop urban qualities of the urban space.

The Influence of Roof Shapes on Solar Energy Efficiency

The effectiveness of the use of solar technology is determined by the global radiation, which depends on the respective geographical location. The efficiency of the energy yield depends, in addition to the type of system, on the orientation and angle of the elements.

A basic distinction must be made between the

- Generation of electricity by means of photovoltaic elements (active solar energy use) and the
- Production of hot water using collectors (passive solar energy use).

The design possibilities in dealing with solar elements are manifold and are constantly increasing due to further developments. A distinction must be made between integrated systems in roof and wall and independent, applied systems. These additional technical elements can be inserted in a restrained manner, but can also be deliberately staged as a creative, technical expression of a building. They can also be used in a sensible interplay of passive and active solar energy use.

On flat roofs the photovoltaic modules or collectors can usually be positioned as desired. Of all pitched roofs, the console form is theoretically the most effective, because it offers a good ratio between yield and utilization of the floor space of the house. However, it is common practice to tilt these roofs so that as much sunlight as possible reaches the living areas. The resulting inclination is unfavourable for solar elements.

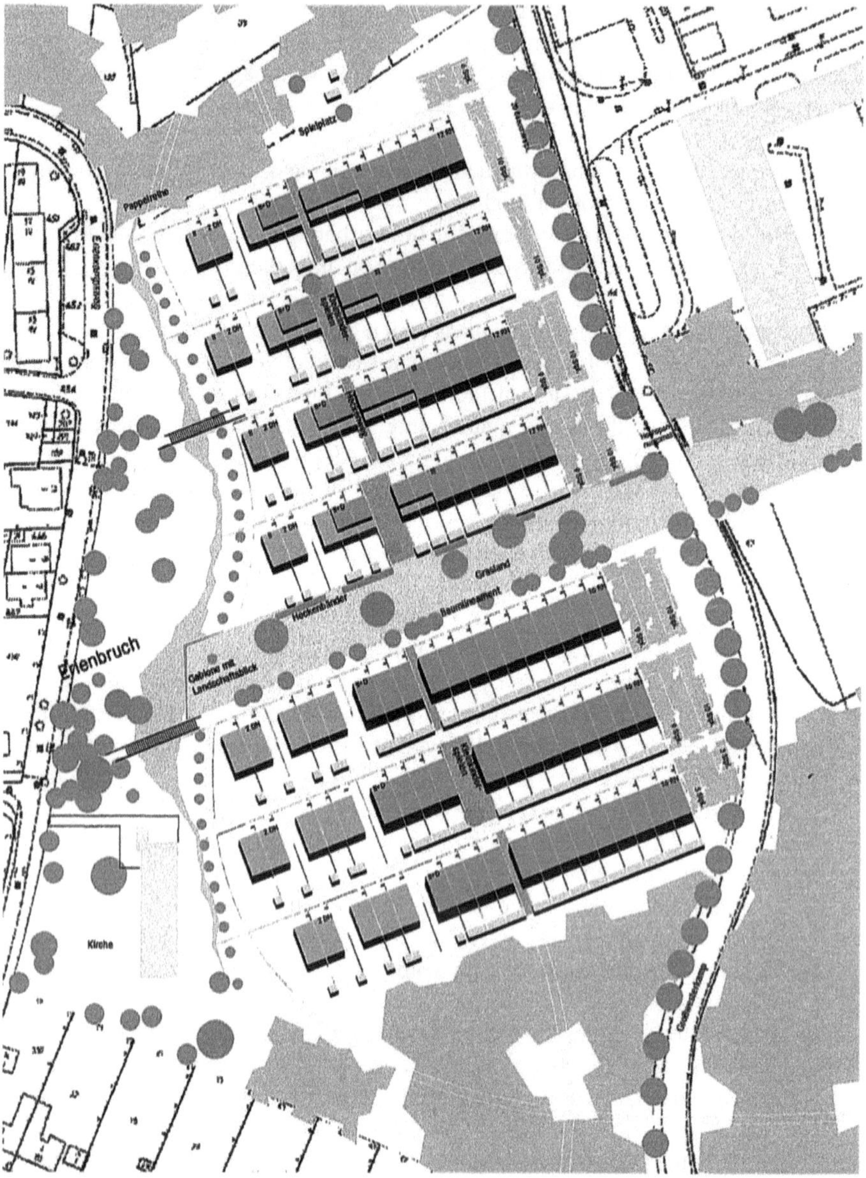

Fig. 12.9 Young residential neighbourhoods Essen, competition entry second prize, RHA architects and urban planners, Aachen

A gabled roof with a 30 degree inclination and facing south, on the other hand, does not look as good as it appears at first glance: it uses only 58% of the floor space of a house with a square floor plan. With gable roofs with a west-east orientation, one can only expect 70–87% of the yield due to the less favourable orientation, but the usable roof area is extremely high in comparison to the floor area, at up to 200%.

With perspectively more favourable production costs for solar systems, the number of roofs in question is growing.

In summary, it can be said that, similar to the building structures, the orientation and shape of the roofs also offer several possibilities with regard to solar energy issues in order to develop urban spatial and design qualities.

> The ecological city of the future is the dense, highly complex city in which distances could be minimized, uses mixed, no expansion of space dispensed with and supply and disposal rationally organized. (Häußermann, Siebel 1993)

Integrated Urban Planning Approach

The Solar City Linz-Pichling is pursuing a holistic approach. With this project, the city of Linz started its attempt to combine different aspects of sustainable building as early as 1995:

- Extensive use of solar energy
- Mix of uses in the new development area
- Addressing social concerns

The aim was to combine technically functional solar technology with social concerns; in the individual solar houses on the one hand and in the urban neighbourhood context on the other. The desire to maximise the winter solar gain and minimise shaded areas is contrasted with the need to achieve a certain density and variety of buildings and a mix of uses, which requires buildings of different types and geographical orientations.

Fig. 12.10 Solar City Linz-Pichling, Austria

The development plan is divided into several compact urban neighbourhood blocks with mixed use. Each block is easily accessible on foot from a central square with public transport stops, which makes the use of cars less attractive. This not only saves energy, but also contributes to social interaction and strengthens social contacts within the neighbourhood. This project is the result of a holistic approach with integrated energy and infrastructure planning.

Fig. 12.11 Brünner Street/Empergasse, Vienna, Marin Treberspurg Consortium, Georg Reinberg, Erich Raith. Despite the high density (FAR 1.8) the buildings are arranged without shading. This is necessary for the superordinate ecological concept with comprehensive urban development and building technology measures: energetically favourable shape and orientation of the building structures, minimisation of heat losses through thermal insulation and heat recovery, passive use of solar energy and use of energy-saving heating systems

2 Dismantling and Conversion

General Conditions

For some years now, there has been an increasing discussion about spatial shrinkage and deconstruction processes. In the spatial context, the term shrinkage initially refers to the loss of population and economic activity. A purely quantitative decrease in population and economic activity is often accompanied by urban and regional processes of decline and decay. Shrinkage is nothing entirely new in the spatial context. Phases of shrinkage have occurred in many epochs of urban development. Nevertheless, the shrinkage tendencies that have been emerging since the 1990s have revealed a new dimension:

- The "growth mentality", that is, the interplay of growth and shrinking phases, is now only valid to a limited extent. The current processes of shrinking appear to be permanent and are self-reinforcing.
- The shrinking processes no longer correspond to the usual clichés of suburbanisation, rural exodus and migration from rural areas. Both the periphery and the centres are affected.
- The phase of spatial redistribution of the population is being replaced by an overall spatial shrinkage.

This new dimension of shrinkage cannot remain without consequences for urban design. Shrinkage can be understood as a transformation process that leads to a change in the content of urban development models. The potential of shrinkage lies in the adaptation of structures and the creation of favourable conditions for new developments. The greater availability of space can be understood as a gain in designable space. The concrete spatial expression of the shrinking processes plays a decisive role for the perception on site and the resulting options for intervention.

> Where cities shrink, emptiness spreads. It disturbs our image of urbanity, but sometimes we consciously seek it out... Empty time is leisure or boredom, longing or anticipation; empty space means desolation, or else. Space for me. (Goerd Peschken)

Urban Management

Conservation areas

Areas are classified as conservation areas if they comprise important structural, spatial and functional, mostly centrally located areas. These can be important historical building structures, but also structures with a cross-location supply function or in the location of infrastructure. The conservation areas must be affected by vacancies as little as possible in order to be able to realise the tasks of these action areas. The ability to identify with the location or the neighbourhood can be maintained by preserving important spatial connections and structures that shape the space.

Restructuring Premises
Restructuring areas are those areas within the town or village that are affected by vacancies and loss of functions. These areas should not be of central importance for the identification of the inhabitants with the place or the neighbourhood. However, these spaces offer, for example, opportunities for the development of a network of public and private open spaces. In many places, the quality of living can be improved by demolition and subsequent use. The aim of the planning intervention in the restructuring areas is to initiate an upgrading and new development in these areas, which are characterised by vacancy and decay, by creating open spaces or even selective new buildings.

Dismantling Areas
Dismantling areas are defined as areas that are affected by vacancies over a large area and are usually located on the edge of a built-up area. They should not have any function or urban planning significance for the entire town or city. In addition, they should not be located directly on the corridors of the technical infrastructure, because deconstruction also involves the long-term abandonment of supply lines.

The aim of the deconstruction of areas of low structural and functional quality and importance is to increase the compactness and efficiency of settlement structures in the long term and to save infrastructure costs. Dismantling can create new, high-quality open spaces in the vicinity of existing buildings. In addition, spatial conflicts can be resolved.

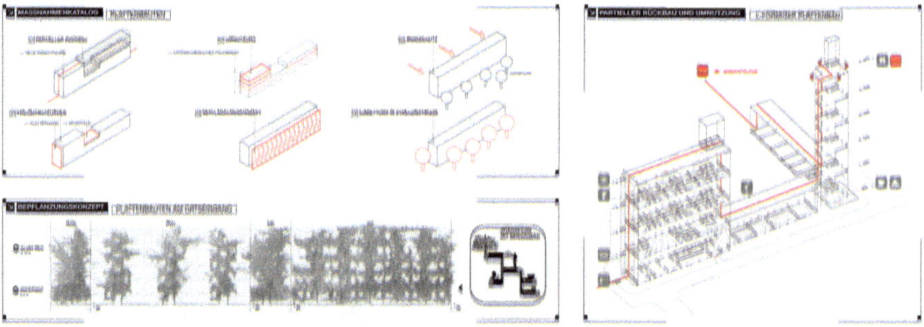

Fig. 12.12 Partial deconstruction and conversion of existing buildings, LEG Prize 2010 competition

Observation Areas
The areas in which only one observation is made are characterised by a comparatively low current need for action. Vacancies and lack of functionality are still low.

Even if the current condition does not yet indicate acute problems, it may be useful to mark observation areas, for example, if they are adjacent to restructuring areas or deconstruction areas where the consequences of development measures cannot yet be fully assessed. Observation areas can also be useful if demographic developments such as the ageing of society make it necessary to adapt the building structure and the living environment in the future.

Fig. 12.13 Action spaces in Aidhausen

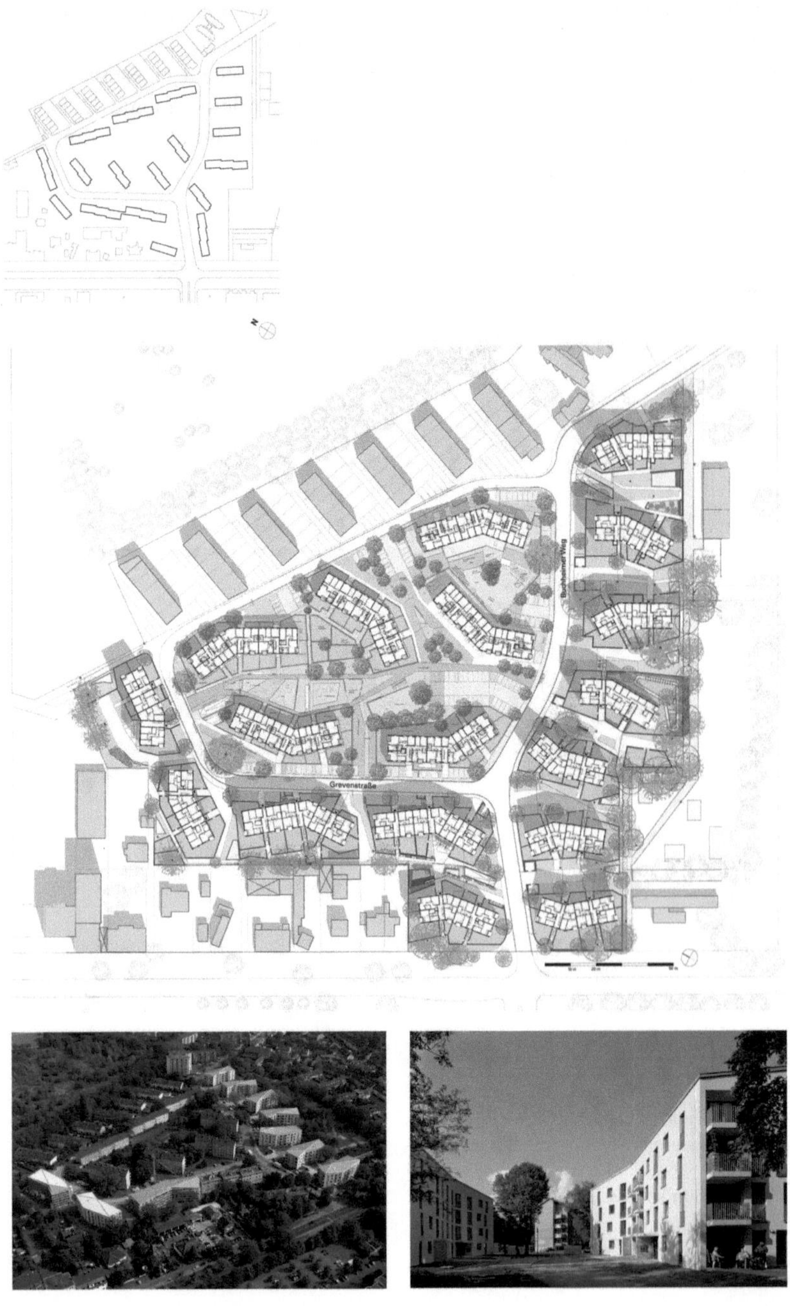

Fig. 12.14 Buchheimer Weg housing estate, Cologne, ASTOC Architects and Planners, Cologne. The Buchheimer Weg/Grevenstraße housing estate is located on the eastern edge of Cologne, 6 km from the centre. It is part of a housing estate from the 1950s, which today occupies about one fifth of the Ostheim district and is typical of post-war urban development in Germany with its terraced

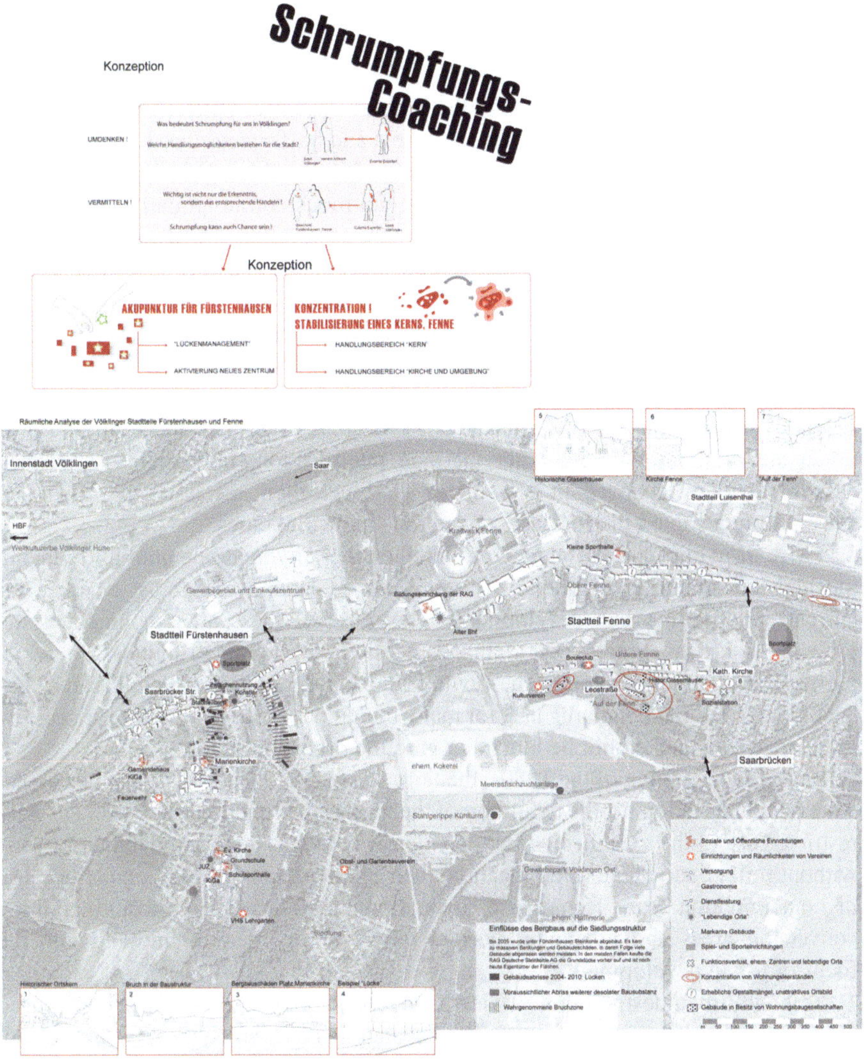

Fig. 12.15 Shrinkage coaching, competition LEG Prize 2010

Fig. 12.14 (continued) houses and open spaces (see map on the left). The existing qualities of the settlement structure of the 1950s were taken over conceptually. In the urban development concept, the existing buildings are gradually replaced by somewhat more voluminous, slightly angled rows. The angular shape of the new buildings, with their spatial arrangement, allows courtyard-like areas between the houses, where the access and other meeting points are located. At the same time, they preserve the "flowing" open space that connects the individual parts of the settlement. With only two building types, their different orientation and positions create a variety and richness of variation both for the exterior spaces and for the apartments themselves

3 Brownfield Development

General Conditions

Every day in Germany, according to the Federal Statistical Office, about 104 ha of building land is newly designated for settlement and traffic areas (see Federal Statistical Office Wiesbaden calculation period 2005–2008). Although a slight decline in land use is recorded in 2010, this is still an enormous consumption of land. At the same time, many areas formerly used for commercial, military or transport purposes have become functionless and lie fallow due to economic and social change.

The structural change first began in the coal and steel sector in Western Europe in the 1960s, soon followed by the manufacturing industry. As a result of this industrial upheaval, especially in the coal and steel industry, but also, for example, in textile processing, many plants were shut down. The areas often located in the narrower urban area fell into disuse. After German reunification, industrial structures in eastern Germany changed in a similar way. Since then, the issue of brownfield sites has preoccupied the formerly prosperous industrial cities.

These brownfields represent a considerable inner city land potential for reducing the high consumption of previously undeveloped land, which must be used for current and future urban development. The reuse of these areas counteracts the process of urban sprawl and at the same time offers opportunities for urban reorganisation and upgrading. The revitalization of underused or derelict areas has become an important planning task for many municipalities. The question arises how these areas can and should be integrated into the organism of the city.

Since the beginning of the 1980s, the term "brownfield land" has been discussed. An essential criterion for the definition of the term is the functionlessness or strongly restricted use of the area. Brownfields are defined as areas that can no longer be used without further ado due to their location, their urban planning conditions or because of contamination from former use. In addition to industrial wasteland, military conversion areas and no longer needed railway properties are moving into the focus of urban development.

From an urban design point of view, conversion areas are areas previously used for military purposes, whose use has been abandoned and which are to be put to civil use. The term conversion area thus already carries the will for rededication in its name, while in the case of brownfield land the focus is still on the discontinued use.

Also, as a result of the restructuring of DB AG, technical modernisation of the railway infrastructure and in particular the transfer of considerable parts of freight transport to the road, real estate previously required for operations is no longer needed for railway operations on a permanent basis. As many of these railway areas are located in the heart of cities, they are important potentials for inner city development. The same applies, perhaps to an even greater extent, to Swiss Post.

Tasks

When an area used for commercial, military or transport purposes, which is usually very closed off, loses its function and becomes a wasteland, a break in the structure of the city occurs. An area that is usually quite large in size lies unconnected in the structure of the city and must be integrated, often combined with major changes in the adjacent neighbourhoods.

In many cases, problems arise due to a lack of building laws, high refurbishment costs, inadequate profit expectations or even a lack of demand for those uses that are desirable from an urban development perspective. Especially the use of existing buildings is often problematic. Contaminated sites are also a frequent problem. Restoring the usability of the sites requires special efforts.

However, these inner city areas also have many locational advantages. These areas often enjoy a distinct advantage in terms of location and are regarded as "prime locations". Also, the already existing infrastructure and development can often be used for the new use and facilitates the realisation.

Urban Design Practice

A reuse of the fallow land makes sense for various reasons despite the restrictions and problems described above:

- From an **ecological point of view**, the new use of inner city land potentials leads to an avoidance of the unchanged high consumption of open spaces. The activation of brownfield sites is an essential approach for an urban inner development.
- The **economic significance** of land activation is demonstrated by the creation of jobs, especially in the case of commercial reuse of brownfield land. The use of existing supply and disposal infrastructure can save considerable costs.
- From a **social perspective**, the fallow of a site often leads to negative effects on the surrounding area. Fallow land carries the risk of becoming a negative image factor for an entire urban area. Reuse can counteract the threat of migration and social weakness.

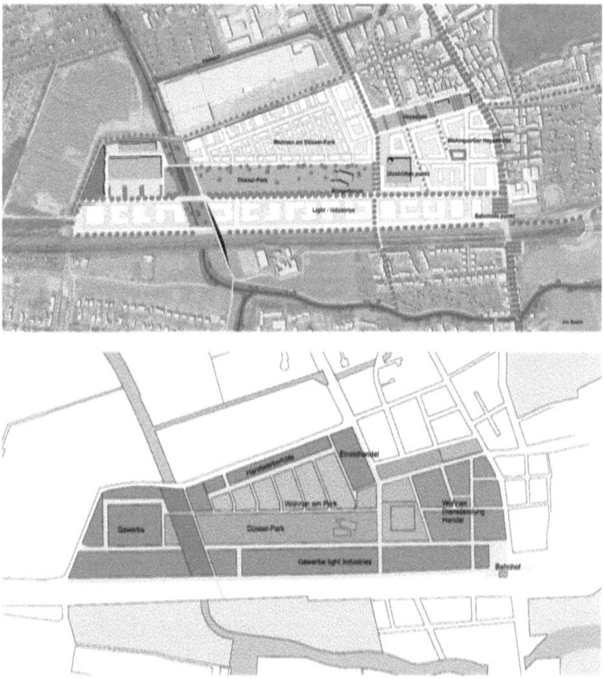

Fig. 12.16 Düsseldorf Gerresheim, RHA Architects and urban planners. Urban development concept and usage zones

Projektkategorien		Städtebau-konzept	Nutzungs-konzept
Östliche Innenstadt Attendorn		eigenständig	eigenständig
Innenstadt Wesseling	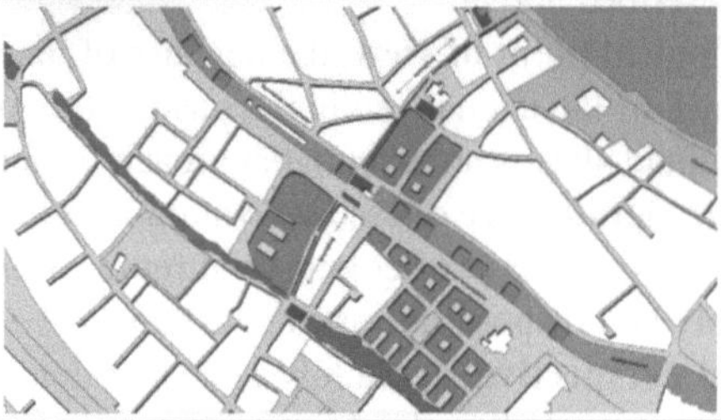	eigenständig	eigenständig
„Güterbahnhof Ost" Dortmund	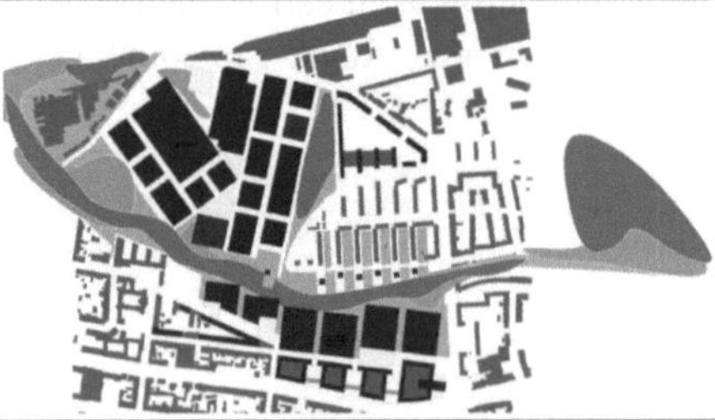	Umgebung wiederspiegelnd	verknüpft + verflochten
Ennepebogen Gevelsberg	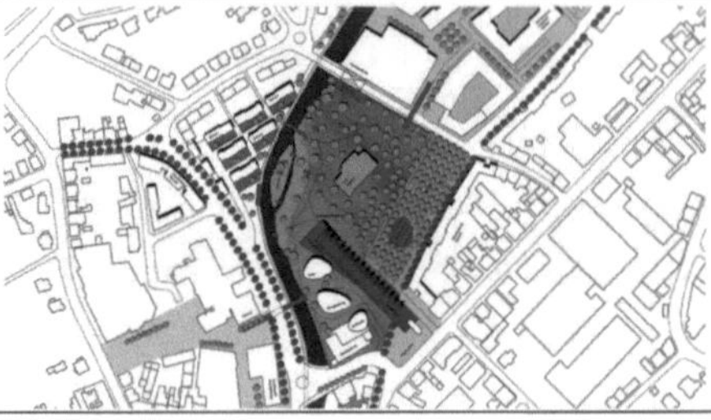	ergänzend + eigenständig	verflochten + eigenständig

Fig. 12.16 (continued)

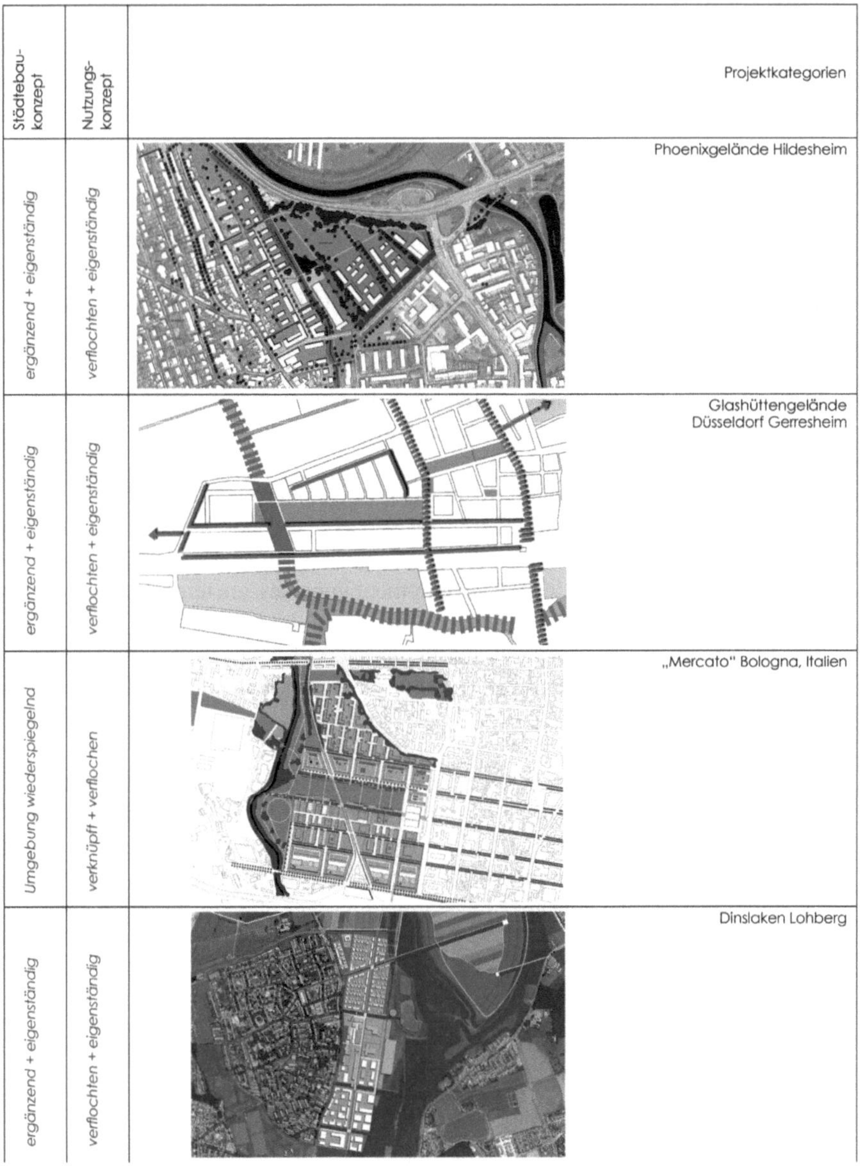

Fig. 12.17 Project categories brownfields

For urban design in dealing with brownfield sites, the examination of two thematic complexes is particularly important:

Identity
How and with which functions, urban development and design means, do you make "old locations" attractive and competitive? How can locations be given an identity of their own and these hitherto mostly closed-off areas be integrated into the urban context?

Flexibility
How and with which urban design means can sufficient flexibility of land development be guaranteed in the case of long development periods and imponderable development conditions?

This means concretely:

- The "inner logic" of the old location decisively determines the new urban development concept.
- At the same time, the urban design concept must give the new location a specific quality, a face and a distinctive character and ensure flexibility for different settlement cases and development processes.
- In all stages of the inevitably incalculable development process in terms of time and content, the interplay of identity and flexibility must be conceptually guaranteed.
- The usually few inner location qualities are to be put in the limelight in terms of urban development and efficiently "in scene", also for necessary marketing strategies.
- The public measures provide the development impulses, are templates and quality standards. They thus shape the level of the overall location.
- The future quality of a location and the success of a site development are also determined to a large extent by the quality of the process design and the bridging of long periods of time.

The resulting urban design concept can be based on different conceptual attitudes. Basically, a differentiation can be made between an independent, complementary, punctual linking and a concept that is interwoven with the surroundings. Which concept is appropriate in each individual case must be worked out on a case-by-case basis. With regard to the concept of use, guidelines are often set by politicians.

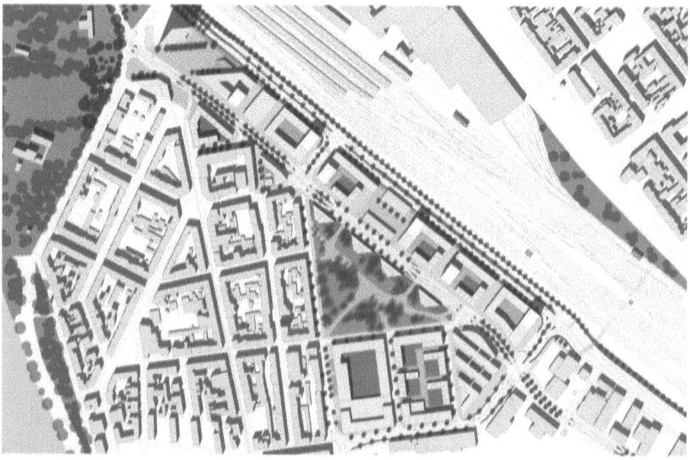

Fig. 12.18 Mannheim 21, ASTOC Architects and Planners, Cologne. Development of available railway areas into a new piece of the city

4 Planning and Building by the Water

General Conditions

It has always urged people to settle by the water or to include water in their settlement. Beyond its practical qualities, for example, as a transport route or fishing ground, water has a wide range of uses and properties: It creates "spatial images" because it can define spaces without creating absolute boundaries. Water delimits areas of use from each other, but at the same time it also connects spaces with each other. Water can appear as a calming or even invigorating element. Depending on its creative use, water contributes to the image of a place. On a small scale, fountains shape market squares, and office buildings use the element of water in the entrance area to show the importance of their institution. On a larger scale, the creation of water areas or the development of attractive areas along the water, which combine this high location quality with urban design qualities and the potential of city-centre locations, is today one of the "highlights" of urban planning.

In urban design, the proximity to water as the basis for attractive habitats has been rediscovered for some years now. In many cities, such as Cologne or Düsseldorf, the Rhine has become an indispensable element of the city, with promenades, attractive buildings and parks in the immediate vicinity of the river. In the current discussion, the trend is clearly towards living, working and living "by the water", so that many areas in direct proximity to the water have already been converted into new urban districts.

Water has always been the basis of life because it provided drinking water, growth on fertile soil or food through fishing. But the most important function of rivers has always been that of trade and shipping routes. In the past, transport developments made rivers increasingly attractive for this purpose: during the Middle Ages, trade led to the foundation of settlements and entire cities all over Europe, preferably at river crossings. Up to our time, water, especially rivers, was only a means of transport and production, only later did it gain great importance as a design element in architecture and urban design. Urban banks become the showcase of cities.

With the advent of industrialisation, river locations became increasingly unattractive because port facilities, railway lines and roads, but also industrial plants that used water as a source of energy or for cooling. Taking possession of the riverbanks. At the latest with the increasing industrial pollution of the waters, the attraction was lost. For this reason, there are many buildings on the Rhine, for example, which turn their backs on the river. Large areas along the waterfront are industrial areas and not open to the public or have an unresolved relationship to the river.

From the 1980s onwards, river basins again became the focus of urban planning. Due to ever scarcer space in inner city locations, attempts have since been made to find new uses for old port or industrial areas that have fallen into disuse as a result of the far-reaching structural change in industrial production. There is the possibility of expanding the inner city by opening it to the water. In this way many rivers are brought back into the cityscape. Various "waterfront projects" such as Duisburg's inner harbour or the Hafencity in Hamburg show that water is a positive location factor today. An attractive address on the waterfront also serves to improve the marketing of plots of land.

Not all locations are suitable for building a settlement on or near the water. Many locations are unsuitable for building use due to ecological or urban planning restrictions or the risk of flooding. Therefore, derelict industrial and port areas offer special potential.

With the numerous settlements that have been built on the waterfront throughout history, three basic forms can essentially be derived, which appear again and again and form the urban planning framework of the settlements: The central harbour, the canal system and the riverside development:

Fig. 12.19 Rheinau harbour Cologne. At the beginning of the twentieth century (top), beginning of the twenty-first century; crane houses under construction (bottom)

Central Port

Ports were always an important crystallisation point of urban life. They were often the starting point of a city's economic development and were therefore very close to the city centre. The structural change of the ports by ever larger ships, the introduction of containers and the departure of the industries attached to them often left behind empty areas and buildings, or entire port areas were no longer used. Thus, an insurmountable wasteland was formed between the city centres and the water. In the 1980s, however, these areas were rediscovered and incorporated into the further development of the cities. Due to the special location of the areas between the city centre and the water's edge, they offer the possibility to orient the city centres with their mix of uses towards the water. Hamburg, London or Duisburg show the importance of reactivating such abandoned, underused urban areas on the waterfront and illustrate the potential for a city precisely in such former port situations.

Canal System

A canal system is a branched network of artificially created waterways starting from the water body. In the historical cities, the primary importance was the development by boat, but also the drainage. All buildings are connected to the waterway, which favoured trade. Today, pure residential or holiday settlements are often built with a canal system to activate the recreational value of the water for the residential environment.

Bank Edge Development

Many settlements have not integrated the adjoining waters into the development, so that the water has been provided with a peripheral development as an unchangeable barrier. The water is, therefore, neither a central point nor an inner city structuring element. Often such settlements did not have a protected harbour for topographical reasons.

Tasks and Objectives

The interaction of water and city is the basis for a lively city life at the water's edge. The combination with experimental forms of living, a mixture of living and working in high urban development quality and networking with leisure activities in the immediate vicinity should be aimed for. In order to prevent riverbanks from being obstructed or harbour areas from being left lying around because it is not clear which uses can be established, principles for sustainable urban development on the waterfront have been developed. The following principles refer to the results of the international EXPO 2000 workshop "Water Cities—An Interim Balance", which took place in Berlin in October 1999 (see Wasserstadt GmbH 2000: p. 194).

- Building on water is always building in the urban fabric. New urban neighbourhoods by the water must be designed as an integral part of the existing city and must be networked with the existing residential neighbourhoods.
- The banks must generally be made publicly accessible. Where existing workplaces are not obstructed, banks should also be physically and visually accessible to visitors. The more quality the space is used, the more it is used. Due to a variety of usage requirements such as leisure, recreation or sports, which often have a high public frequency, the banks must be designed in a high quality.

- Buildings (especially monuments) that create identity should be used to give the new neighbourhood a specific character. The preservation of parts of old port and industrial areas, such as storage buildings worthy of preservation or structural facilities such as cranes and thus contemporary witnesses of the industrial past, should be an integral part of sustainable development, because they embody the change in urban life and work.

Water is an important urban development element. Within urban settlement structures it forms a natural element of arrangement, it enlivens the urban or settlement structure. In addition, water surfaces are guidelines for the development systems and represent an urban development connection with the surrounding area. In terms of urban design, a water surface integrated into a settlement is always a point of identification in the surroundings, since water is used to set orientation points that counteract the uniformity of a development. This special situation in a residential area can usually be perceived from a greater distance, because there is a particular contrast between the water surface and the built elements.

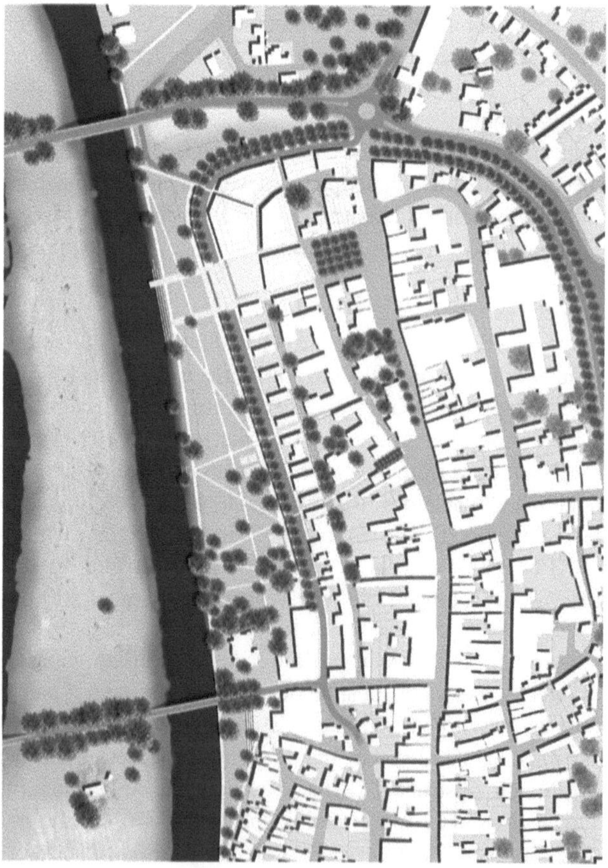

Fig. 12.20 Allerufer Verden, urban planning competition, kellner schleich wunderling architects + urban planners GmbH, Irenelohauspetercarl Landscape Architecture

Urban Planning Practice

Interlocking

How the elements of water, public space and development are brought into a relationship has to be worked out individually in each urban design concept. A close visual relationship between water and development as an urban design approach makes it possible to integrate water very strongly into the planning process. Suitable design elements are U-shaped buildings open to the water, but also differentiated height developments between the bank development and the "second row" or staggered building structures.

Proportions

It is important for urban design to include the proportions of the water areas in the design. The linear expansion of for example, harbour basins must not be disregarded. The height development must be adapted to the environment and the intended urban character. How does the design look from a distance, which scale is appropriate? Especially the development along the banks of the river plays a special role, because if the buildings are too low, connections can be lost, and structures along the banks that seem too large can overwhelm the existing buildings and appear to be out of scale. There is no preference for closed or open development. However, it is undisputed that protrusions or recesses, different heights or different architectural languages, consciously used, enliven the image of the development.

Public Space

The promenade is an essential design element in the design of the public space on the waterfront. Even today, strolling is a popular leisure activity. The visual relationship with the water from the promenade is important. In the design of the promenade, classical elements are staircases running out to the water with seating facilities and pontoons permanently installed on the water with a wide variety of uses ranging from sports facilities to catering establishments. The most important quality of public spaces on the water is the use of the water itself. Leisure uses on or on the water make these areas attractive for visitors: marinas, marinas for sport boat harbours and gastronomic or cultural uses determine the special flair of water locations.

Green and Open Space

Green and open spaces on the waterfront in the direct residential environment are a central element of local recreation. If these areas are open to the public, they enhance a neighbourhood in a special way. These areas must, therefore, be designed in such a way that accessibility is always guaranteed, and the stay is attractive.

The project development of waterfront sites can also bring with it a variety of problems:

- **Contaminated sites from previous use:** Soil contamination is to be expected, particularly in the case of old port areas.
- **Security measures:** In flood areas, measures against flooding and for the rescue of persons must be provided, and bank reinforcements are often costly.

- **Access situation:** Many locations on the waterfront are cut off from the surrounding development by brownfields or traffic arteries, so that these barriers must first be overcome in order to develop these areas.
- **Immissions:** Remaining industries or even only small businesses sometimes lead to pollution which, for example, has a massive impact on housing.
- **Specialist planning:** The interests of flood protection or waterway protection must be included at an early stage.
- **Ownership:** The discussion about future uses should not be initiated too early. Otherwise owners could use this knowledge to their advantage.

In order to turn riparian zones and areas along the waterfront into attractive engines of urban and regional development with a special appeal, they must first be made accessible to everyone. In addition, areas for recreational use or areas for landscape development must be available alongside areas to be built on. Attractive water and shore use can be a city's calling card, a location factor and an important image carrier. It is therefore important to make targeted use of the potential of rivers, canals or lakes and to integrate water into urban design concepts so that places are created that contribute to a specific identity.

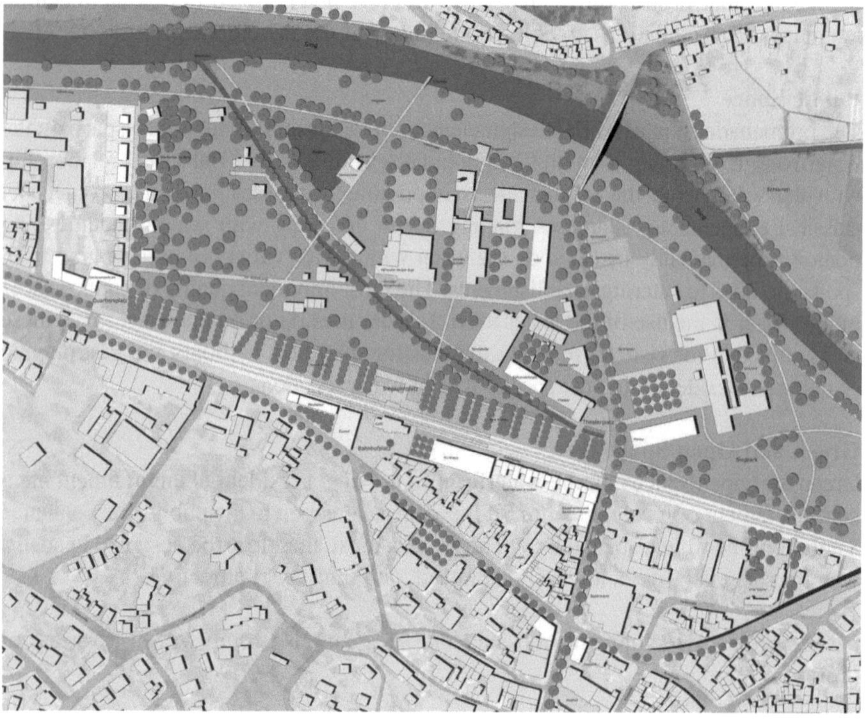

Fig. 12.21 Eitorf, RHA Reicher Haase Architects and Urban planners with Club L 94. Urban and landscape planning concept "Leap to victory"

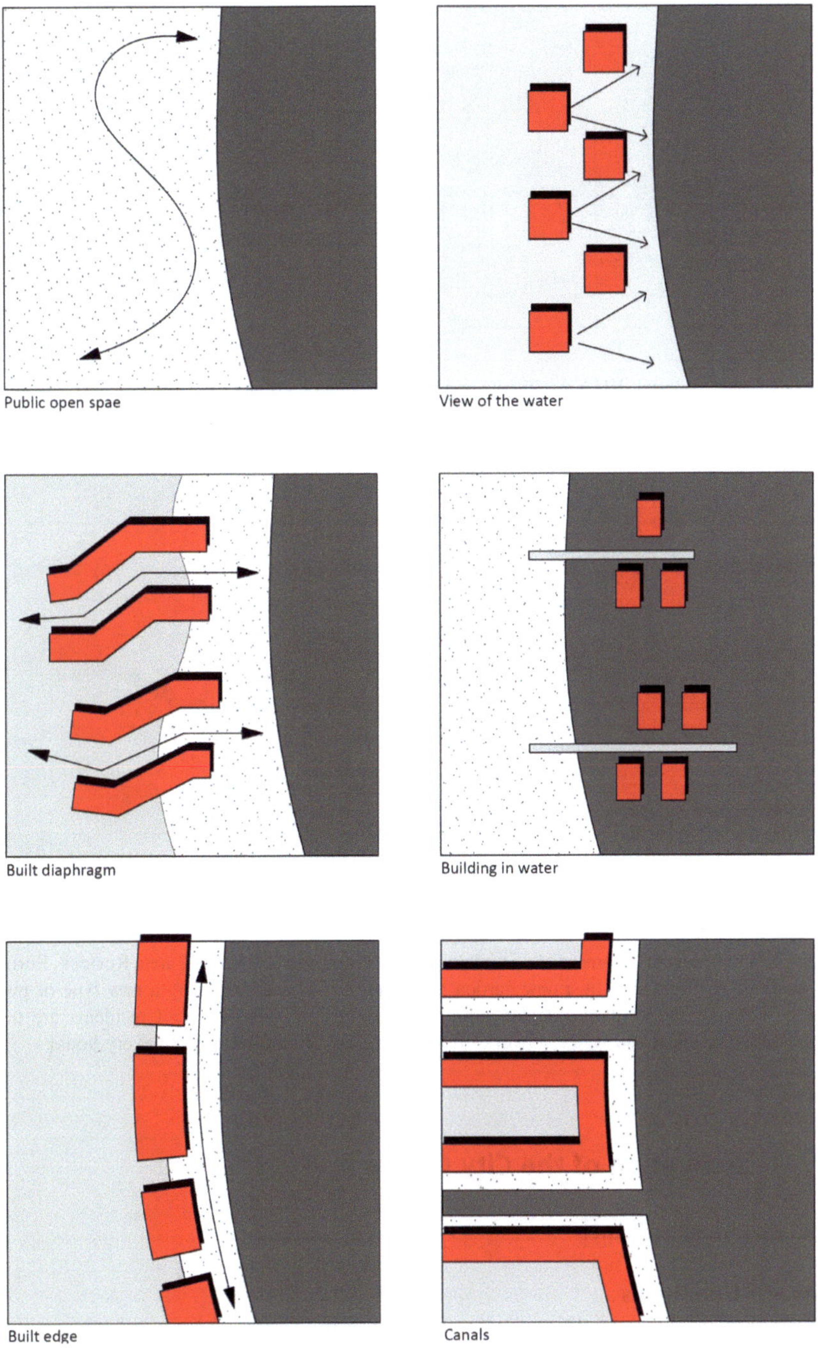

Fig. 12.22 Typologies of urban design at the waterfront

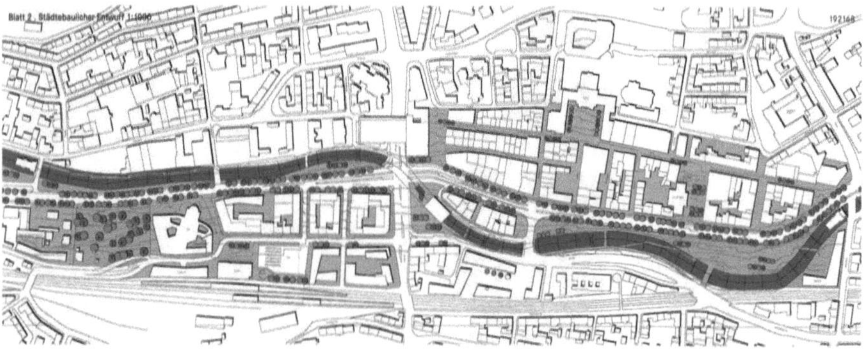

Fig. 12.23 Wuppertal, RHA Architects and Urban planners, competition entry

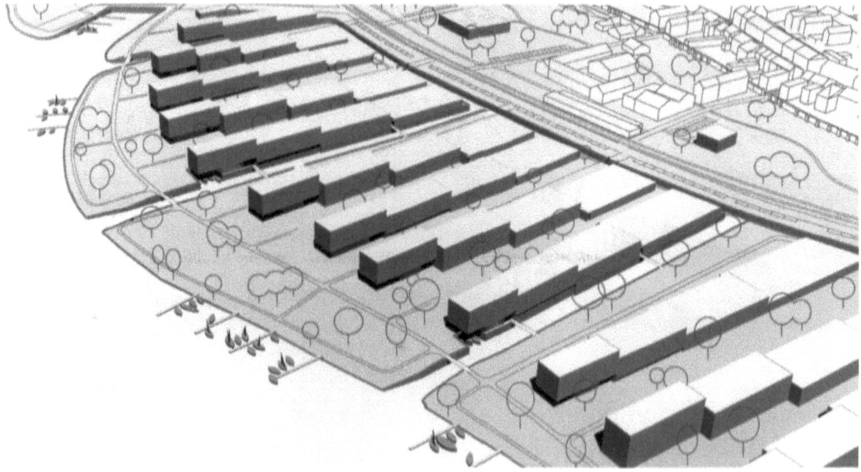

Fig. 12.24 Urban development of an underused residential and commercial area, Rostock, Europan competition 9. Formation of a new garden and water city characterized by a new type of public space. Through the moderate densification and the mix of uses, different residents are to be addressed. The canal and water network offer direct access to the water from every house

5 Activation of the City Centre

with Ilka Mecklenbrauck

General Conditions

The face of a city is traditionally sought in the inner city. As a symbolic core, the inner city represents the city outwards and inwards, that is, for the city's inhabitants themselves. The city is identified through the inner city or its city centre. This is where the importance of the city, its culture and its economic and social "climate" can be seen.

With regard to the city centre, very different terms are used: the city centre, the city, the central supply area, the historical city centre, the old town and, especially in the case of larger cities or regional agglomerations, the city and district centres or sub-centres. Common to all terms is the demarcation of the symbolic core of the city or district, regardless of the size of the city and its spatial location. In the urban fabric, the city centre has a variety of functions as the spatial centre, the cultural centre, the political centre, the centre of trade and services and the inner city as a residential location.

Consequently, the inner city is the place to be:

- local history and cultural heritage,
- of politics and power, of political and social debate,
- of local democracy,
- of civic engagement,
- social and community contacts and the public,
- local exchange of ideas and innovation,
- of cultural creation and education,
- the collective as well as the individual self-presentation,
- of diversity, liveliness and multifunctionality,
- the variety of participation possibilities,
- market and exchange, trade and services (see Netzwerk Innenstadt NRW 2010).

Fig. 12.25 Cologne city centre, juxtaposition of historic and new buildings

History

The inner cities have always been places of action, culture, social and political life. They are not rigid structures, but have changed over the course of different architectural periods. Even today, the social and cultural reality of many epochs can still be read in the cities. This presence of history distinguishes the European city: It is the place where modern society was born. By walking through a European city, you can make sure of your own history.

In the city of ancient times, the urban centre is characterised by trade on the central square, the "agora" in Greece or the "forum" in Roman city foundations. In addition to trade, people also lived and worked in the city, all in close proximity to each other. The city of the Middle Ages is also characterised by trade in the centre. Frequently settlements were founded at the crossroads of important trade routes—for example, along the Hellweg—or at harbours or fords, typical, for example, for the bourgeois governed Hanseatic cities. In addition to traffic routes and trade, the proximity to church institutions and secular rulers is often a decisive factor in the founding of cities. Urbanisation also takes place where monasteries, palaces or castles have their seat. During the period of absolutism, the city centres there were often transformed into representative centres of the city, streets were laid out in a rational grid or in relation to the centre of the city—the seat of power. The main streets, which are oriented towards representative buildings, are particularly magnificently designed.

The development of transport and production technology from the middle of the nineteenth century onwards brought about profound changes in the structure of the city. Industrial production in factories is associated with a great need for production areas and space for the immigrant labour force, which the city cannot meet in its present size. A condition for this industrialisation in all cities is the great progress in the development of transport by rail. The railway stations, initially often built outside the old city centres, are soon caught up by the rapid growth of the cities and often develop into the centre of the inner city infrastructure—even today. At the end of the nineteenth century, this development leads to an explosive growth of cities and a densification of the city centres.

At the beginning of the twentieth century, functionalism asserted itself as a planning model and finally propagated the separation of urban functions with the Charter of Athens. Initially, this primarily affects the further development of cities on the periphery. Only in the context of the reconstruction of the cities destroyed in the Second World War did this model also take effect in the city centres. In terms of the separation of functions, the inner cities are increasingly being developed into places of administration and consumption.

The type of large department store that was already established in the big cities before the war is now being realised in all city centres. In many places, the remains of the historical building structure have to make way for this. The scale of the inner cities is changing: the new buildings are much larger and higher, and the plot structure that has dominated the picture up to now is being pushed back. Even more significant for the change in scale is the often-excessive expansion of street space, which is believed to be able to cope with the rapid increase in individual traffic in the 1960s and 1970s. In return, much of the historical structure is being destroyed in favour of the new traffic arteries.

As a result of this development, the residential function of city centres is losing importance. Living space is being converted into commercial or office space. The quality of living decreases with the desolation of the inner cities due to this monofunctional orientation and the associated traffic load. Living in the city centre loses its attractiveness, which encourages people to move to the countryside or to the outskirts of the cities.

The establishment of pedestrian zones and the construction of bypasses in the 1970s and 1980s eased this situation and gave new impetus to the city centres. In the pedestrian zones, the functions of retail trade and also services are particularly concentrated. After initial criticism, they are well accepted and are now a characteristic feature of the inner city in many cities, especially when a historical urban structure forms the framework.

Since the beginning of the 1990s, almost all cities have developed new competition to the city centre in the form of large-scale retail centres on greenfield sites outside the city gates, which advertise with favourable transport connections and free parking spaces. The return of the retail trade to the inner cities heralds a new trend. This welcome development raises the question of the scale and integration capacity of these buildings into the mostly small-scale inner city structure. In addition, for some years now, inner cities have also been becoming more preferred residential locations again: the "renaissance of the centre" has become a common concern of urban society.

Fig. 12.26 Masterplan Cologne, excerpt, Albert Speer and Partners. Supplements to the urban structures, new spatial edges and connections between the urban areas on the right and left bank of the Rhine.

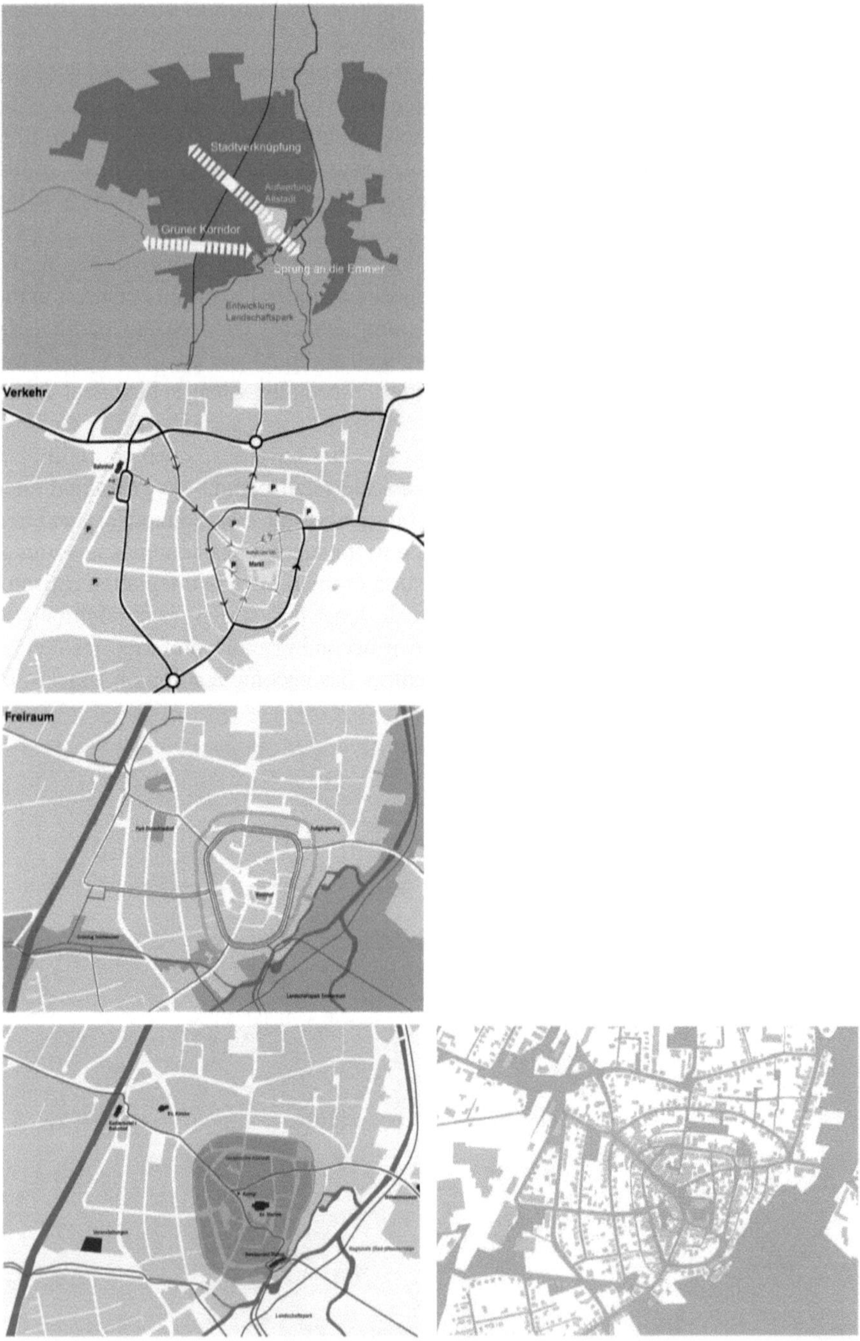

Fig. 12.27 City centre concept Steinheim, RHA architects and urban planners

Spatially, the concept for the public space in Steinheim's inner city is based on the following guidelines:

- Upgrading of the old town: The various problems of the old town are to be addressed by upgrading the public space. "Jump to the Emmer": The Emmer floodplains are to be upgraded in the form of a landscape park and by improving access and development.
- Better connection between the city centre and the western part of the city: The barrier of the railway line is to be overcome by urban development measures.
- Creation of the Green Corridor Pond Water: The strengthening of the green corridor at the pond water should have a positive effect on the conversion of western brownfield sites.
- Development of a landscape park: A landscape area of high quality and attractiveness is created via the pond water in the immediate vicinity of the settlement area.

Goals

The goals concretise the mission statement in the various development-related issues. Constant feedback between the content and the formulated goals leads to a constant specification in the overall process and thus to a jointly supported basis.

Traffic

- Support of the ring system: outer ring + inner ring
- Traffic calming on Market Street by one-way traffic guidance in the direction of the ring
- More pedestrian-friendly design of the road spaces
- Preserving the opening of the old town for individual motorized forms of mobility
- Creation of a public transport thoroughfare through the old town
- stronger enforcement of the ban on trucks

Free Space

- Creation of a landscape park for city dwellers and (bicycle) tourists
- Equipment with numerous leisure/local recreation possibilities
- Preservation of the floodplain as a retention area (information boards in the park)
- Networking the old town and inner city green spaces with the floodplain
- Design and hierarchisation of street spaces through trees
- Creating pedestrian-friendly public spaces and connections in the city

Tourism

- Profiling as a "creative" furniture city
- Profile as a historic city on the waterfront
- "Bringing the past to life"
- Networking of existing (cycle) hiking trails
- Strengthening as a destination for cyclists through special infrastructure (including a cyclist hotel)
- Starting point for water hikers (canoe, etc.)
- Installation of a wayfinding system/information steles

Challenges of the Inner City

Current considerations on inner city design are increasingly oriented towards the concept of mixed use. They pursue the model of a city of short distances. The impulses for this rethinking are manifold:

- The demographic development and social change of an urban society that is becoming "less, older and more colourful" is leading to changed lifestyles and new demands on housing.
- The economic changes in the retail trade (concentration processes, chain stores, increase in sales space and changes in shopping behaviour) have an impact on the functional structure of inner cities.
- Urban space, especially public space, has been increasingly shaped by the phenomena of commercialisation and privatisation.

New concepts are needed for the inner city and its spaces that deal constructively with these challenges. With regard to open space and especially public space, obvious contrasts and conflicts arise:

Urbanity Versus Housing Needs

The expectation of a qualified urban density with the characteristics of physical and social density, mix of uses and urban design often contradicts the desire for tranquillity, especially in the residential environment.

Permanent Versus Temporary

Areas fallow due to abandoned uses and thus not only influence the structural appearance, but also have a negative impact on public space and open spaces, with enormous negative consequences. The opportunities that lie in a temporary activation of unused or underused areas are not yet sufficiently recognized and utilized today.

Exploitation Interests Versus Opportunities for Appropriation

The exploitation interests of public space in connection with adjacent uses often conflict with the appropriation and use interests of the population, residents and neighbouring residents.

Conflicts and contradictions are also evident with regard to the urban layout and architectural design of the buildings in the inner cities.

Reconstruction Versus Contemporary Architecture

The desire for a historical appearance that follows a traditional urban layout and traces lost facades is in full swing. This goes hand in hand with the danger of setting "wrong tracks" in the history of urban development.

Large-Scale Use Versus Small-Scale Structure

The return of large-scale uses to the inner city, such as retail, is often at odds with the smaller-scale structure.

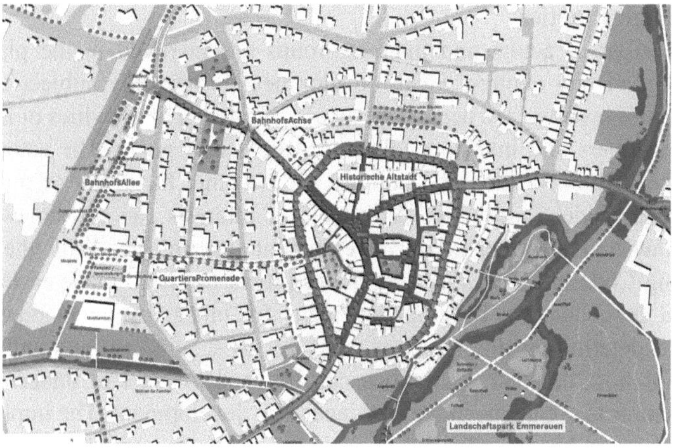

Fig. 12.28 Overall urban development concept, RHA architects and urban planners

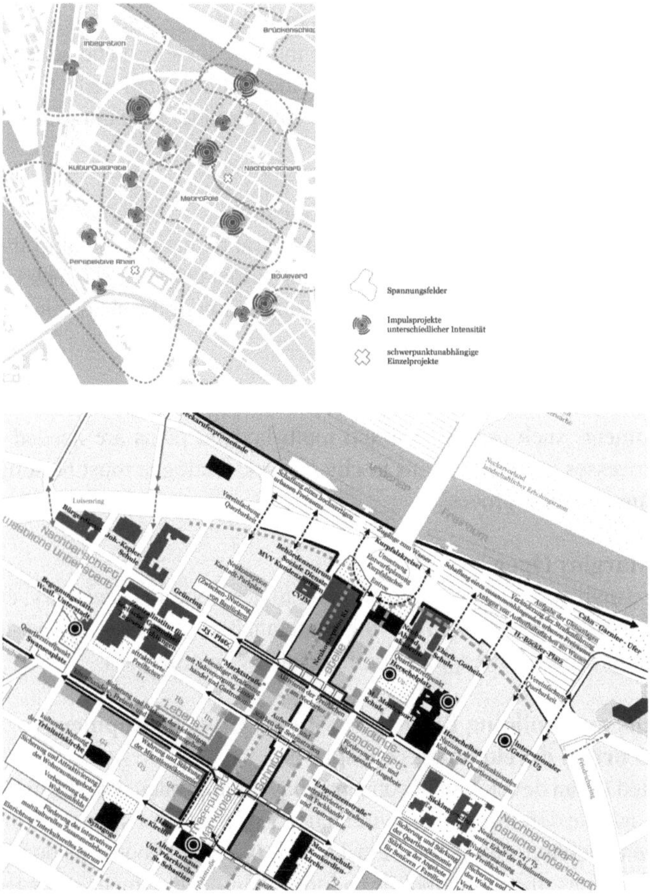

Fig. 12.29 Development concept inner city Mannheim, ASTOC Architects and Planners, Cologne. Impulse projects within the different spatial areas of conflict control the development of the inner city. Spatial measures are initiated within the differentiated spatial focuses—the neighbourhoods.

Urban Design Practice
A large number of actors and interest groups are involved in the planning and development of city centres. Since development is basically subject to planning law, the parties involved are dependent on cooperation. The activities of landowners, project developers and investors follow private sector goals. The actors in urban development, administration and politics, on the other hand, are oriented towards overall urban models and goals. A balance of interests can be achieved through planning law and informal urban development instruments. With the update of the Building Code in 2007, inner city development has been made easier. Both the accelerated procedure for development plans for inner city development and the methods of determining the types of measures to secure central supply areas are examples of how detailed regulations can be used to react to undesirable urban development or to create incentives for inner city development. The implementation of concepts must be preceded by a clarification of the guidelines for inner city development.

Guidelines for a Holistic Inner City Development
Strengthening and revitalisation of core functions
The traditional images of the city centre must be reviewed and adjusted. The city centre will continue to play a major role as a location for retail trade. Complementary uses such as culture, housing and services can provide a meaningful rounding off, beyond consumption.

Maintenance and Vitalization of the Cityscape
City centres embody the cultural and architectural diversity typical of the European city. The existing qualities, especially the individual character of each city, must be cultivated and valued. In this context, the examination of the shape and use of public space plays a central role.

Promotion of Creative Processes
New instruments such as informal and multi-layered plans are needed to promote creative processes and innovations in city centres. Dialogue must be sought through active communication processes.

Keeping Margins Open
Due to the usually strong economic pressure to develop, it is necessary, especially in the inner cities, to leave room for the unplannable, to secure spaces that are not subject to any superficial exploitation interests by interim uses.

Instruments and Building Blocks of an Inner City Development
Integrated urban development concepts
Integrated urban development concepts play an important role: from the setting of objectives and the definition of individual fields of action for strengthening inner city areas to the concrete formulation of tasks with the allocation of responsibilities and financial resources, the municipalities adopt urban development strategies.

Many factors make the inner city location interesting for major investors in the retail trade and real estate industry:

- Traffic development
- High audience frequency
- Tourist appeal
- Historically grown places as background of the production and
- Revitalization of neighbouring neighbourhoods as preferred residential locations

In the integrated urban development concept, the various factors must be brought into a convincing balance.

Real Estate and Location Communities (ISG)
A real estate and location community should bring together business people in a clearly defined area (usually a business area) with the aim of developing measures to maintain the value of the neighbourhood. The focus is on marketing and optimising the design of the business district.

Shared Space Concepts
In city centres, the planning philosophy of "shared space" as a space shared by different road users is becoming increasingly relevant. By dispensing with traffic signs, signal installations and road markings, the traffic space is once again becoming a recreational space.

Civic Involvement
The vibrancy and development of the inner cities is hardly conceivable without the city society and the citizens. Planning processes and procedures must understand active citizenship as a resource.

> Vitality, atmosphere, quality of public spaces, cultural diversity, an authentic cityscape, attractive shopping, supply and leisure facilities in the inner cities and local centres are the 'capital' that cities and municipalities must continue to strive for. In order to stabilise the centres in a targeted manner, it will be important to maintain and consolidate a sufficient variety of uses. The urban centre can only have a future as a multifunctional place. (Christina Simon-Philipp)

As diverse and individual as the inner cities are, the solutions for a sustainable further development are just as complex. Depending on the location, the objectives of the planning and the implementation process, mixes are possible vertically and horizontally, fine-grained and coarse-grained, in buildings, in blocks or in neighbourhoods. Mixes can be made with shops, offices and service providers, even with non-disturbing craft businesses. Demographic and social developments will continue to drive the trend towards urban lifestyles: Older people, new lifestyles and the increase in global migration trends will increase the demand for central residential locations.

Strategies and Fields of Action

In dealing with the activation of the inner city, a variety of strategies and fields of action can be deduced from this

- **Mix and diversity**: promote a consistent mix and density of uses, with the aim of optimising synergies between the individual uses
- **Accessibility and networking**: networking functional and urban priorities of the city centre
- **Concentration**: Developing strategies for the spatial and functional concentration of uses within the inner city
- **Quality of design**: Commitment to high quality structural and spatial design and high building culture
- **Reclaiming urban space:** deconstruction of oversized infrastructure such as roads and railways
- **Urban open space**: landscape as an equal element in the urban fabric of the city centre
- **Urban community communication**: Enabling informal appropriation of public spaces
- **Use**: Use of the city centre by promoting events
- **Holistic city marketing**: combining planning, strategic and advertising aspects of city marketing into a holistic approach
- **Public-Private-Partnership**: Establishing cooperation and strategic alliances between municipal and private actors
- **Willingness to engage in dialogue and moderation**: improving communication processes, trust and dispute culture

The planning task is to ensure a balanced variety of uses with functions typical for a centre: Retail trade, services, crafts, administration, culture, educational institutions, social and church facilities, communication, gastronomy, leisure and sophisticated living for young and old. The upgrading and activation of the public space in its various manifestations—from the pedestrian zone to the city square—and its various possibilities of use and appropriation is a central component of an inner city concept. A high degree of multifunctionality, combined with high-quality, varied and sophisticated spaces and buildings, must be the goal of all urban planning.

> The world of experience and the quality of life in our inner cities are decisively influenced by their evolved urban structures and the design of public space. Cities that have a historical architectural heritage, a formative architecture and/or a lively and diverse culture of squares with green zones and parks continue to prove to be magnets for visitors and tourists. Especially in view of the competition from the 'green meadow', there are decisive urban planning advantages here. (Ulrich Hatzfeld)

Literature

1. **Solar and energy efficient urban development**
2. **Dismantling and conversion**
3. **Brownfield development**

Dosch, Fabian 1994: Gewerbebrachen als Baulandreserven. In: Bestand, Bedarf und Verfügbarkeit von Baulandreserven–Materialien zur Raumentwicklung Heft 64. Bundesforschungsanstalt für Landeskunde und Raumordnung.

Feldkeller, Andreas (Hrsg.) 2001: Städtebau: Vielfalt und Integration–Neue Konzepte im Umgang mit Stadtbrachen. Stuttgart, München: DVA.

4. **Planning and building by the water**

Wasserstadt GmbH 2000: Wasser in der Stadt–Perspektiven einer neuen Urbanität. Berlin: Transit Buchverlag.

5. **Activation of the city centre**

Brühl, Hasso; Echter, Claus-Peter; Jekel, Gregor: Frölich von Bodelschwing; Franciska 2005: Wohnen in der Innenstadt–eine Renaissance? Dlfu-Belträge zur Stadtforschung 41. Berlin.

Monheim, Heiner; Monheim, Rolf: Innenstädte zwischen Autoorientierung, Verkehrsberuhigung, Shopping Centern und neuen Steuerungsmodellen. In: Monheim, Heiner, Zöpel, Christoph (Hrsg.) 2008: Raum für Zukunft. Zur Innovationsfähigkeit von Stadtentwicklungs- und Verkehrspolitik. Essen.

Annexure

A.1. Housing Typologies

Single family house types

Sample sizes of detached single family houses

Sample sizes row houses

Types of apartment buildings

Vertical development types / "Spanner" types

- Single-unit type
- Two-unit type
- Three-unit type
- Four- and more unit type

Horizontal development types / "pergola" types

Internal and external access

A.2. Street Typologies

Building set

- Pavement accompanying the road
- Roadside bicycle path
- Separating strips
- Parking
- Separation and mixing principle

Exemplary road cross sections

- Travelling
- Residential street
- Assembly line

C. Reicher, *Urban Design*, https://doi.org/10.1007/978-3-658-34370-5

- Neighbourhood street
- Local shopping street

Main shopping street
Pedestrian zone
Commercial street

A.3. Housing Typologies

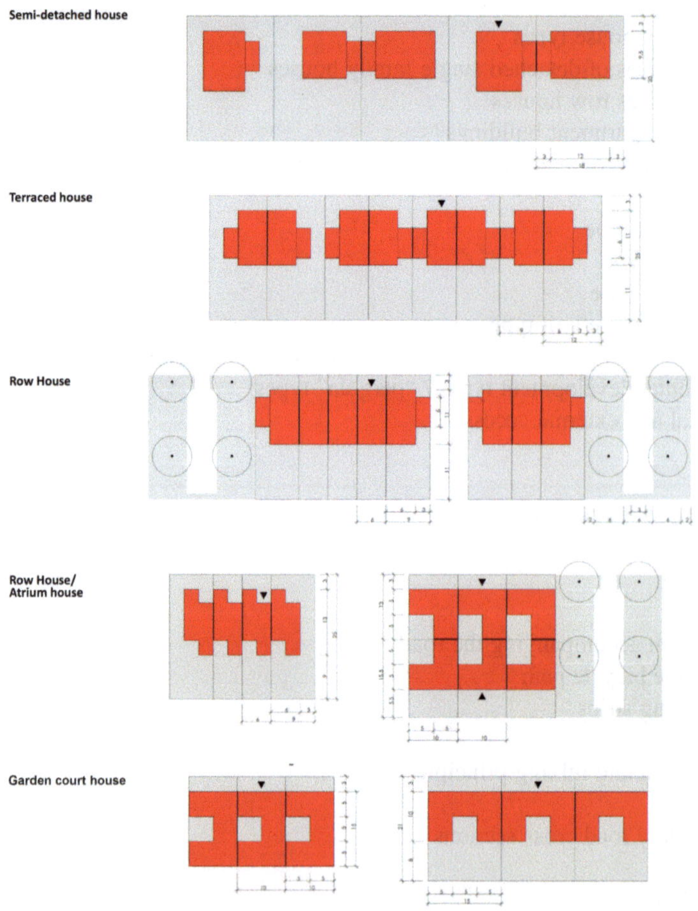

Fig. A.1 Single-family house types. Freestanding. Semi-detached house. Terraced house. Row house Atrium house. Garden court house

m/sq m	BUILDING WIDTH / DEPTH	BUILDING PLOT WIDTH* / DEPTH	BUILDING PLOT AREA	GROUND FLOOR	GRZ	NOTE
MIN	7 X 11 m	13 X 25 m	325 sq m	77 sq m	0.24	Strongly limited variability, structure only offset
MID	9 X 13 m	15 X 25 m	375 sq m	117 sq m	0.31	Minimum level for unrestricted variability
MAX (+)	8-13 X 13 m	20 X 30 m	600 sq m	bis ca. 145 sq m	0.24	Sizes open at the top

* 3m minimum distance between the main building and the neighbouring property (BauO NRW)

Detached family house

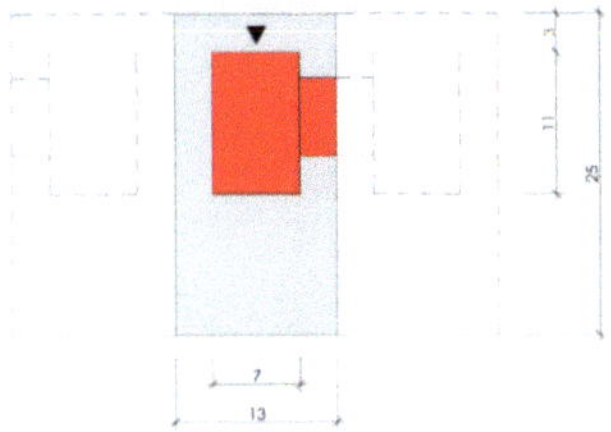

Angle type

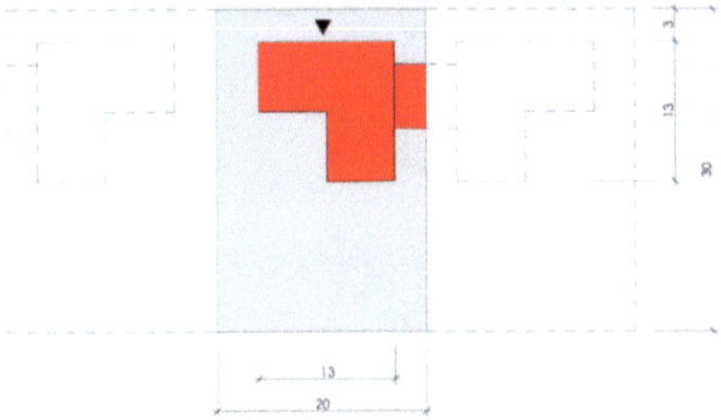

Detached family house

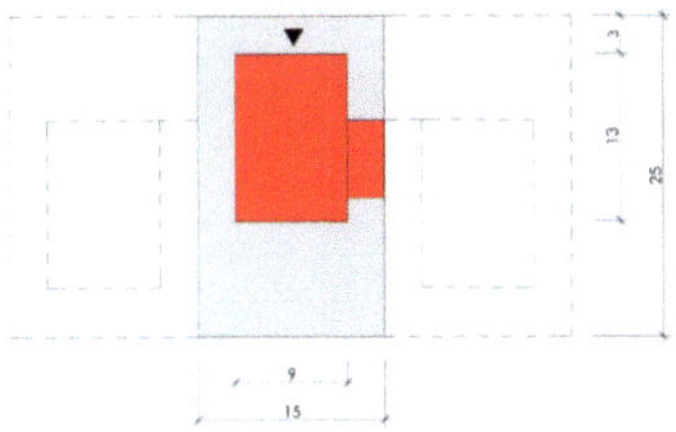

Bungalow (1-storey)

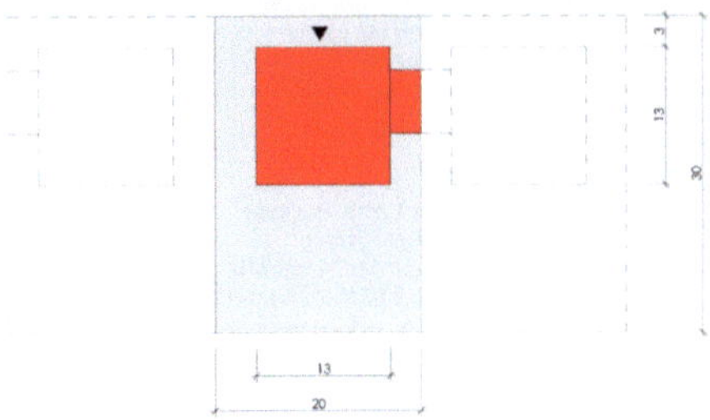

Fig. A.2 Sample sizes of detached single family houses. *3 m minimum distance between the main building and the neighbouring property (BauO NRW). Detached family house. Angle type. Detached family house. Bungalow (1-storey)

m/sq m	BUILDING WIDTH / DEPTH	BUILDING WIDTH / DEPTH	BUILDING PLOT AREA	GROUND FLOOR	GRZ	NOTE
MIN	5.25 x 11 m	28.5* m	150 sq m	57.85 sq m	0.38	Min. possible width 4.30 m. *Depth of building plot with regard to permissible GRZ
MID	6.50 x 10 m	35 m	200 sq m		0.32	
MAX (*)	9.5 x 13* m	35 m	332.5 sq m	123.5 sq m	Approximately. 0.38	*For special shapes, e.g., offset building parts

* 3m minimum distance between the main building and the neighbouring property (BauO NRW)

Narrow row house

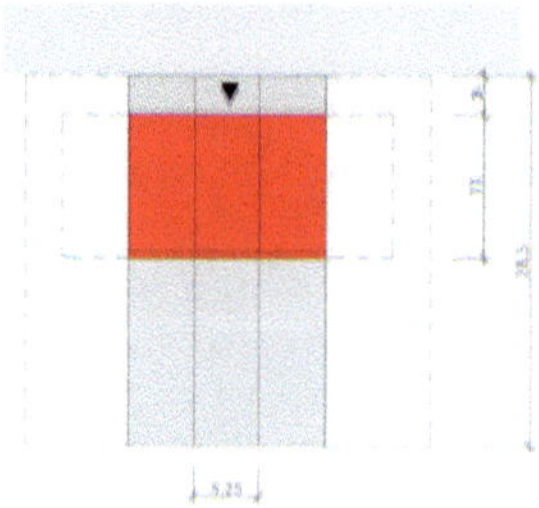

Row house normal

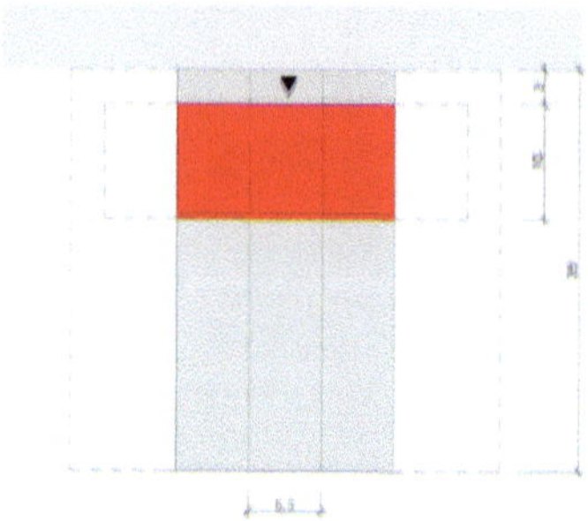

Garden

The depth of the garden should, among other things, react to the direction of the sun.
In the case of north gardens, these should be designed deeper, as the own building shades the garden area close to the building.

Terraced house wide

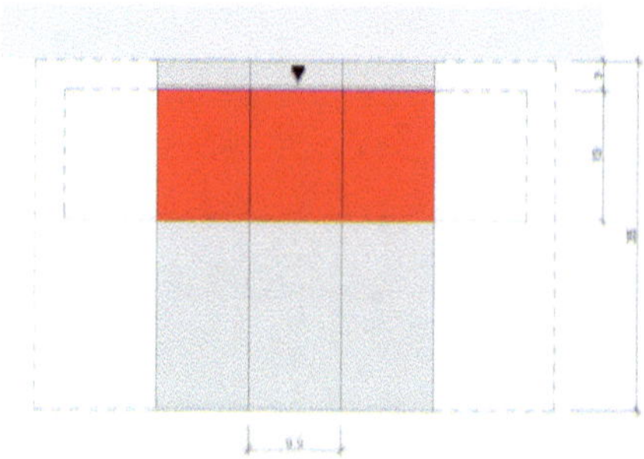

Front garden

If available at least 1.50m, standard 3.00m as narrow front garden.
From 3.00 m you can park lengthwise to the house, from 6.00 m crosswise to the house.
Even with parks in front gardens, the design of the greenery must always be taken into account (directly visible to the public).

Fig. A.3 Sample sizes of row houses. * 3 m minimum distance between the main building and the neighbouring property (BauO NRW). Narrow row house. Row house normal. Garden. The depth of the garden should, among other things, react to the direction of the sun. In the case of north gardens, these should be designed deeper, as the own building shades the garden area close to the building. Front garden. If available at least 1,50 m, standard 3,00 m as narrow front garden. From 3,00 m you can park lengthwise to the house, from 6,00 m crosswise to the house. Even with parks in front gardens, the design of the greenery must always be taken into account (directly visible to the public). Terraced house wide

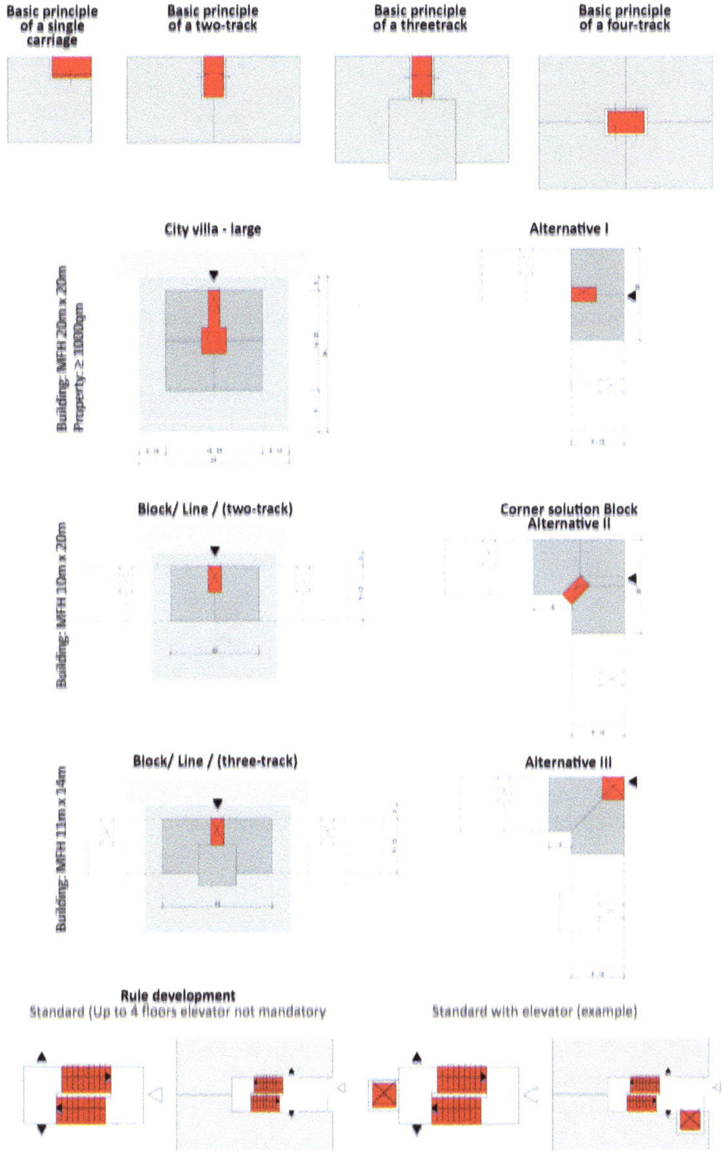

Fig. A.4 (a) Principle of multi-family house/trailer. Basic principle of a one-unit type. Basic principle of a two-unit type. Basic principle of three-unit type. Basic principle four-unit type. City villa—large. Alternative I. Block/line/ (Spanner). Corner solution Block Alternative II. Block/ line/ (three horses). Alternative III. Rule development. Standard (Up to 4 floors elevator not mandatory). Standard with elevator (example) (b) Principle of multi-family house/aisle types. Pergola house. Timber-framed buildings. Depending on the length of the corridor, fire safety measures must be observed during construction

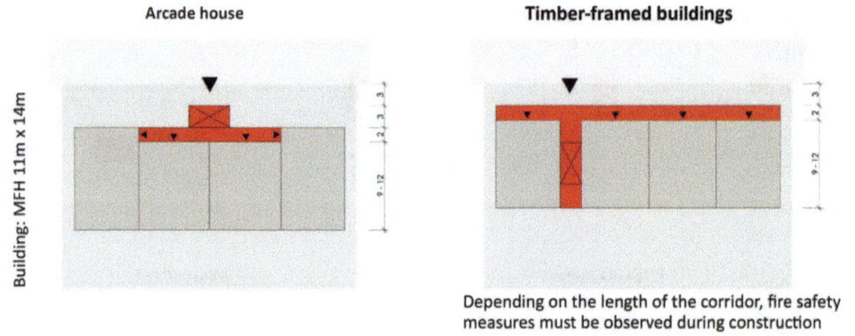

Fig. A.5 Vertical development type: Single unit. Basic principle. Type of development: Single-unit

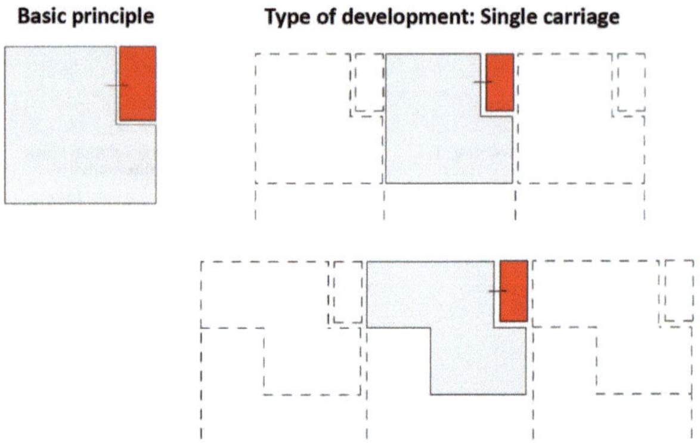

Fig. A.6 Vertical access type: Two-track. Basic principle. Type of development: Two-unit

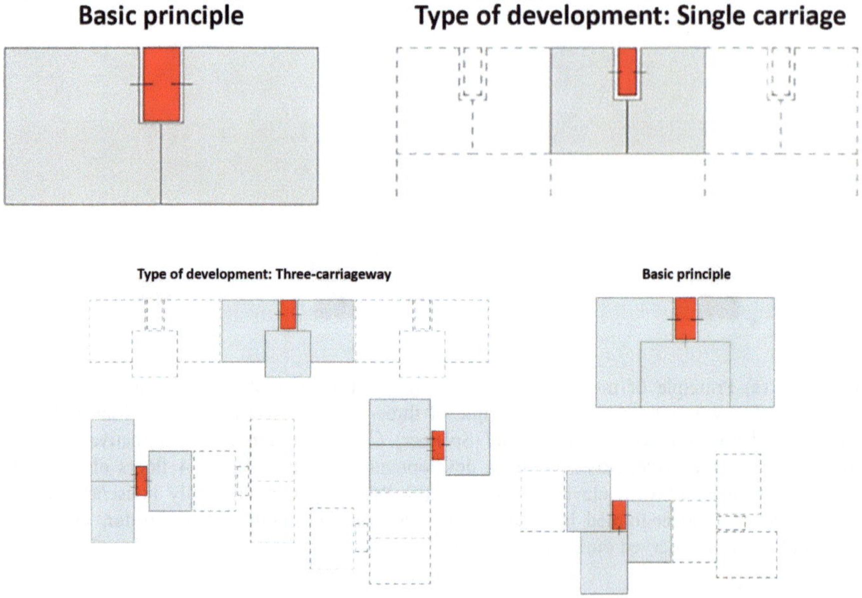

Fig. A.7 Vertical access type: three-car. Type of development: Three unit. Basic principle

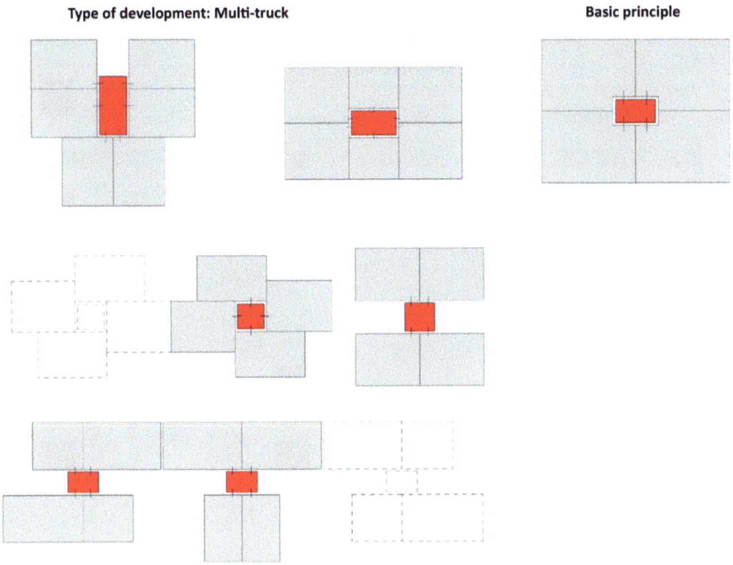

Fig. A.8 Vertical development type: Four and more units. Type of development: Multi-unit. Basic principle

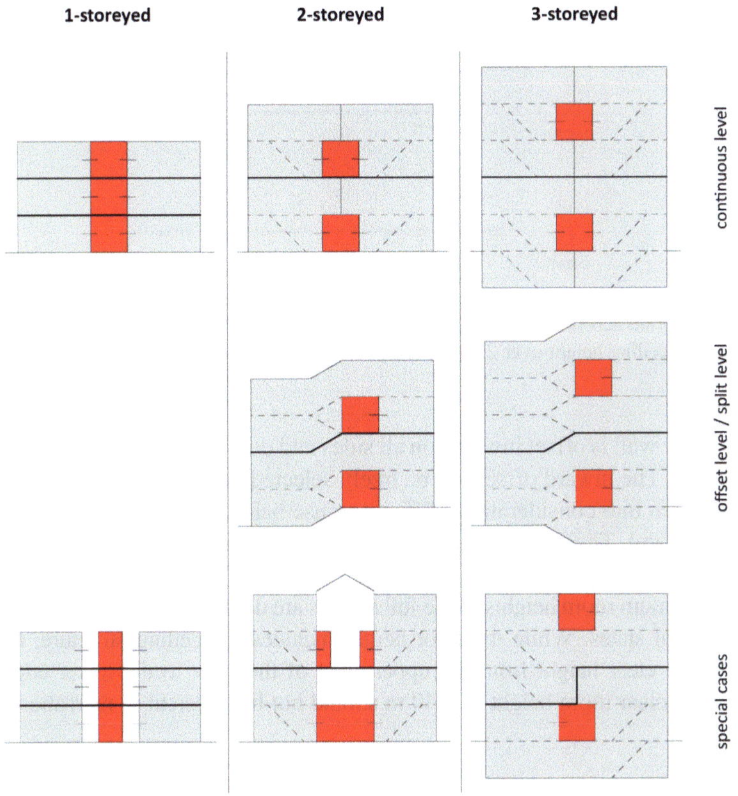

Fig. A.9 Possibilities of inside aisle opening cut

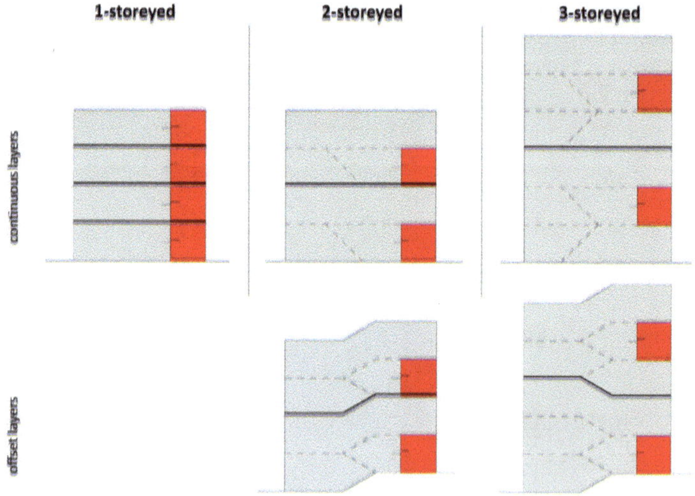

Fig. A.10 Possibilities of external access cut

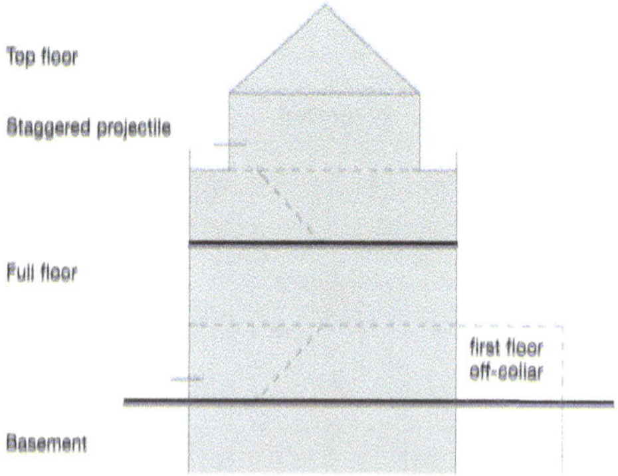

Fig. A.11 External access on every second floor (maisonette). Top floor: An attic floor has 1/3 less usable space (ceiling height over 2.40 m) compared to a full floor

Relay:

The outer wall is offset inwards on all sides and does not load the outer wall of the floor below. The inward offset can be freely selected, but the usability of the terrace must be taken into consideration (difficult to use below 1 m terrace depth, optimum 2.00 m or more). The ceiling height corresponds to the ceiling height of a full storey.

Full floor:

The minimum room heights of the full storeys are defined in the building regulations of the federal states. While the floor height includes the ceiling structure, the room height is the clear height from the upper edge of the floor to the lower edge of the ceiling. The clear room height of 2.40 m should not be undercut in recreation rooms.

The average floor heights in residential construction are approx. 3.00 m, for uses such as offices or practices approx. 3.50 m–4.00 m.

In order to enable a mixture of uses in the ground floor zones and passages, a higher floor height or greater depth should be selected.

The floor and room heights show which demands are made on the use, location and technical equipment of the respective building. Currently there is a clear tendency towards higher room and floor heights.

Basement:

The basement/basement may only project up to 1.60 m above the surface (BauO NRW).

A.4. Road Typologies

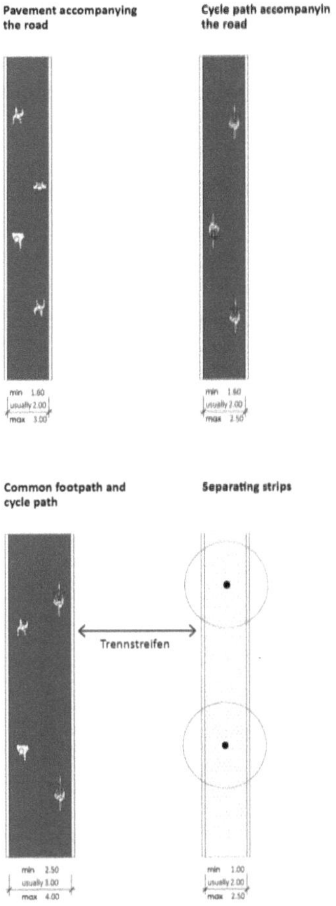

Pavement accompanying the road. Cycle path accompanying the road. Common footpath and cycle path. Separating strips

1. Sidewalk accompanying the road

A sidewalk accompanying the road should not be narrower than 1.50 m, in many cases a wider version is appropriate. Especially in the vicinity of schools, shopping centres, leisure facilities and similar uses, a wider design is desirable.

Social demands—stay, play, etc.—and demands for barrier-free design vary greatly depending on the use of the surroundings, the location and the importance of the street space within the pedestrian network. In principle, these demands must be taken into account when designing sidewalks accompanying the road.

2. Cycle path accompanying the road

A minimum width of 1.60 m should not be undercut when planning a road-accompanying cycle path.

However, the exact width can be derived from the importance and location of the street within the city's overall street and bicycle network. In principle, a wider cycle path offers greater riding comfort and greater safety on the road.

3. Common footpath and cycle path

Common footpaths and cycle paths are particularly suitable for a small number of pedestrians/cyclists and often also outside built-up areas. The width should not be less than 2.50 m. If side room widths of more than 3.50 m are available, it is more sensible to provide separate facilities for pedestrians and cyclists.

4. Separating strips

The width of separating strips varies between a minimum width of 1 m and a maximum of 2.50 m. They are generally used on busy, multi-lane roads to separate the two directions of travel more clearly. Dividing strips must be planted with vegetation. If they are wide enough, they can also serve as parking strips in residential and collective streets and thus act as a separation between bicycle/foot traffic and the roadway.

4a. Parking parallel to the carriageway

In the case of a parking space arrangement parallel to the carriageway, a minimum width of 2.00 m and a length of 6.00 m must be assumed for a passenger car. This type of parking is a space-saving variant and is particularly recommended for narrow roads.

Parking parallel to the carriageway

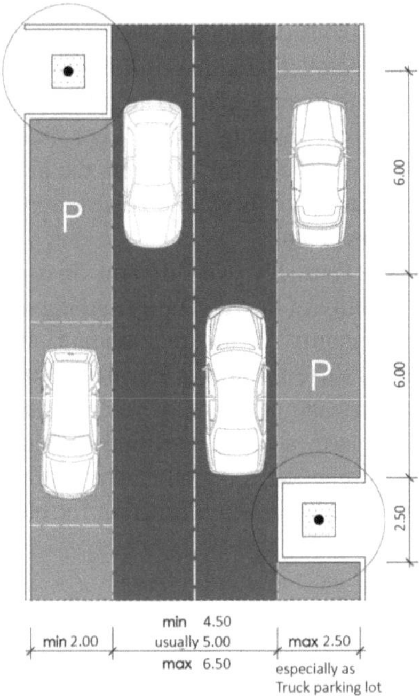

Parking vertical to the road

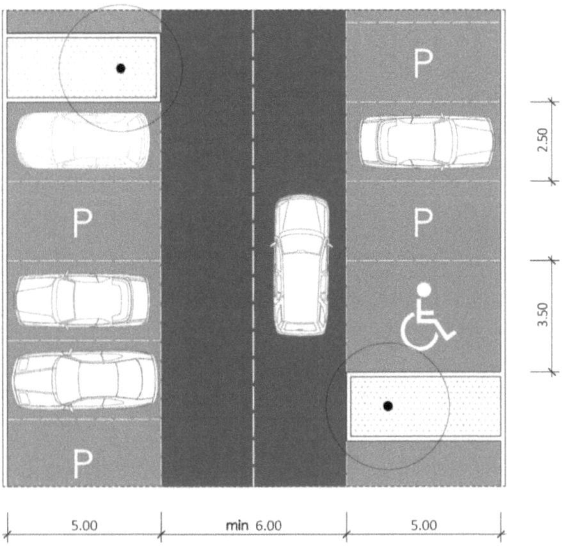

Parking parallel to the carriageway. Parking vertical to the road

4b. Parking perpendicular to the carriageway

Parking perpendicular to the carriageway is associated with a higher shunting effort. At the same time, however, it offers the possibility of realizing a larger number of parking spaces. The dimensions of a parking space start at 2.30 m width × 5.00 m length.

Pitches for physically handicapped persons (wheelchair users etc.) should not be less than 3.50 m wide. Both vertical and parallel parking positions must be provided with greenery in line with the design of the street space. The following applies: A tree stands after approximately every fifth parking space. Each tree requires a minimum unsealed root area of 4.00 sq. m. The larger this area, the better. For this reason, it should also, if possible, be wider or connected to adjacent green spaces.

5a. Lane separation principle: Access road

The width of access roads with two lanes for two different driving directions varies greatly depending on the functionality of the road or its ranking in the overall urban context. As a rule, widths of approx. 5 m are common. It should be remembered that not only passenger cars use the roads, but that a roadway must also be able to accommodate buses, trucks and other motor vehicle traffic.

The term "access road" is a generic term and includes both service roads and service roads that have a development function for motor vehicle traffic. Access roads can certainly fulfil an important main connection function for bicycle traffic. Speed 30 zones are common here.

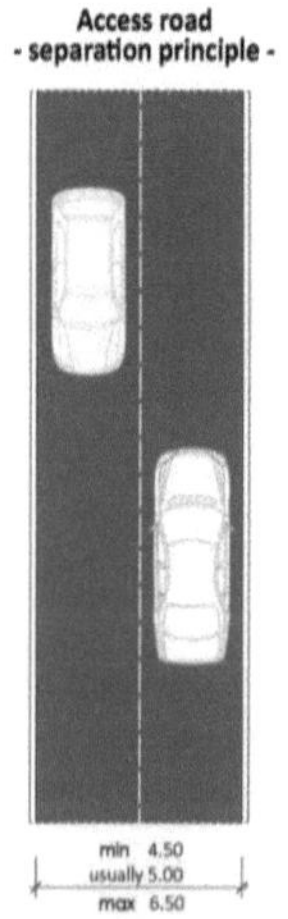

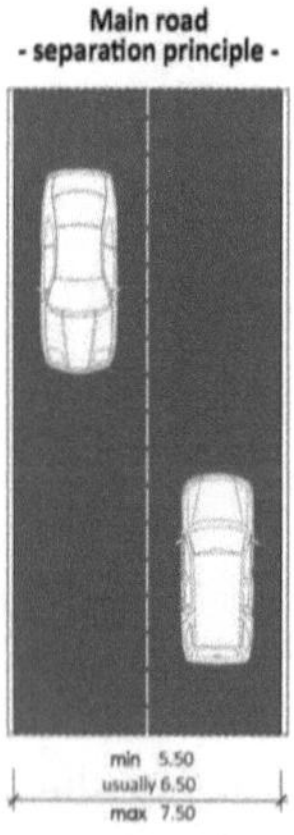

Access road—separation principle. Main road—separation principle

5b. Lane separation principle: Main road

A main road usually has a width of 5.50 m to 6.50 m and can be both cultivation-free and developable.

The difference to access roads is that these usually collect traffic and lead it to the main roads to direct it in different directions from there. A speed limit of 50 km/h or higher is common here.

6. Lane mixing principle

Traffic-calmed roads (colloquially "play streets") are the typical form of a lane mixing principle. Here, the pavement and roadway are neither separated from each other by different types of paving nor by paint stripes, but are available to all road users. However, a minimum pure roadway width of 3.00 m must be guaranteed even with this principle. Motor vehicle traffic must not travel faster than walking speed here.

Mixed traffic area

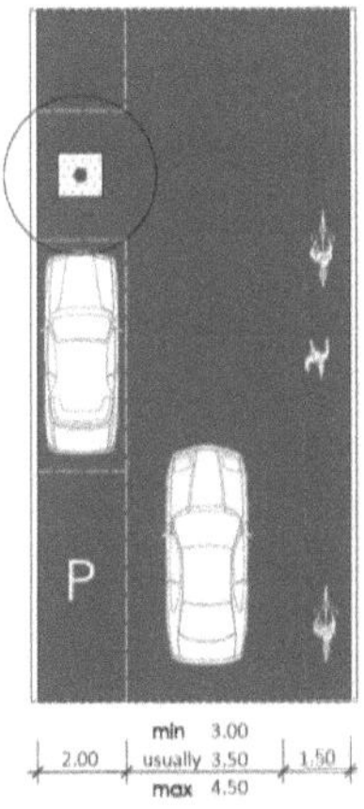

Mixed traffic area (so-called 1-2-3 principle)

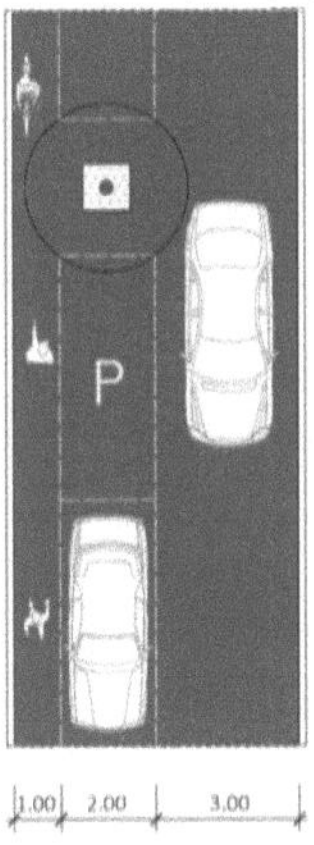

Mixed traffic area. Mixed traffic area (so-called 1–2-3 principle)

1. Residential way

The residential path is one of the most traffic-calmed down forms of access roads. Only private plots of land with residential buildings are developed from it, so that the planning should focus on a particularly high quality of stay. Here, above all, the lane mixing principle is applied. Usually, there are residential buildings at the edges of the road with associated semi-private areas such as front gardens, garages, carports, etc. These are connected by a traffic area which is shared by cars, bicycles and pedestrians. Often a small-scale paving is provided here. This influences the car driver to slow down his speed. Trees/planting beds planted on the left or right have the same effect. It should also be noted that in the event of a dead end, a turning area must be provided for refuse collection vehicles and emergency vehicles.

When determining the dimensions of all urban spaces, it should be noted that a ratio of 5:1 between street width and building height represents the absolute maximum in order to be able to perceive the space as such from a human perspective.

2. Residential street

A residential street has more the character of a public space. From here, only residential buildings are developed, but in a higher density than in a residential street. Special demands are made not only on the quality of stay, but also on traffic and parking.

A semi-private green strip is optionally attached in front of the development. This can be followed (again optionally) by a separate sidewalk and a strip of green and parking areas. The roadway in the middle should usually allow two cars to meet. In any case, the bicycle traffic shares the traffic area with the car traffic. It should be noted that alternative routes must be provided for cars and refuse collection vehicles to meet.

Housing paths (mixed area)

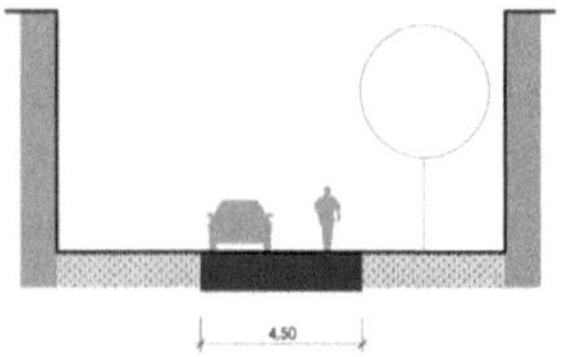

Residential way

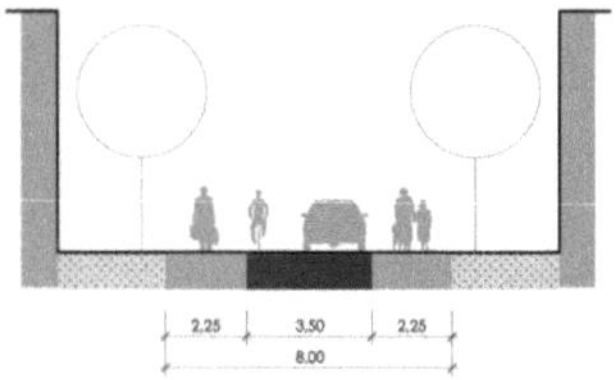

Residential street

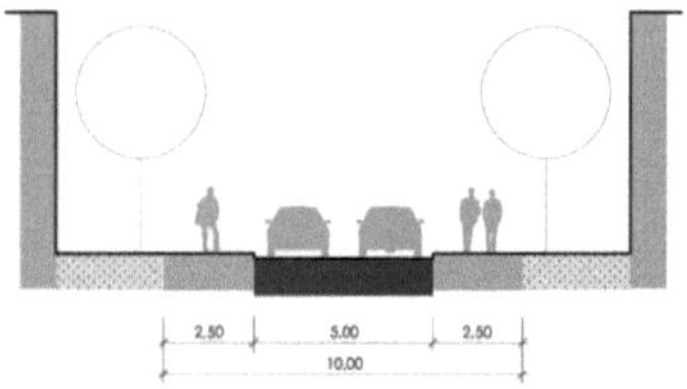

Residential street (longitudinal parking)

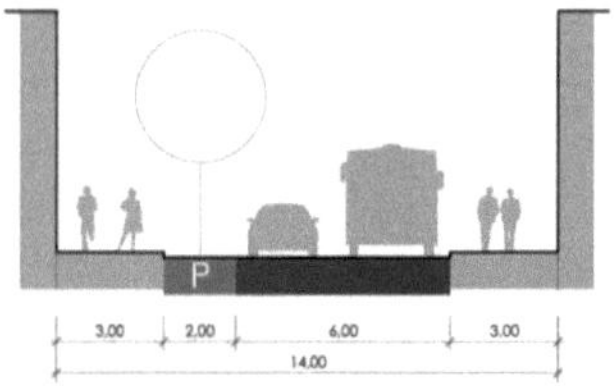

Residential street (cross parking)

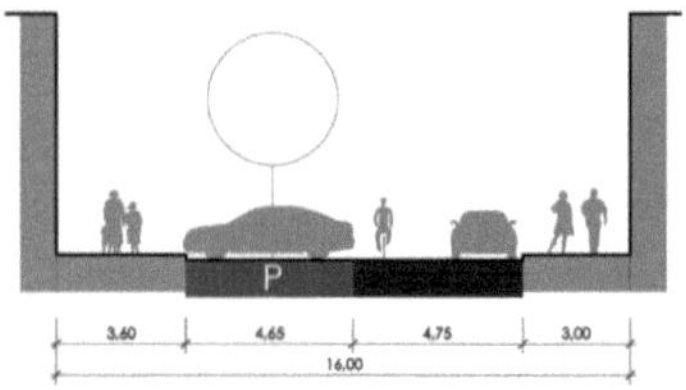

Housing paths (mixed area). Travelling. Residential street. Residential street (longitudinal parking). Residential street (cross parking)

3. Assembly line

A collection road provides access to residential areas as well as to individual shops and public amenities. The residential density here is higher than in a residential street. A service road must above all be suitable for pedestrian traffic (longitudinal and cross connections), bicycle traffic and, to some extent, regular bus traffic. If the central reservation is wider, or if it does not include areas for pedestrians or cyclists, it can also be equipped with trees or rows of trees as avenues.

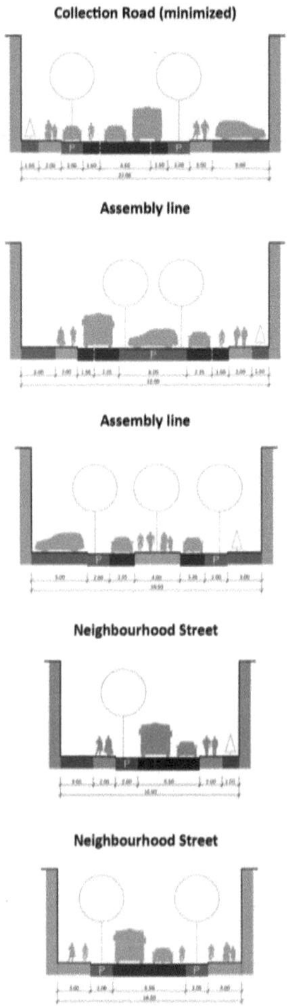

Collecting line (minimized). Assembly line. Assembly line. Neighbourhood Street. Neighbourhood Street

4. District street

A neighbourhood street is characterised by dense, often closed development. Mixed uses of residential, commercial and service areas predominate here.

Typical problems resulting from the high density of use are reflected in a high demand for parking space in street space. At the same time, the quality of stay must be taken into account. This is possible, for example, by connecting green/tree and parking strips.

A street cross-section of a neighbourhood street can be very similar to that of a residential street, except that here the pedestrian must be given a separate area, especially since neighbourhood streets can also fulfil the function of Plauptstraße. It is also important here to provide pedestrians with clear crossing possibilities.

5a. Local shopping street (as access road)

Local shopping streets are usually located in district centres. They are often characterised by a busy shopping population on the ground floors and dense construction. Here, very diverse usage requirements come together: longitudinal and lateral pedestrian traffic must be taken into account as well as areas for delivery and collection of the ground-floor business areas. Due to the mixed uses, a large number of parking spaces for residents and the target traffic of the shops must also be guaranteed.

In cross-section, a road could be constructed as follows: In front of the buildings there is often a large pedestrian area. This should be followed by parking facilities, for example parking bays divided by individual trees. Here, care should be taken to ensure that the parking/green strip is not too narrow and is possibly loosened up by areas without parking so that pedestrians have the opportunity to change sides of the street spontaneously.

Set in the middle, the roadway should not only provide space for cars but also for public buses and delivery vehicles. It is also possible to create a crossable central reservation, which should serve as a crossing aid for pedestrians. Since local shopping streets often have a speed limit of 30 km/h, cycling can be easily integrated into the traffic flow with cars.

Local shopping street. Local shopping street (with central reservation)

5b. Local shopping street (as main street)

In the case of a local shopping street, which is designed as a main road, more facilities for pedestrians to cross the street should definitely be provided, as speed limits of 50 km/h are not uncommon here.

Special situations can arise when trams are also to be used. It should be noted that the more road users share the road, the safer it is for cyclists to be given a separate cycle lane.

6. Main shopping street

Main shopping streets are located in the centres of large and medium-sized cities. A dense commercial population on the ground floors, but also on the upper floors is the rule here; living is the exception.

Some of them are designated as pedestrian zones, in which delivery traffic must have access in the morning hours. However, cities that do not have a purely pedestrian zone must meet special requirements in the planning and design of their main shopping streets:

Particular attention must be paid here to the quality of the stay as well as pedestrian cross and longitudinal connections. Generously proportioned shop forecourts with opportunities to linger (e.g., benches), broken up by a row of trees or flowerbeds are desirable to generate a high quality of stay. At the same time, the street space must meet the requirements of the public transport system, provide space for delivery and collection, and at the same time offer (at least short-term) parking facilities. The extent to which cycle traffic is directed onto the road or over a separate lane depends on the volume of vehicle traffic on the road.

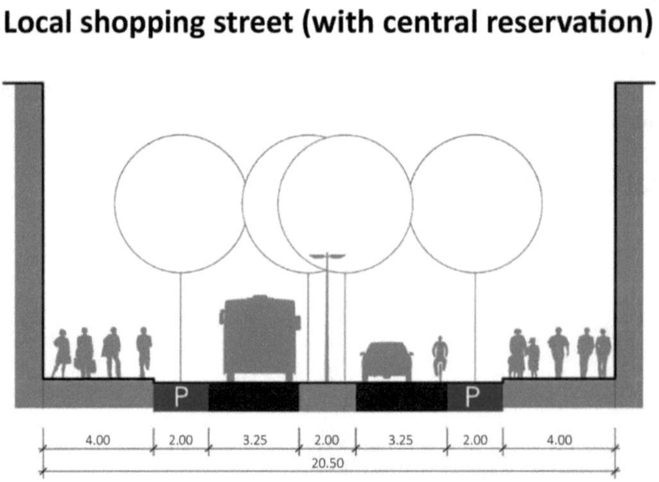

Main shopping street/ boulevard

7. Pedestrian zone

A pedestrian zone is a street that is generally closed to vehicles and may only be used by pedestrians. Depending on the location, suppliers and often cyclists are exempt from the blockade, although they are usually only allowed to enter the pedestrian zone at certain times. The street width of a pedestrian zone is not fixed.

Furnishing elements such as benches, waste baskets and lighting should be integrated, for example in the form of furniture islands.

Pedestrian zone (narrow). Pedestrian zone with zoning

8. Commercial street

A commercial road primarily develops large parceled plots of land with individual buildings and associated parking areas. On these sites, uses such as retail, office or leisure are realized. Here, the traffic area must above all meet the requirements of motor vehicle and delivery traffic. For the development of large commercial plots, a separate turning lane should be considered for safe access and exit.

Despite car-related structures, the connection of a commercial road with public transport and the creation of bicycle lanes should not be neglected.

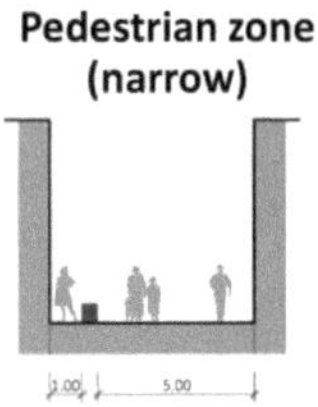

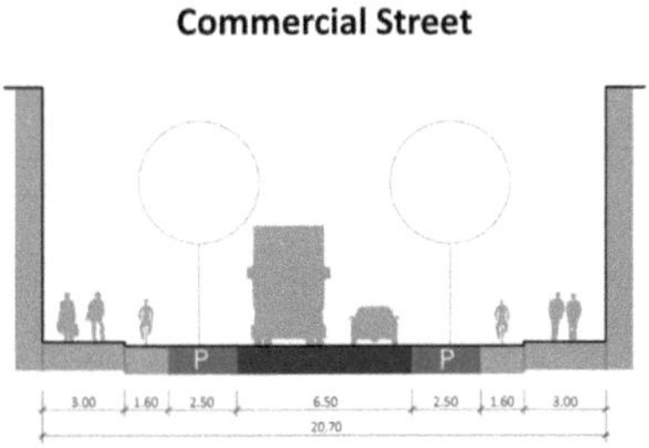

Commercial Street

MIX
Papier aus verantwortungsvollen Quellen
Paper from responsible sources
FSC® C105338

If you have any concerns about our products,
you can contact us on
ProductSafety@springernature.com

In case Publisher is established outside the EU,
the EU authorized representative is:
Springer Nature Customer Service Center GmbH
Europaplatz 3, 69115 Heidelberg, Germany

Printed by Libri Plureos GmbH
in Hamburg, Germany